Springer Complexity

Springer Complexity is an interdisciplinary program publishing the best research and academic-level teaching on both fundamental and applied aspects of complex systems—cutting across all traditional disciplines of the natural and life sciences, engineering, economics, medicine, neuroscience, social and computer science.

Complex Systems are systems that comprise many interacting parts with the ability to generate a new quality of macroscopic collective behavior the manifestations of which are the spontaneous formation of distinctive temporal, spatial or functional structures. Models of such systems can be successfully mapped onto quite diverse "real-life" situations like the climate, the coherent emission of light from lasers, chemical reaction-diffusion systems, biological cellular networks, the dynamics of stock markets and of the internet, earthquake statistics and prediction, freeway traffic, the human brain, or the formation of opinions in social systems, to name just some of the popular applications.

Although their scope and methodologies overlap somewhat, one can distinguish the following main concepts and tools: self-organization, nonlinear dynamics, synergetics, turbulence, dynamical systems, catastrophes, instabilities, stochastic processes, chaos, graphs and networks, cellular automata, adaptive systems, genetic algorithms and computational intelligence.

The three major book publication platforms of the Springer Complexity program are the monograph series "Understanding Complex Systems" focusing on the various applications of complexity, the "Springer Series in Synergetics", which is devoted to the quantitative theoretical and methodological foundations, and the "Springer Briefs in Complexity" which are concise and topical working reports, case studies, surveys, essays and lecture notes of relevance to the field. In addition to the books in these two core series, the program also incorporates individual titles ranging from textbooks to major reference works.

Editorial and Programme Advisory Board

Understanding Complex Systems

Future scientific and technological developments in many fields will necessarily depend upon coming to grips with complex systems. Such systems are complex in both their composition–typically many different kinds of components interacting simultaneously and nonlinearly with each other and their environments on multiple levels–and in the rich diversity of behavior of which they are capable.

The Springer Series in Understanding Complex Systems series (UCS) promotes new strategies and paradigms for understanding and realizing applications of complex systems research in a wide variety of fields and endeavors. UCS is explicitly transdisciplinary. It has three main goals: First, to elaborate the concepts, methods and tools of complex systems at all levels of description and in all scientific fields, especially newly emerging areas within the life, social, behavioral, economic, neuro- and cognitive sciences (and derivatives thereof); second, to encourage novel applications of these ideas in various fields of engineering and computation such as robotics, nano-technology, and informatics; third, to provide a single forum within which commonalities and differences in the workings of complex systems may be discerned, hence leading to deeper insight and understanding.

UCS will publish monographs, lecture notes, and selected edited contributions aimed at communicating new findings to a large multidisciplinary audience.

More information about this series at http://www.springer.com/series/5394

Mark Edelman · Elbert E.N. Macau
Miguel A.F. Sanjuan
Editors

Chaotic, Fractional, and Complex Dynamics: New Insights and Perspectives

Springer

Editors
Mark Edelman
Stern College for Women
Yeshiva University
New York, NY
USA

Miguel A.F. Sanjuan
Departamento de Fisica
Universidad Rey Juan Carlos
Móstoles, Madrid
Spain

Elbert E.N. Macau
INPE/LAC
Instituto Nacional de Pesquisas Espaciai
São José dos Campos, São Paulo
Brazil

ISSN 1860-0832 ISSN 1860-0840 (electronic)
Understanding Complex Systems
ISBN 978-3-319-88550-6 ISBN 978-3-319-68109-2 (eBook)
https://doi.org/10.1007/978-3-319-68109-2

Printed on acid-free paper

This Springer imprint is published by Springer Nature
The registered company is Springer International Publishing AG
The registered company address is: Gewerbestrasse 11, 6330 Cham, Switzerland

Contents

Contributors

Luis F. C. Alberto Sao Carlos Engineering School, University of Sao Paulo, Sao Carlos, SP, Brazil

Fabiolo M. Amaral Federal Institute of Bahia-College Eunápolis, Eunápolis, BA, Brazil

Alvar Daza Nonlinear Dynamics, Chaos and Complex Systems Group, Departamento de Física, Universidad Rey Juan Carlos, Móstoles, Madrid, Spain

J. L. Deneubourg Center for Nonlinear Phenomena and Complex Systems, Brussels, Belgium

Mark Edelman Department of Physics, Stern College, Yeshiva University, New York, NY, USA; Courant Institute of Mathematical Sciences, New York University, New York, NY, USA; Department of Mathematics, BCC, CUNY, Bronx, NY, USA

Joana G. Freire Instituto de Altos Estudos da Paraíba, João Pessoa, PB, Brazil; Departamento de Física, Universidade Federal da Paraíba, João Pessoa, Brazil; Faculdade de Ciências, Instituto Dom Luiz, Universidade de Lisboa, Lisbon, Portugal

Marcia R. Gallas Instituto de Altos Estudos da Paraíba, João Pessoa, PB, Brazil; Departamento de Física, Universidade Federal da Paraíba, João Pessoa, Brazil; Complexity Sciences Center, Surfside, FL, USA

Jason A. C. Gallas Instituto de Altos Estudos da Paraíba, João Pessoa, PB, Brazil; Departamento de Física, Universidade Federal da Paraíba, João Pessoa, Brazil; Complexity Sciences Center, Surfside, FL, USA

Bertrand Georgeot Laboratoire de Physique Théorique, IRSAMC, Université de Toulouse, CNRS, UPS, Toulouse, France

Anandamohan Ghosh Indian Institute of Science Education and Research—Kolkata, Mohanpur, India

Josaphat R. R. Gouveia Jr. Federal Institute of Bahia-College Eunápolis, Eunápolis, BA, Brazil

J. M. V. Grzybowski UFFS - Federal University of Fronteira Sul, Erechim, RS, Brazil

David Guéry-Odelin Laboratoire Collisions, Agrégats, Réactivité, IRSAMC, Université de Toulouse, CNRS, UPS, Toulouse, France

Tomasz Kapitaniak Division of Dynamics, Lodz University of Technology, Lodz, Poland

Vladimir Klinshov Institute of Applied Physics of the Russian Academy of Sciences, Nizhny Novgorod, Russia

J. Kurths Institut für Physik, Humboldt-Universität zu Berlin, Berlin, Germany; Potsdam Institut für Klimafolgenforschung, Potsdam, Germany; Institute for Complex Systems and Mathematical Biology, University of Aberdeen, Aberdeen, UK; Department of Control Theory, Nizhny Novgorod State University, Nizhny Novgorod, Russia

António M. Lopes Faculty of Engineering, UISPA–LAETA/INEGI, University of Porto, Porto, Portugal

Albert C. J. Luo Department of Mechanical and Industrial Engineering, Southern Illinois University Edwardsville, Edwardsville, IL, USA

Leonhard Lücken German Aerospace Center (DLR), Institute of Transportation Systems, Berlin, Germany

Elbert E. N. Macau INPE—National Institute for Space Research, São José dos Campos, SP, Brazil

J. A. Tenreiro Machado Department of Electrical Engineering, Institute of Engineering, Polytechnic of Porto, Porto, Portugal

Vladimir Nekorkin Institute of Applied Physics of the Russian Academy of Sciences, Nizhny Novgorod, Russia

Lev A. Ostrovsky University of Colorado, Boulder, USA; University of North Carolina, Chapel Hill, USA

Elisa Ramalho Chemical Engineering Department, CIETI/DEQ/ISEP, Centre of Innovation on Engineering and Industrial Technology, Institute of Engineering, Polytechnic of Porto, Porto, Portugal

Gokul PM Indian Institute of Science Education and Research—Kolkata, Mohanpur, India; Division of Dynamics, Lodz University of Technology, Lodz, Poland

G. M. Ramírez-Ávila Instituto de Investigaciones Físicas, Universidad Mayor de San Andrés, La Paz, Bolivia

Miguel A. F. Sanjuán Nonlinear Dynamics, Chaos and Complex Systems Group, Departamento de Física, Universidad Rey Juan Carlos, Móstoles, Madrid, Spain; Department of Applied Informatics, Kaunas University of Technology, Kaunas, Lithuania; Institute for Physical Science and Technology, University of Maryland, College Park, MD, USA

Lea F. Santos Department of Physics, Yeshiva University, New York, NY, USA

Yury A. Stepanyants University of Southern Queensland, Toowoomba, Australia

Vasily E. Tarasov Skobeltsyn Institute of Nuclear Physics, Lomonosov Moscow State University, Moscow, Russia

Valentina V. Tarasova Lomonosov Moscow State University Business School, Lomonosov Moscow State University, Moscow, Russia

E. Jonathan Torres-Herrera Instituto de Física, Benemérita Universidad Autónoma de Puebla, Puebla, Mexico

Alexandre Wagemakers Nonlinear Dynamics, Chaos and Complex Systems Group, Departamento de Física, Universidad Rey Juan Carlos, Móstoles, Madrid, Spain

Siyuan Xing Department of Mechanical and Industrial Engineering, Southern Illinois University Edwardsville, Edwardsville, IL, USA

Serhiy Yanchuk Technical University of Berlin, Institute of Mathematics, Berlin, Germany

T. Yoneyama ITA–Aeronautics Institute of Technology, São José dos Campos, SP, Brazil

New Insights and Perspectives in Chaotic, Fractional, and Complex Dynamics

Mark Edelman, Elbert E.N. Macau and Miguel A.F. Sanjuán

1 Introduction

Chaotic dynamics is mainly focused on the analysis of the behavior of dynamical systems that are extremely sensitive to initial conditions. Nowadays it constitutes a very fruitful interdisciplinary field of research, where beyond the apparent randomness of chaotic dynamical systems, many structures and patterns can be observed and analyzed through the numerous methods and techniques developed in the past few

M. Edelman
Department of Physics, Stern College, Yeshiva University,
245 Lexington Ave, New York, NY 10016, USA
e-mail: edelman@cims.nyu.edu

M. Edelman
Courant Institute of Mathematical Sciences, New York University,
251 Mercer St., New York, NY 10012, USA

M. Edelman
Department of Mathematics, BCC, CUNY, 2155 University Avenue,
Bronx, NY 10453, USA

E.E.N. Macau
Laboratório Associado de Computaçao e Matemática Aplicada,
Instituto Nacional de Pesquisas Espaciais, Av. Dos Astronautas, 1758 - C. Postal 515,
São José Dos Campos - SP 12227-010, Brazil
e-mail: elbert.macau@inpe.br

M.A.F. Sanjuán (✉)
Nonlinear Dynamics, Chaos and Complex Systems Group,
Departamento de Física, Universidad Rey Juan Carlos,
Tulipán s/n, 28933 Móstoles, Madrid, Spain
e-mail: miguel.sanjuan@urjc.es

M.A.F. Sanjuán
Department of Applied Informatics, Kaunas University of Technology,
Studentu 50-415, 51368 Kaunas, Lithuania

M.A.F. Sanjuán
Institute for Physical Science and Technology, University of Maryland,
College Park, MD 20742, USA

M. Edelman et al. (eds.), *Chaotic, Fractional, and Complex Dynamics: New Insights and Perspectives*, Understanding Complex Systems,
https://doi.org/10.1007/978-3-319-68109-2_1

years. Fractional dynamics also constitutes an interdisciplinary field of study aiming at investigating the behavior of dynamical systems that are described by differentiation of fractional orders, using methods of the fractional calculus. Fractional-order systems are dynamical systems that can be described by fractional differential equations containing derivatives of non-integer order. They are used to model non-local systems and systems with power-law memory in sciences and engineering. Complex systems appear in many areas of natural sciences, social sciences, engineering, and mathematical sciences. They are often composed of large numbers of interconnected and interacting units, whose behavior is typically difficult to predict due to the relationships and interactions between their components. As a consequence of these relationships and interactions, these systems possess some properties, such as nonlinearity, emergence, spontaneous order, adaptation, and feedback loops. There are many examples of applications including transportation, communication and social networks, ecological and biochemical systems, engineering systems, computational, mechanical and electrical, social and economic systems, besides many basic examples from basic sciences. Chaotic, fractional, and complex dynamics are also deeply connected and, as a whole, they constitute a rich and fruitful field of research in physics, mathematics, biology, engineering and social sciences.

The current book provides a collection of chapters with contributions from leading scientists working in nonlinear dynamics, chaotic dynamics, and complex systems, including fractional dynamics and networks as well. The main goal is to provide an overview of cutting-edge research in different topics, including both fundamental and applied research in this flourishing area of science.

The purpose of this chapter is to review the contributions of the authors to this book. Among the different chapters we can find contributions of different kinds. Some contributions are more fundamental and other are more related to applications. Admitting that a classification always has its risks, we have considered that we can classify all the chapters in three main groups: nonlinear, chaotic dynamics and applications; fractional dynamics and applications; and complex dynamics and applications. The first group of chapters is of a more basic nature and focuses on more fundamental aspects of nonlinear dynamics, such as predictability of dynamical systems by observing their basins of attraction [1], different ideas related to the common problem of synchronization [2, 3] that affects many disciplines, time-delay effects on periodic motions of nonlinear oscillators [4], generation of mix-mode oscillations in an ecological model [5], and the analysis of bifurcations and stability of dynamical systems [6]. Another group is focused on various aspects of fractional dynamics [7–9], of basic or applied nature, as well. And a final group is more oriented towards applications in the field of acoustics [10] and nonequilibrium dynamics [11], including oscillations in neuron networks [12] and power grids modeled using complex networks [13].

Needless to say, with this selection of chapters all we can do is give an overview showing the current progress, and how active and rich the research field is. Furthermore, we believe that the book can provide new ideas and inspiration for further original results in these active fields of scientific endeavour.

2 Nonlinear, Chaotic Dynamics and Applications

One fundamental concept in science and dynamical systems in particular, is the prediction of the final state of a dynamical model. In [1] Daza et al. describe the concept of basin entropy as a measure of the final state unpredictability and they apply it to the chaotic scattering of cold atoms. The concept of basin of attraction plays an important role in nonlinear dynamics. It represents the set of initial conditions leading to an attractor or, in general, to a final state. Actually, we can have basins of attraction when there are attractors and escape basins when there are no attractors, as in the case of open Hamiltonian systems. Furthermore, in multistable systems where there are several attractors, the boundaries separating different basins may be of a fractal or a smooth nature. The predictability of a given dynamical system strongly depends on the structure of these basins. The authors introduce the concept of basin entropy to quantify the final state unpredictability associated with the basins. Several paradigmatic examples of nonlinear dynamics, such as the Duffing Oscillator, the Hénon-Heiles Hamiltonian and the map that computes the complex roots of the unity using the Newton method have been used to explain it. Besides the fundamental and basic nature of the problem, the authors apply these ideas to experiments with cold atoms, where they use the basin entropy to detect directly from experimental measurements the appearance of fractal structures in phase space.

Synchronization continues to be a topic of interest in nonlinear dynamics, and precisely in [2] the authors study synchronous flashing in fireflies as one of the observed natural phenomena displaying synchronization of a large ensemble. Definitely, the fireflies constitute an example of synchronous collective behavior and nowadays it is considered as a paradigmatic example of synchronization. They explain biological aspects related to fireflies flashing and their functionality. By using a model based on electronic fireflies, they illustrate the response to synchronization observed in some firefly species. Other interesting aspects are explored as the consequences of the firefly courtship as a whole process, including the males synchronization and the corresponding females response. They also point other interesting aspects, such as models inspired by fireflies for communication networks, and the use of firefly synchronization concept in mobile networks and other devices. Finally, they explore the actual applications inspired in fireflies synchronicity and also the perspectives both in biomedical issues and in technological systems including robotics.

The synchronization behavior of discrete and continuous cyclic systems is discussed in [3], where the authors show the emergence of the mixed synchronization states. In these states both the synchronized and the anti-synchronized behaviors can coexist. The mixed synchronized states are shown to depend only on the initial conditions and the basin of attraction is numerically estimated.

Time-delay effects on periodic motions in a time-delayed, hardening Duffing oscillator are analyzed by Luo and Xing in [4]. One of the results that they show is bifurcation trees of periodic motions to chaos varying with time-delay. Furthermore, they simulate numerically the time-delayed oscillator and show that the time-delay effects on period-1 motions to chaos in nonlinear dynamical systems are strongly related to the distributions and quantity levels of harmonic amplitudes.

In [5] Freire et al. discuss the recently detected chaos-mediated mixed-mode oscillations as a kind of complex oscillations supported by a prey-predator model including dormancy, a strategy to avoid extinction. They show that as the carrying capacity grows, surprisingly wide phases of nonchaos-mediated mixed-mode oscillations appear before the onset of chaos in the system. Furthermore, they show that nonchaos-mediated cascades display spike-adding sequences while chaos-mediated cascades show spike-doubling. In addition, they find a host of exotic periodic phases embedded in a region of control parameters dominated by chaotic oscillations of the prey-predator populations. They describe these complicated phases and show how they are interconnected and how their complexity unfolds as control parameters change. The new nonchaos-mediated phases are stable and large even at low values of the carrying capacity.

The problem of the bifurcations and stability regions of nonlinear dynamical systems is analyzed in the chapter by Alberto et al. [6]. Stability regions of nonlinear dynamical systems may suffer drastic changes as a consequence of parameter variation. These changes are triggered by local or global bifurcations of the vector field. In particular, they analyze these changes for two types of local bifurcations on the stability boundary: saddle-node bifurcations and Hopf bifurcations on the stability boundary. They also characterize the local and the global characterizations of the stability boundary at the bifurcation points and the changes of the stability boundaries and stability regions at these bifurcations.

3 Fractional Dynamics and Applications

Among the group of contributions devoted to fractional dynamics, the chapter by Edelman [7] analyzes the universality in systems with power-law memory and fractional dynamics. The main goal of the chapter is to introduce the new features of universality, including cascades of bifurcations on single trajectories, which appear in fractional nonlinear dynamical systems. Regular nonlinear dynamical systems can be extended by introducing power-law memory or considering fractional differential equations in the case of continuous systems or difference equations in the case of discrete systems, instead of integer ones. This extension allows the introduction of families of nonlinear dynamical systems converging to regular systems in the case of an integer power-law memory or an integer order of derivatives or differences. The author considers, by using various examples, the phenomenon of transition to chaos through a period-doubling cascade extending it to fractional maps, which are maps with power-law memory.

After this interesting and fundamental chapter on fractional dynamics, Tarasova and Tarasov [8] describe an application of the fractional calculus to factor analysis of dynamical systems in economy. They suggest some basic concepts and methods for a better understanding of the effects of memory and nonlocality in the deterministic factor analysis. These methods give a quantitative description of the influence of individual factors on the change of the effective economic indicator. They also suggest two methods of fractional integro-differentiation of non-integer order for the deterministic factor analysis of economic processes, where apparently they can give more exact results than the standard methods of factor analysis based on differentiation and integration of integer orders.

Finally, in [9] Lopes et al. provide an interesting chapter on the application of fractional calculus, to a fractional-order model of wine. They use the electrical impedance spectroscopy to characterize several types of wine and they compare the experimental results with those of chemical analysis. The electrical impedance of wine is measured and modeled by means of fractional transfer functions. They compare the model with standard chemical analysis, showing strong correlation between the two distinct descriptions. In order to analyze and visualize the relationships embedded in the data, they adopt hierarchical clustering. As an interesting result, they demonstrate that fractional models describe wine adequately with a reduced number of parameters.

4 Complex Dynamics and Applications

A final block of chapters is more related to applications. Among them, we have the chapter by Ostrovsky and Stepanyants [10], where the authors analyze the complex dynamics of microparticles and gaseous bubbles in an acoustic field caused by the period-averaged radiation force. They also discuss some recent results concerning the effects of particles concentration and mixing in plane and cylindrical resonators, and furthermore they compare the theory with the available experimental data. Their results could be also applied to other scientific fields, in particular they briefly outline among others, some modern biomedical applications.

In the chapter by Santos and Torres-Herrera [11], the authors review some results for the nonequilibrium quantum dynamics of isolated many-body quantum systems described by one-dimensional spin-1/2 models. The evolution of these systems depends on the initial state and the strength of perturbation that takes them out of equilibrium; on the Hamiltonian, whether it is integrable or chaotic; and at the onset of multifractal eigenstates that takes place in the vicinity of the transition to a many-body localized phase. Furthermore, they unveil different behaviors at different time scales, and also discuss how information about the spectrum of a many-body quantum system can be extracted by the sole analysis of its time evolution. This approach is useful for experiments that routinely study dynamics, but have limited or no direct access to spectroscopy, such as experiments with cold atoms, trapped ions, and nuclear magnetic resonance.

The multi-jittering instability in oscillatory systems with pulse coupling and other recent results are reviewed by Klinshov et al. [12], where they also discuss its main features and applications. The multi-jitter instability may provoke a destabilization of the regular spiking regimes of oscillatory systems with pulse coupling. At the bifurcation point numerous so-called jittering regimes with distinct interspike intervals emerge simultaneously. Such regimes were first discovered in a single oscillator with delayed pulse feedback and later were found in networks of coupled oscillators.

Finally, in [13] Grzybowski et al. study the stability and robustness of the power grids as complex networks. Power grids are well-known in engineering and are composed of many interconnected subsystems. A recent line of research in complex systems has considered these systems as complex networks. Several attempts have been made to approach the key problem of the stability and robustness of power grids, in spite of decades of intensive research on power grids to uncover these problems. The authors review a number of recent results on the topic of robustness and stability in power grids, that have been developed within the framework of the theory of complex networks, especially those where a second-order Kuramoto model has been used.

5 Conclusions

In this section we summarize the topics that have been covered in this book. On the one hand a group of chapters deals more directly with fundamental problems in nonlinear, chaotic dynamics and applications: the new tool of basin entropy for the predictability of dynamical systems associated with their basins; synchronization phenomena, including the fireflies as a paradigm and the emergence of mixed synchronization states; time-delay effects on periodic motions in a time-delayed nonlinear oscillator; the recently detected chaos-mediated mixed-mode oscillations as a kind of complex oscillations supported by a prey-predator model including dormancy and the analysis of bifurcations and stability regions of nonlinear dynamical systems.

The second group of chapters deals with fractional dynamics and applications. The universality in systems with power-law memory and fractional dynamics; an application of the fractional calculus to factor analysis of dynamic systems in economy; and an application of fractional calculus to a fractional-order model of wine.

Finally, a third group is devoted to complex dynamics and applications. The dynamics of microparticles in an acoustic field; the nonequilibrium quantum dynamics of isolated many-body quantum systems; the multi-jittering instability in oscillatory systems with pulse coupling; and the stability and robustness of the power-grids as complex networks.

All this offers us a rich perspective of the current state of the research fields dealt with in this book. We also hope that this will contribute to a new vision and stimulus to attack future scientific problems. They can be analyzed and investigated under the research techniques and tools that are shared by the fields of chaotic, fractional, and complex dynamics providing new insights and perspectives.

Acknowledgements ME acknowledges support by the Stern College at Yeshiva University. EENM acknowledges the financial support of the São Paulo Research Foundation (FAPESP) grant

#2015/50122-0 and the Brazilian National Scientific Research Council (CNPq). MAFS acknowledges financial support by the Spanish Ministry of Economy and Competitiveness under Project No. FIS2013-40653-P and by the Spanish State Research Agency (AEI) and the European Regional Development Fund (FEDER) under Project No. FIS2016-76883-P. Furthermore, he also acknowledges the jointly sponsored financial support by the Fulbright Program and the Spanish Ministry of Education (Program No. FMECD-ST-2016).

References

1. Daza, A., Wagemakers, A., Georgeot, B., Guéry-Odelin, D., Sanjuán, M.A.F.: Basin entropy, a measure of final state unpredictability and its application to the chaotic scattering of cold atoms. In: Edelman, M., Macau, E.N.N., Sanjuán, M.A.F. (eds.) Chaotic, Fractional, and Complex Dynamics: New Insights and Perspectives. Springer, Cham (2018)
2. Ramirez-Avila, G.M., Kurths, J., Deneubourg, J.L.: Fireflies: a paradigm in synchronization. In: Edelman, M., Macau, E.N.N., Sanjuán, M.A.F. (eds.) Chaotic, Fractional, and Complex Dynamics: New Insights and Perspectives. Springer, Cham (2018)
3. Gokul, P.M., Kapitaniak, T., Ghosha, A.: Mixed synchronization in the presence of cyclic chaos. In: Edelman, M., Macau, E.N.N., Sanjuán, M.A.F. (eds.) Chaotic, Fractional, and Complex Dynamics: New Insights and Perspectives. Springer, Cham (2018)
4. Luo, A.C.J., Xing, S.: Time-delay effects on periodic motions in a duffing oscillator. In: Edelman, M., Macau, E.N.N., Sanjuán, M.A.F. (eds.) Chaotic, Fractional, and Complex Dynamics: New Insights and Perspectives. Springer, Cham (2018)
5. Freire, J.G., Gallas, M.R., Gallas, J.A.C.: Nonchaos-mediated mixed-mode oscillations in a prey-predator model with predator dormancy. In: Edelman, M., Macau, E.N.N., Sanjuán, M.A.F. (eds.) Chaotic, Fractional, and Complex Dynamics: New Insights and Perspectives. Springer, Cham (2018)
6. Alberto, L.F.C., Amaral, F.M., Gouveia Jr., J.R.R.: Bifurcations and stability regions of nonlinear dynamical systems. In: Edelman, M., Macau, E.N.N., Sanjuán, M.A.F. (eds.) Chaotic, Fractional, and Complex Dynamics: New Insights and Perspectives. Springer, Cham (2018)
7. Edelman, M.: Universality in systems with power-law memory and fractional dynamics. In: Edelman, M., Macau, E.N.N., Sanjuán, M.A.F. (eds.) Chaotic, Fractional, and Complex Dynamics: New Insights and Perspectives. Springer, Cham (2018)
8. Tarasova, V.V., Tarasov, V.E.: Fractional deterministic factor analysis of economic processes with memory and nonlocality. In: Edelman, M., Macau, E.N.N., Sanjuán, M.A.F. (eds.) Chaotic, Fractional, and Complex Dynamics: New Insights and Perspectives. Springer, Cham (2018)
9. Lopes, A.M., Tenreiro Machado, J.A., Ramalho, E.: Fractional-order model of wine. In: Edelman, M., Macau, E.N.N., Sanjuán, M.A.F. (eds.) Chaotic, Fractional, and Complex Dynamics: New Insights and Perspectives. Springer, Cham (2018)
10. Ostrovsky, L.A., Stepanyants, YuA: Dynamics of particles and bubbles under the action of acoustic radiation force. In: Edelman, M., Macau, E.N.N., Sanjuán, M.A.F. (eds.) Chaotic, Fractional, and Complex Dynamics: New Insights and Perspectives. Springer, Cham (2018)
11. Santos, L.F., Torres-Herrera, E.J.: Nonequilibrium quantum dynamics of many-body systems. In: Edelman, M., Macau, E.N.N., Sanjuán, M.A.F. (eds.) Chaotic, Fractional, and Complex Dynamics: New Insights and Perspectives. Springer, Cham (2018)
12. Klinshov, V., Lücken, L., Yanchuk, S., Nekorkin, V.: Multi-jittering instability in oscillatory systems with pulse coupling. In: Edelman, M., Macau, E.N.N., Sanjuán, M.A.F. (eds.) Chaotic, Fractional, and Complex Dynamics: New Insights and Perspectives. Springer, Cham (2018)
13. Grzybowski, J.M.V., Macau, E.E.N., Yoneyama, T.: Power-grids as complex networks: emerging investigations into robustness and stability. In: Edelman, M., Macau, E.N.N., Sanjuán, M.A.F. (eds.) Chaotic, Fractional, and Complex Dynamics: New Insights and Perspectives. Springer, Cham (2018)

Basin Entropy, a Measure of Final State Unpredictability and Its Application to the Chaotic Scattering of Cold Atoms

Alvar Daza, Alexandre Wagemakers, Bertrand Georgeot, David Guéry-Odelin and Miguel A.F. Sanjuán

1 Introduction to Basin Entropy

Dynamical systems describe quantities evolving in time according to deterministic rules, and towards some asymptotic behavior depending on the initial conditions and on the specific choice of parameters. Basins of attraction link a given set of initial conditions to its corresponding final states. This notion appears in a broad range of applications where several outcomes are possible, which is a common situation in neuroscience, economy, astronomy, ecology and many other disciplines. Depending on the nature of the basins, prediction can be difficult even in systems that evolve under deterministic rules. From this respect, a proper classification of this unpredictability is clearly required. To address this issue, we introduce the basin entropy, a measure to quantify this uncertainty.

A. Daza (✉) · A. Wagemakers · M.A.F. Sanjuán
Nonlinear Dynamics, Chaos and Complex Systems Group, Departamento de Física, Universidad Rey Juan Carlos, Tulipán s/n, 28933 Móstoles, Madrid, Spain
e-mail: alvar.daza@urjc.es

A. Wagemakers
e-mail: alexandre.wagemakers@urjc.es

M.A.F. Sanjuán
e-mail: miguel.sanjuan@urjc.es

B. Georgeot
Laboratoire de Physique Théorique, IRSAMC, Université de Toulouse, CNRS, UPS, Toulouse, France
e-mail: bertrand.georgeot@irsamc.ups-tlse.fr

D. Guéry-Odelin
Laboratoire Collisions, Agrégats, Réactivité, IRSAMC, Université de Toulouse, CNRS, UPS, Toulouse, France
e-mail: dgo@irsamc.ups-tlse.fr

M.A.F. Sanjuán
Institute for Physical Science and Technology, University of Maryland, College Park, MD 20742, USA

M. Edelman et al. (eds.), *Chaotic, Fractional, and Complex Dynamics: New Insights and Perspectives*, Understanding Complex Systems,
https://doi.org/10.1007/978-3-319-68109-2_2

1.1 Final State Unpredictability in Dynamical Systems

Imagine a rainy day. A droplet falls in the ground and runs until it eventually reaches a river. Another droplet falls nearby, but in its run towards the lowest point, it ends in a different mass of water, a lake for example. If we were to determine the fate of each falling droplet, we should study the regions leading to each mass of water, that is, we should study their different basins. This picture explains perfectly the origin of the term basin in nonlinear dynamics: a basin is the set of initial conditions leading to a particular region of phase space [1]. Dynamical systems specify the evolution of some magnitudes in time according to deterministic rules, in a similar way as the profile of the ground determines the path of streams of water. In dissipative systems, the final destination is typically an attractor, in open Hamiltonian systems, the particular region of phase space usually refers to an exit.

If a given dynamical system has only one attractor or exit, then the fate of any initial condition is clearly determined. However, dynamical systems often present several possible final outcomes and, in these cases of multistability, elucidating which orbits tend to which attractor becomes a fundamental question. For instance, if a system has two attractors, then two basins exist separated by a basin boundary. This basin boundary can be a smooth curve or can be instead a fractal curve. The study of these basins can provide much information about the system since their topology is deeply related to the dynamical nature of the system. For example, systems with chaotic dynamics usually display basins of attraction with fractal structures [2].

In order to give an intuitive picture of our problem we may look at Fig. 1a and b. The figures show the escape basins of the Hénon-Heiles Hamiltonian for two different values of the energy E above the critical energy that separates bounded motions from unbounded motions. Most initial conditions leave the region through one of the three different exits to infinity for any E above this critical energy. The colors represent points that taken as initial conditions leave the region through a specific exit. With this in mind, we may intuitively understand that it is harder to predict in advance which will be the final destination of an orbit in Fig. 1a than in Fig. 1b.

The problem is that even though, we can have an intuitive notion that Fig. 1a is more uncertain than Fig. 1b, there is no quantitative measure to affirm this. Moreover, this is not easy to assess when we compare two figures of basins corresponding to close values of the energy.

This is precisely the idea of uncertainty or unpredictability which we are considering here. This remark is important since we are aware that these terms are polysemic and consequently its use in the literature might be confusing. Here we refer to unpredictability or uncertainty as the difficulty in the determination of the final state of a system, that is, to which attractor the initial conditions will tend to. Note that we speak about attractors for simplicity, though the discussion is identical for open Hamiltonian systems, where there are no attractors. This notion of unpredictability strongly differs from others used in nonlinear dynamics, like the Kolmogorov-Sinai entropy [3, 4], the topological entropy [5], or the expansion entropy [6], which refer to the difficulty of predicting the evolution of the trajectories. All these quantities

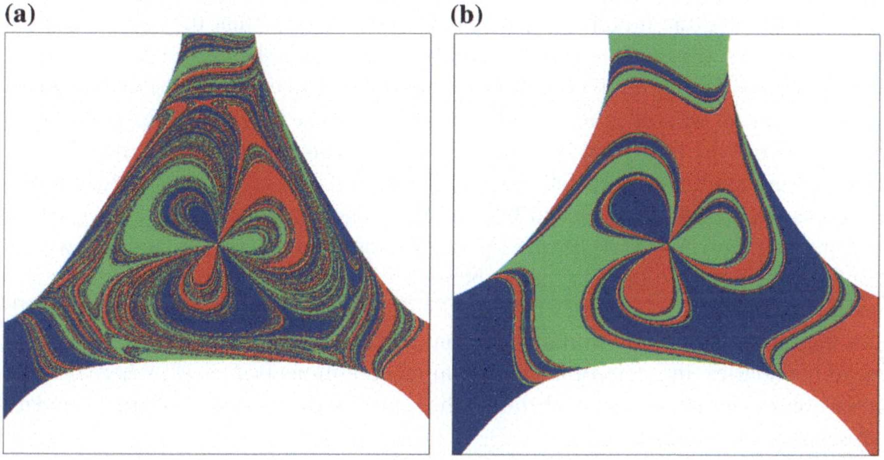

Fig. 1 **Comparison between basins**. Escape basins for the Hénon-Heiles system but different energies. They represent which exit will take each initial condition. It is clear that determining the final destination of the trajectories in the case (**a**) is harder than in the case (**b**)

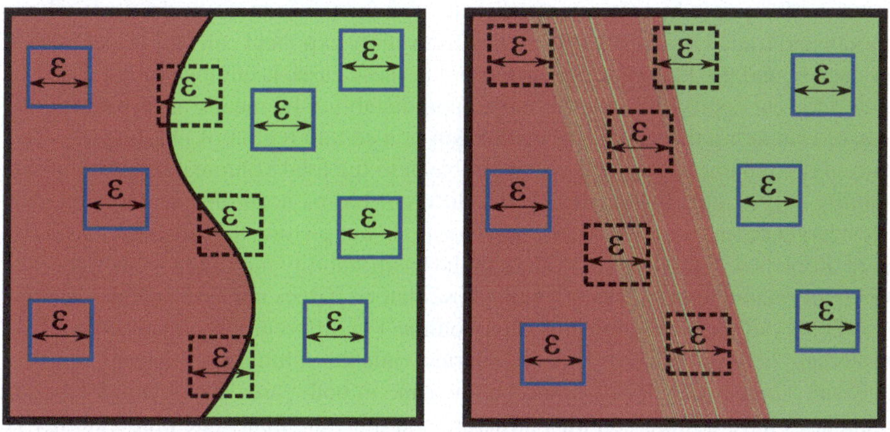

Fig. 2 **Uncertainty in the basin boundaries**. Given some uncertainty in the initial conditions ε, the final state of a dynamical system depends on the structure of its basins. In the left panel, two basins separated by a smooth boundary. In the right panel, two basins separated by fractal boundaries. The black-dashed boxes are for uncertain initial conditions (boxes intersecting the basin boundary)

are related to the topology of the trajectories, whereas our aim here is to develop an entropy depending on the topology of the basins.

A first approach to study the final state uncertainty in dynamical systems was investigated by Grebogi et al. [7]. Given two attractors, they studied how the predictability of the system depends on the fractal or smooth nature of the basin boundaries. Let us describe their methodology by looking at Fig. 2. In the picture, we can see two different basins (red and green) leading to two different final destinations.

Now consider that the initial conditions have some error or that they suffer a small perturbation of size ε, a situation which is always unavoidable in practice. Therefore, instead of considering initial conditions as points, we rather consider them as boxes of linear size ε. If we study the evolution of many of these boxes of initial conditions, we can find out the ratio f of boxes whose future is uncertain, in the sense that not all the initial conditions within the box will end in the same attractor. These boxes are represented by a black dashed line in Fig. 2, and we would have $f = 3/10$ for the left panel and $f = 5/10$ for the right one. Coming back to our analogy, we are counting how many droplets fall in the boundary between basins in this rain of initial conditions. If we vary the size of the boxes ε, we can find that for smooth boundaries the ratio of uncertain initial conditions f grows linearly with ε. However, for fractal boundaries, the ratio of uncertain initial conditions is $f \sim \varepsilon^{\alpha}$, where α is the dimension of the phase space D minus the capacity dimension d of the boundary that separates both basins

$$\alpha = D - d. \tag{1}$$

This quantity α is called the uncertainty exponent, and we have $\alpha = 1$ for smooth boundaries whilst $\alpha < 1$ for basins with fractal boundaries. The closer α gets to zero the more difficult it becomes to predict the system as we zoom in. In cases where smooth and fractal basins are mixed, the uncertainty exponent can still be calculated for each boundary. However, the procedure in these cases is cumbersome [8].

A different approach to measure the unpredictability by means of its basins consists of evaluating the volume of each basin in a certain region of phase space. The ratio of the volume occupied by a single basin to the total volume defines the basin stability [9]. It aims at classifying the different basins according to their relative sizes: larger basins are considered more stable (in our picture, the larger the basin the more droplets will fall into it). Although the basin stability may shed some light into problems related to networks of coupled oscillators, it does not take into account the morphology of the basins, but only their volume. For different sets of parameters, two basins can show smooth or fractal boundaries while the volume of each basin remains constant. The basin stability would be the same in both cases but obviously fractal boundaries have a more complex structure and thus, the final state predictability is not the same.

Figure 3 reveals the limitations of both basin stability and the uncertainty exponent α. The four basins have the same basin stability (the proportion of red and green is the same in the four pictures) although they are clearly different. The uncertainty exponent also fails to capture the uncertainty associated to these basins: it cannot distinguish among different smooth boundaries or among different riddled boundaries [10–12]. However, the basin entropy [13] takes increasing values for each basin, matching our intuition. In the following we introduce the mathematical definition of the basin entropy and a method for its computation.

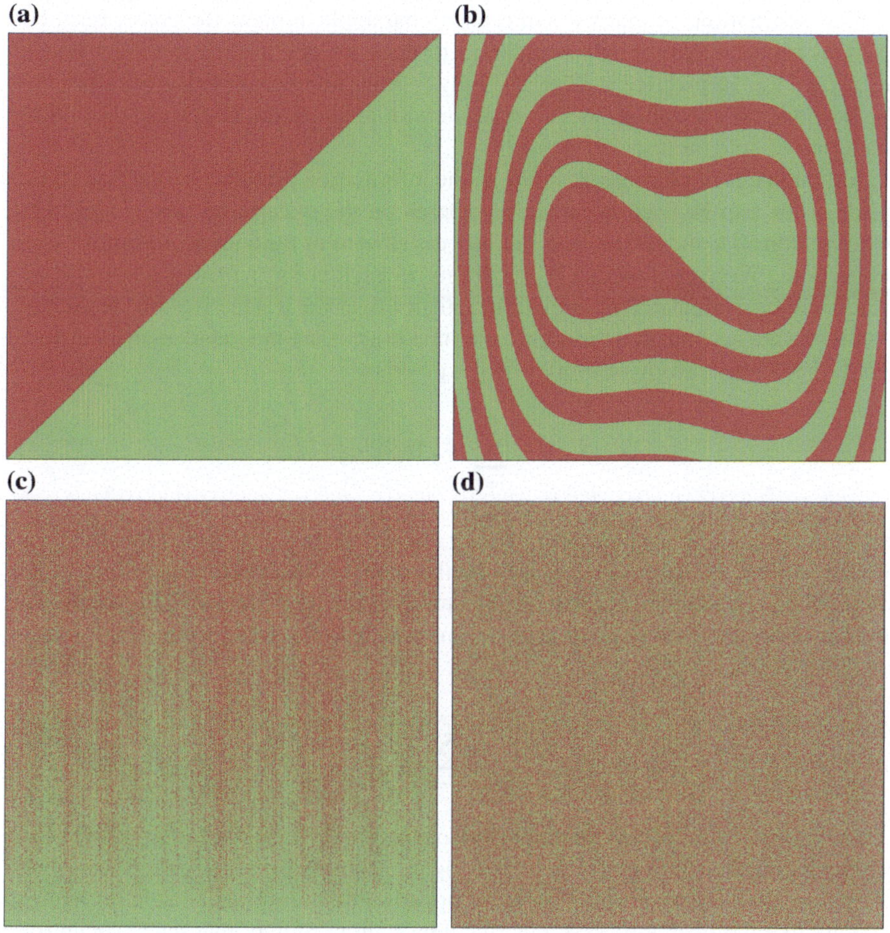

Fig. 3 Comparison of basins and the methods to characterize them. The figure shows different basins obtained from well-known dynamical systems with two attractors. In the two upper panels, the uncertainty exponent is $\alpha = 1$ since both boundaries are smooth, while for the two lower panels $\alpha = 0$ since both of them are riddled basins. The basin stability is equal to 1/2 for the four basins. However, the basin entropy is able to distinguish the four cases and provides a method to measure quantitatively the unpredictability in increasing order from (**a**) to (**d**)

1.2 Definition and Computation of the Basin Entropy

Suppose we have a dynamical system with N_A possible final outcomes for a choice of parameters in a certain region Ω of the phase space. We can discretize Ω via a finite number of boxes covering it, or sample Ω randomly using a sufficient number of boxes. Here we study two-dimensional phase spaces, so that we cover Ω with boxes of linear size ε. Now we build an application relating each initial condition to its final destination, so that we will refer to that application as the *color*. Each box contains in principle infinitely many trajectories, each one leading to a color labeled from 1 to N_A. In practice we can use only a finite number of trajectories per box. We will discuss this point later in relation with an experimental setup used in the scattering of cold atoms.

Although ε is our limiting resolution, the information provided by the trajectories inside a box can be used to make hypotheses on the uncertainty associated to the box. We consider the colors into the box distributed at random according to some proportions. We can associate a probability to each color j inside a box i as $p_{i,j}$ which will be evaluated by computing statistics over the trajectories inside the box.

Taking into account that the trajectories inside a box are independent in a statistical sense, the Gibbs entropy of every box i is given by

$$S_i = -\sum_{j=1}^{m_i} p_{i,j} \log\left(p_{i,j}\right), \tag{2}$$

where $m_i \in [1, N_A]$ is the number of colors inside the box i, and the probability $p_{i,j}$ of each color j is determined simply by the number of trajectories leading to that color divided by the total number of trajectories in the box. Finally, using a sufficient number of boxes N we can define the basin entropy as the mean value of the entropy for those boxes:

$$S_b = \sum_{i=1}^{N} \frac{S_i}{N}. \tag{3}$$

An interpretation of this quantity is associated to the degree of uncertainty of the basin, ranging from 0 (a sole attractor) to $\log N_A$ (completely randomized basins with N_A equiprobable attractors). This latter upper value is in practice seldom realized even for extremely chaotic systems. It is important to remark that the basin entropy depends on the scaling box size ε, i.e., the basin entropy is an extensive property. Therefore, to make quantitative comparisons of different basins we must fix ε.

The procedure for the calculation of the basin entropy is quite similar to the procedure used for the determination of the uncertainty exponent. However, there are important differences. The first one is that when we compute the basin entropy we use the information contained in the boxes. We do not just label the boxes as certain or uncertain like for the uncertainty exponent, but we study the probabilities of the different outcomes through the proportions of different colors inside each box. The second important difference is that, as we will show later, we do not need different scales ε to compare the uncertainty of different basins. Even more, we can detect fractal structures using only one scale. This is fundamental to study the final state unpredictability in experimental systems with finite resolution, like the beam splitter for cold atoms described later.

1.3 What Does the Basin Entropy Measure?

At this point, we can delve deeper into the consequences of this definition by considering a simple hypothesis, which is to assume that the colors inside a box are equiprobable, thus $p_{i,j} = 1/m_i, \forall j$. If we add the entropy of all the trajectories in a box, then we recover the Boltzmann expression for the entropy $S_i = \log(m_i)$, where m_i are the different colors inside a box (the accessible microstates of the Boltzmann entropy). Then the equiprobable total entropy becomes $S = \sum_{i=1}^{N} S_i = \sum_{i=1}^{N} \log(m_i)$. Furthermore, if we have a grid on a given region of phase space, many boxes will have an equal number of colors. That is, many boxes will be in the interior or lie near the boundary between two or more basins. Then we can say that there are N_k equal boxes (in the sense that they have the same number of colors), where $k \in [1, k_{max}]$ is the label for the different boundaries. Boxes lying outside the basin boundaries do not contribute to the entropy as they only have one color. In other words, what matters is what happens at the basin boundaries. Then, the basin entropy reads

$$S_b = \sum_{k=1}^{k_{max}} \frac{N_k}{N} \log(m_k). \tag{4}$$

By following the method of the box-counting dimension D_k [14], by which we compute fractal dimensions of basin boundaries, the number of boxes that contains a boundary grows like $N_k = n_k \varepsilon^{-D_k}$ where n_k is a positive constant. In the case of smooth boundaries, the equation $D_k = D - 1$ holds, D being the dimension of the phase space. For fractal boundaries D_k can be larger, but obviously we always have $D_k \leq D$. On the other hand, the number of boxes in the whole region of phase space, grows as $N = \tilde{n}\varepsilon^{-D}$, where $\tilde{n}$ is a positive constant. Substituting these expressions for N_k and N in Eq. 4, and recalling that $\alpha_k = D - D_k$ is the uncertainty exponent [7] for each boundary, we get

$$S_b = \sum_{k=1}^{k_{max}} \frac{n_k}{\tilde{n}} \varepsilon^{\alpha_k} \log(m_k). \tag{5}$$

This last expression reveals important information. The basin entropy has three components: the term $n_k/\tilde{n}$ is a normalization constant that accounts for the boundary size which is independent of ε; the term of the uncertainty exponent α_k, is related with the fractality of the boundaries and contains the variation of the basin entropy with the box size; finally there is a term that depends on the number of different colors m_k. All these terms depend on the dynamics of the system, while the scaling box size ε depends only on the geometry of the grid.

Equation 5 sheds light into some interesting questions. First, we can compare smooth boundaries ($\alpha_k = 1$) and fractal boundaries ($\alpha_k < 1$). For both of them, smooth and fractal basins, we get $S_b \to 0$ when $\varepsilon \to 0$, but it converges faster in the smooth case. That is, it is more difficult for the basin entropy to decrease its value in a system with fractal boundaries. Despite other important factors, fractal boundaries introduce a larger uncertainty than the smooth ones. Furthermore, if $\alpha_k = 0$ then $S_b > 0$ no matter the scaling box size (this might happen in riddled basins [10–12]).

These ideas can be successfully applied for Wada basins. Basins exhibiting the Wada property have only one boundary that separates all the basins [15, 16]. We can argue that increasing the number of colors in the boundary boxes increases the basin entropy and therefore its uncertainty. In particular, having all possible colors in every boundary box is a unique situation found only in Wada basins. Nevertheless, Eq. 5 also reveals that some non-Wada basins can show larger basin entropy than others exhibiting the Wada property. This can be the case when a system has the Wada property but there is one basin which occupies most of the phase space. Other factors like the number of attractors and the boundary size also play a role in the uncertainty according to the basin entropy formulation. Therefore the Wada property increases the uncertainty under the basin entropy perspective, but each case must be carefully studied.

1.4 Numerical Examples to Help Understand the Basin Entropy

Here we illustrate the main features of basin entropy with several examples of dynamical systems, showing how its dependence on the boundary size $n_k/\tilde{n}$, the uncertainty exponent α_k and the number of attractors N_A.

The term $n_k/\tilde{n}$ corresponds to an estimate of the size of the boundary, since it normalizes the number of boxes containing the boundaries divided by the total number of boxes covering Ω:

$$\frac{N_k}{N} = \frac{n_k}{\tilde{n}} \varepsilon^{\alpha_k}. \tag{6}$$

To study the contribution of this term, we consider the damped Duffing oscillator given by

$$\ddot{x} + \delta \dot{x} - x + x^3 = 0. \tag{7}$$

This equation describes the motion of a unit mass particle in a double well potential with dissipation. This system presents two attractive fixed points in $(\pm 1, 0)$ of the $(x, \dot{x})$ phase space, which correspond to the minima of the double well potential function. The higher the damping coefficient δ the faster the orbits tend to the fixed points and, as a consequence, the basin of attraction appears more deformed for smaller values of δ (Fig. 4a–c). The damped Duffing oscillator is bistable, $N_A = 2$, and has a smooth boundary with uncertainty exponent $\alpha = 1$.

Observing the basins of attraction corresponding to the three different values of δ, it is noticeable that the basin of Fig. 4c has a much simpler structure than the basin in Fig. 4a. The outcome of an initial condition within an ε-box would be more difficult

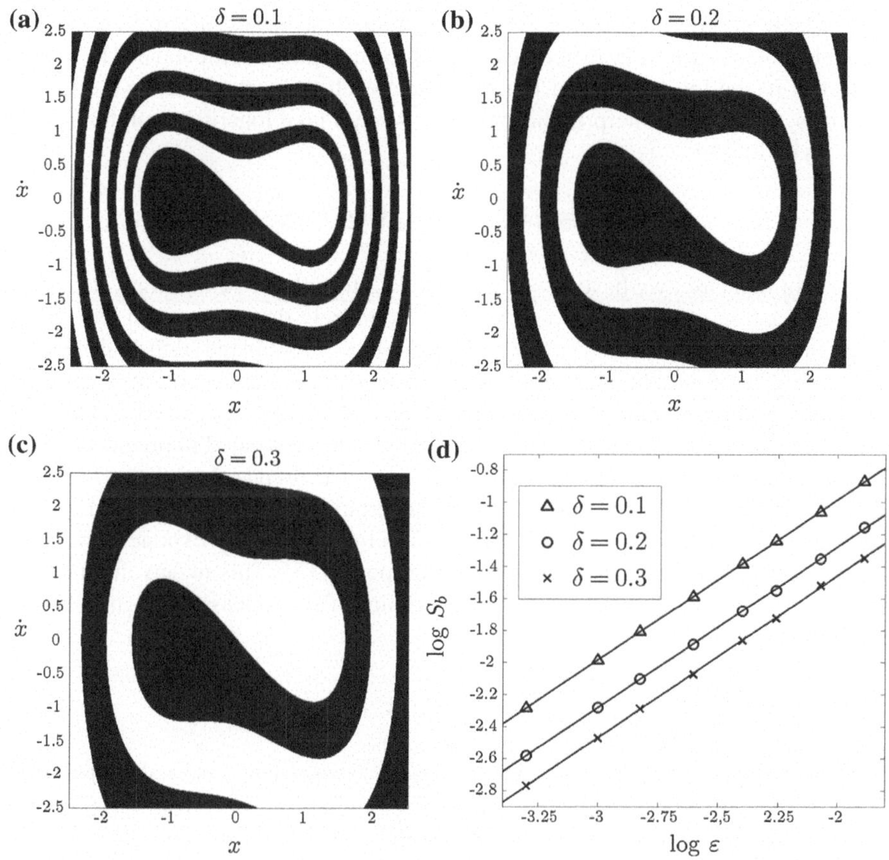

Fig. 4 **Basin entropy dependence on the boundary size.** **a–c** Basins of attraction of the damped Duffing oscillator (Eq. 7) for different values of the damping coefficient δ. As the damping increases the boundary occupies a smaller region of the phase space. Although the boundary is always smooth ($\alpha = 1$), the uncertainty in basin (**a**) is larger than in basin (**c**) no matter the scaling box size ε. **d** A log-log plot of the basin entropy versus the scaling box size for values of the damping coefficient $\delta = 0.1$ (triangles), $\delta = 0.2$ (circles) and $\delta = 0.3$ (crosses). The three fits have the same slope $\alpha = 1$ within statistical error. However, the basin entropy is different for each value of the parameter δ, reflecting the different uncertainty associated to each basin

to predict in the second case. Nevertheless, both basins have the same uncertainty exponent $\alpha = 1$ since in both cases the boundary is smooth. The differences in the values of the basin entropy originates from the differences in the region of discretized phase space occupied by the boundary, that is, the boundary size, which is reflected by the term $n/\tilde{n}$ (indices have been dropped since now there is only one boundary).

In order to highlight this effect, we have computed the basin entropy S_b versus the scaling box size[1] ε for three different values of the damping coefficient δ. The results are shown in the log-log plot of Fig. 4d, where each fit corresponds to a different value of δ. In order to interpret these results, we can take logarithms on both sides of Eq. 5 yielding to

$$\log(S_b) = \alpha \log(\varepsilon) + \log\left(\log(N_A)\frac{n}{\tilde{n}}\right). \tag{8}$$

Since in this case, we have $\alpha = 1$ and $N_A = 2$ for all our simulations, it is clear that the variation of the basin entropy with δ is entirely due to the term $n/\tilde{n}$. Most importantly, we have obtained values of the slope $\alpha = 1$ within the statistical error for all the fits. Therefore, although all these basins have the same uncertainty exponent, they have a different basin entropy for a given value of ε. The basin entropy is sensitive to their different structure and is able to quantify their associated unpredictability.

The fractal dimension of the boundaries also plays a crucial role in the formulation of the basin entropy. This is reflected in the uncertainty exponent α_k [7] of Eq. 5. In order to highlight the effects of the variations in the uncertainty exponent, we have chosen a model that can display the Wada property [17]. This means that there is only one fractal boundary separating all the basins. The model is the Hénon-Heiles Hamiltonian [18],

$$H = \frac{1}{2}(\dot{x}^2 + \dot{y}^2) + \frac{1}{2}(x^2 + y^2) + x^2 y - \frac{1}{3}y^3, \tag{9}$$

which describes the motion of a particle in an axisymmetrical potential well that for energy values above a critical one, the trajectories may escape from the bounded region inside the well and go on to infinity through three different exits. For this Hamiltonian system, we define escape basins in a similar way to the basins of attraction in dissipative systems, i.e., an escape basin is the set of initial conditions that lead to a certain exit. If we vary the energy from $E = 0.2$ to $E = 0.22$, the fractal dimension of the boundaries is modified with E, though the Wada property is preserved [19] (see Fig. 5a–c). The proportion of red, blue and green remains as a constant for these three basins, leading to constant values of the basin stability. However, the basin entropy accounts for their different structures.

As we compute the basin entropy for different scaling box sizes, we observe that the main effect of varying the parameter E is a change of the slope in the log-log plot of Fig. 5d. Equation 8 relates these changes in the slope to the uncertainty exponent α of the boundary. Smaller energies lead to smaller uncertainty exponents, since the boundaries have a more complex structure and consequently the slopes in the log-log plot also decrease. Obviously the offset also varies for the different values of the energy. This is related to changes in the boundary size $n/\tilde{n}$ which in this case cannot be completely separated from the changes in α. This example shows that the

[1] In this work we have normalized the region of the phase space, so that the values of the scaling box size ε in the plots are the inverse of the number of pixels used as a grid.

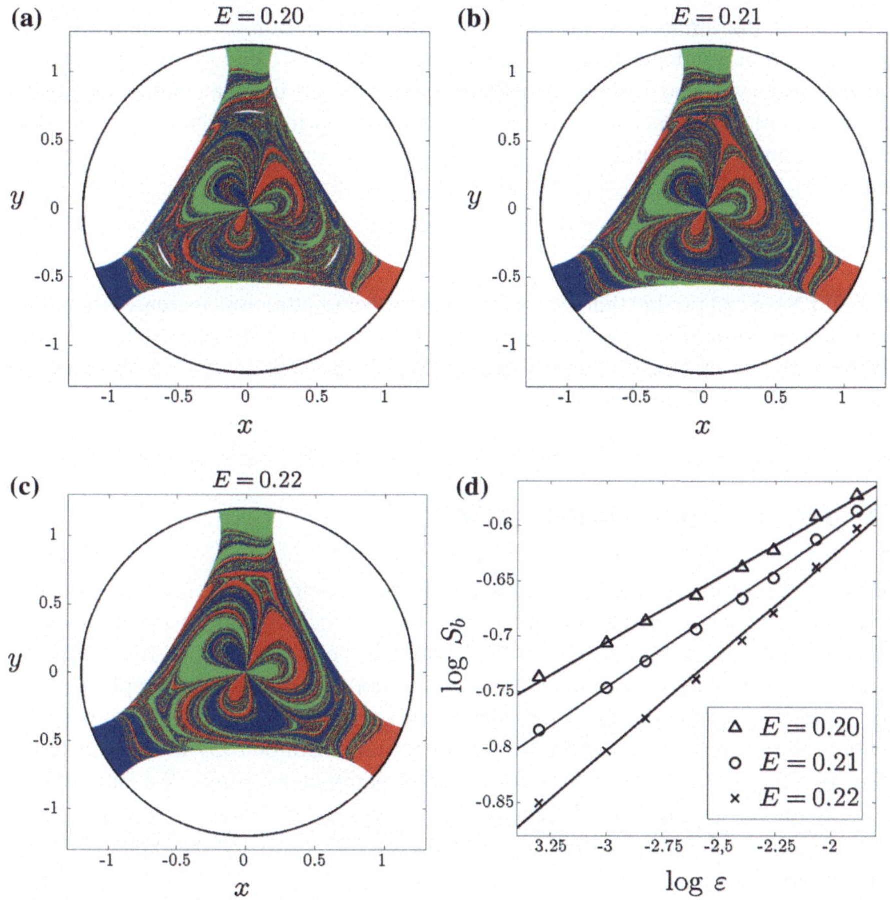

Fig. 5 **Basin entropy dependence on the uncertainty exponent. a–c** Escape basins of the Hénon-Heiles Hamiltonian (Eq. 9) for different values of the energy E. Inside the circles the proportion of red, blue and green boxes is always equal to 1/3. However, as E increases the boundary becomes less uncertain, so that we can intuitively see that basin (**a**) is more unpredictable than basin (**c**). This intuition is confirmed quantitatively by the computation of the basin entropy in the log-log plot of panel (**d**). The most remarkable effect observed in the fits is that the slopes change because of the different dimensions of the boundaries, as expected. This effect cannot be isolated since the offsets also vary. Finally, for coarse-grained basins the basin entropy is almost equivalent

scaling of the basin entropy with box size directly reflects the fractal dimension of the basin boundaries. For small box sizes this effect dominates and the largest fractal dimensions of the basins gives the largest basin entropies, even though the offsets are different (see Fig. 5).

The last factor that contributes to the basin entropy, according to Eq. 5, is the number of attractors N_A. In general, as the number of attractors increases, the uncertainty increases too, and so does the basin entropy. Furthermore, it is impossible to isolate the effect of the number of attractors from the contribution of the boundary size, since they are not independent: if a new attractor emerges while tuning a parameter, a new boundary is also created. We illustrate these effects using a simple map where the number of attractors can be tuned. This map comes from the Newton method to find the complex roots of unity $z^r = 1$ [20], and can be written as

$$z_{n+1} = z_n - \frac{z^r - 1}{r z^{r-1}}. \tag{10}$$

From Eq. 8 we can predict that increasing the number of attractors increases the offset in the log-log plot of the basin entropy versus the box size. This can be observed in Fig. 6, where an increasing number of attractors leads to an increasing value of the basin entropy for all the ε considered.

1.5 Basin Entropy Parameter Set

One of the most interesting applications of the basin entropy is to use it as a quantitative measure to compare different basins of attraction. We propose an analogy with the concept of *chaotic parameter set* [21], which is a plot that visually illustrates in a parameter plane when a dynamical system is chaotic or periodic by simply plotting the Lyapunov exponents for different pairs of parameters. Here, first we choose a given scaling box size ε, and then we evaluate the basin entropy associated to the corresponding basins of attraction for different parameter settings. We call the plot of the basin entropy in a two-dimensional parameter space *basin entropy parameter set*. To illustrate the possibilities of this technique, we study the periodically driven Duffing oscillator

$$\ddot{x} + \delta\dot{x} - x + x^3 = F \sin \omega t, \tag{11}$$

whose dynamics can be very different depending on the parameters. We vary the forcing amplitude F and the frequency ω of the driving, and for each basin we compute its corresponding basin entropy. We have used a resolution of 200×200 boxes ($\varepsilon = 0.005$) with 25 trajectories per box (a million trajectories per basin) to compute the basins of attraction and the same region of the phase space $\Omega = [-2.5, 2.5] \times [-2.5, 2.5]$ for all the pairs (F, ω).

The result is presented in Fig. 7a, which is a color-code representation of the basin entropy in the parameter plane (F, ω) for different values of the forcing amplitude and frequency. The *hot* colors indicate higher values of the basin entropy, while the white pixels are for zero basin entropy. The set of parameters with zero basin entropy indicates that the basin of attraction has only one attractor. Although there is no uncertainty about the final attractor of any initial condition, trajectories may still be very complicated if the attractor is chaotic. This is actually the case for Fig. 7b, where there is only one chaotic attractor.

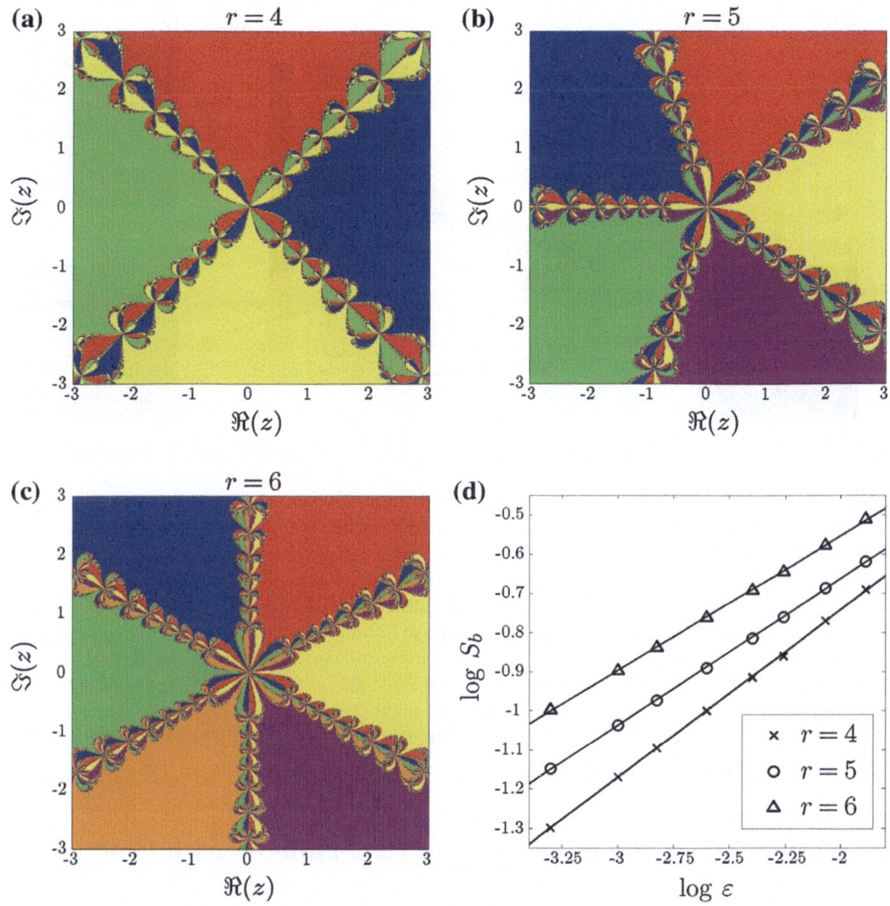

Fig. 6 Basin entropy dependence on the number of attractors. **a–c** The basins of attraction indicate the initial conditions that lead to the complex roots of unity using the Newton method described by $z_{n+1} = z_n - \frac{z^r - 1}{r z^{r-1}}$. Here we plot the cases $r = 4, 5, 6$. The log-log plot of panel (**d**) shows that the basin entropy increases when the number of attractors increases, leading to larger values in the intercepts of the fits as predicted. Nevertheless, the effect of the increasing number of attractors is impossible to separate from the other contributions to the basin entropy, since the boundaries change with the number of attractors

The *hottest* point of the basin entropy parameter set corresponds to the basin of attraction shown in Fig. 7c with eight different attractors whose basins are highly mixed. The reason for having this high value of the basin entropy lies at a combination of a high number of attractors and the uncertainty exponent associated to the boundaries that makes basins of attraction more unpredictable. In Fig. 7d, we can see a basin of attraction with extremely mixed basins, but it has only three attractors so its basin entropy is lower than for Fig. 7c. The converse situation arises in Fig. 7e, where there are sixteen different attractors but the boundaries are not very intricate.

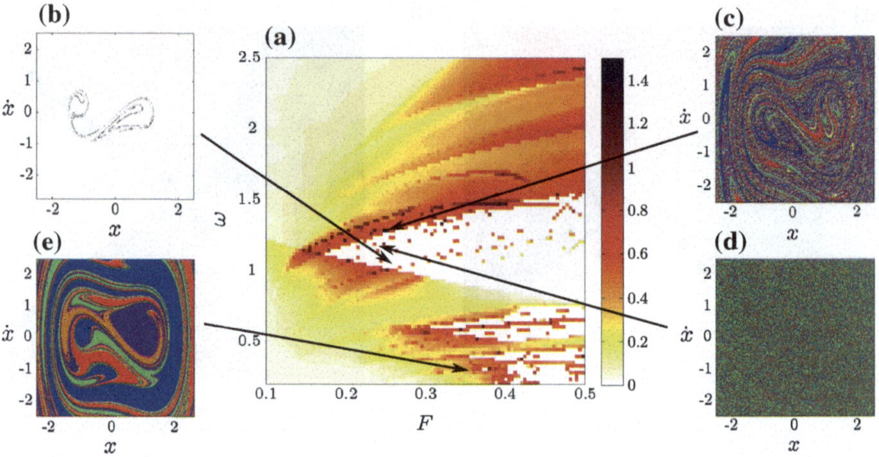

Fig. 7 Basin entropy parameter set. **a** Basin entropy parameter set for the periodically driven Duffing oscillator (Eq. 11). It is a color-code map of the basin entropy for different values (F, ω) of the forcing amplitude and frequency, where we have fixed the scaling box size $\varepsilon = 0.005$ and the damping coefficient $\delta = 0.15$. We have used a color code where the *hot* colors represent larger values of the basin entropy. **b** Example of a basin of attraction with zero basin entropy because there is only one attractor, actually a chaotic attractor (whose Poincaré section is plotted in black), for the parameters $F = 0.2575$ and $\omega = 1.075$. **c** Basins of attraction corresponding to the highest value of the basin entropy in this parameter plane, for $F = 0.2495$ and $\omega = 1.2687$. **d** Basins of attraction with three attractors and a very low uncertainty exponent corresponding to $F = 0.2455$ and $\omega = 1.1758$. **e** Basins of attraction with sixteen different attractors for the parameters $F = 0.3384$ and $\omega = 0.2929$

2 Application of Basin Entropy to Experiments with Cold Atoms

In this section, we show how the basin entropy can be used to characterize the chaotic dynamics of the system and to demonstrate the presence of fractal structures in phase space. All the results presented here are numerical experiments, but we also describe how to perform such experiments in real settings.

2.1 *Cold Atoms and the Crossed Beam Configuration*

In the past few years, beam splitters for guided propagating matter waves were thoroughly investigated in the thermal regime [22–26]. More recently, the Bose-Einstein condensate regime was explored using optical waveguides [27, 28]. Despite the quantum nature of these systems, some results could be understood using classical mechanics. For instance, a classical approximation was employed to unveil the chaotic dynamics underlying the experimental results in Ref. [28]. Positive Lyapunov

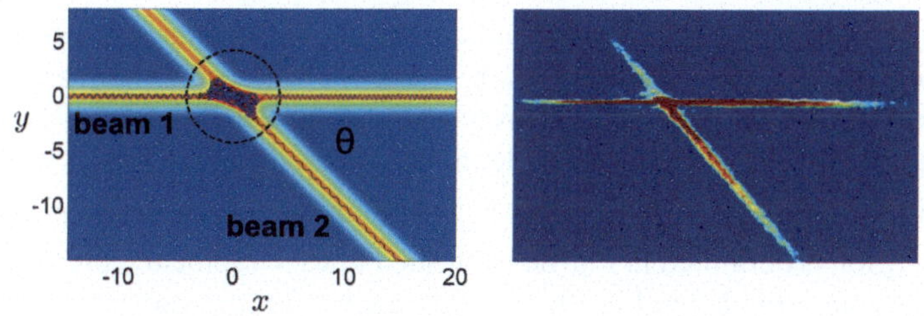

Fig. 8 **The crossed beam configuration**. On the left, two classical trajectories of the Hamiltonian defined by Eq. 12 with parameters $\alpha_1 = \alpha_2 = \beta_1 = \beta_2 = 1, \theta = 45°$. The dashed circle represents the scattering region $\sqrt{x^2 + y^2} < 3\sigma$, with $\sigma = \sqrt{\frac{2}{\beta_i}}$. On the right, an experimental absorption image of the crossed beam configuration in the splitter regime ($\alpha_2/\alpha_1 \approx 1$). Hot colors indicate atoms escaping through the four different exits

exponents, a hallmark of chaos, were also estimated from experiments with ultracold atoms [27]. Here we exploit the concept of basin entropy to characterize this kind of systems and, more importantly, we explain how to use basin entropy in real experiments with cold atoms.

For concreteness, we focus in the configuration experimentally investigated in Refs. [25–28]. In these experiments, atoms are moving into two crossed waveguides (see Fig. 8). For the sake of simplicity, we shall use a two-dimensional model that captures the main features of the experimental system [28, 29]. Using some general assumptions [30], the motion of the particles in the Gaussian potential of the laser beams can be described by the following Hamiltonian,

$$H = \frac{1}{2}\left(\dot{x}^2 + \dot{y}^2\right) - \alpha_1 e^{-\beta_1 y^2} - \alpha_2 e^{-\beta_2 (x \sin\theta + y \cos\theta)^2}. \tag{12}$$

The features of each laser beam are condensed into two characteristic parameters: α, related to the depth of the potential and β, related to the laser waist. Along this work, we will use $\alpha_i = \beta_i = 1$.

Figure 8a shows two examples of a classical trajectory of this Hamiltonian. In Fig. 8b we can see an absorption image where a cloud of atoms is scattered, and we can measure the population of atoms in each branch. The coupling of the longitudinal and the transverse degrees of freedom that occurs at the crossing region is responsible for the complex dynamics. Surprisingly, as shown in [28], the classical description can account for the experimental results. The physical reasons are twofold: (1) the typical scale of variation of the potential is large compared to the de Broglie wavelength associated with the incoming velocity, and (2) interference effects were marginal because of the relatively short time that the wave packet spends in the scattering region, and the 3D dynamics limiting the overlap of the packet with itself. In the following, we shall investigate the fractal properties of this system using a tiling of the classical phase space. The results presented here remain pertinent for the experiments once the phase space cells considered for the statistical analysis are significantly larger than $\hbar$.

If we fix the shooting distance x_0, and consider v_{x_0} and θ as parameters we can analyze the dynamics in terms of (y_0, v_{y_0}). The set of initial conditions (y_0, v_{y_0}) that yields an escape through a given exit is referred to as an escape basin [17]. Given the Gaussian profile of the potentials, we define unbounded trajectories as those going further than $3\sigma_i$ of each laser beam $i = 1, 2$, with $\sigma_i = \sqrt{\frac{2}{\beta_i}}$. An example of such a scattering region is delimited in Fig. 8a by dashed lines. Graphical representations of escape basins are provided in Fig. 9b–d, where each color represents an exit according to the color code of Fig. 9a. White pixels are for trajectories associated to atom losses (because of the finite depth of the potential) and also for what we call sticky trajectories, i.e., that spend more than 2×10^6 time steps without escaping. These two kinds of trajectories will not be considered for the calculations of the basin entropy due to their negligible influence. Their corresponding basin is however interwoven with the other basins but it is only important for extremely low values of v_{y_0} and large initial transverse positions y_0. In the following, we will restrict our study to the parameter ranges $v_{y_0} \in [-1.5, 1.5]$, $y_0 \in [-1.5, 1.5]$.

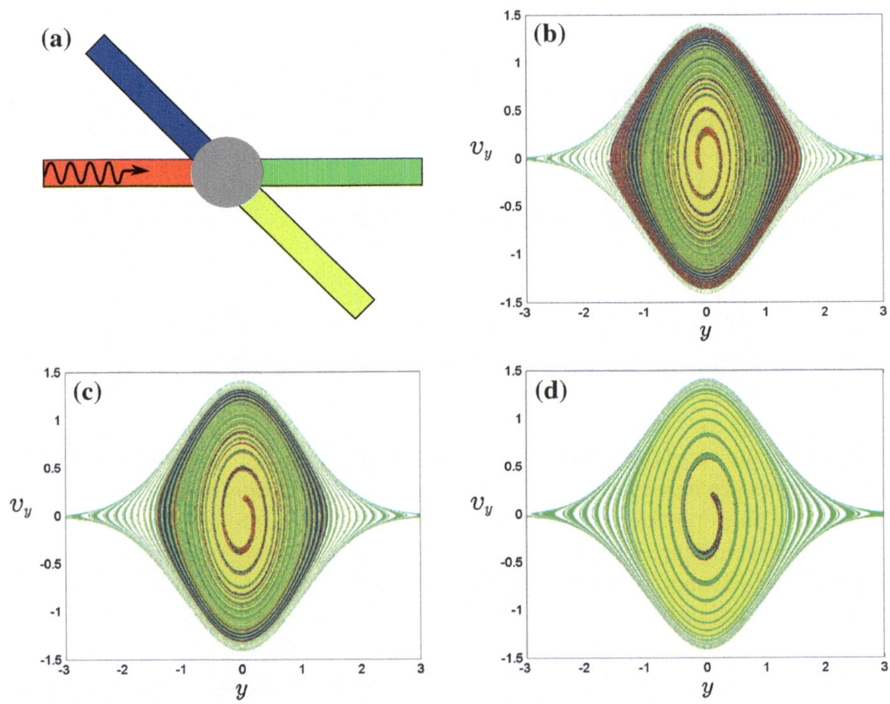

Fig. 9 Escape basins for the crossed beam configuration. The parameters for these basins are $\theta = 45°$, $x_0 = -50$, $\alpha_1 = \alpha_2 = \beta_1 = \beta_2 = 1$ in Eq. 12. **a** Color code for the escape basins. **b–d** Escape basin for a shooting speeds $v_{x_0} = 0.2, 0.5, 0.8$ respectively. Basins are *less fractalized* (they have smaller basin entropy) as the shooting speed v_{x_0} increases

The presence of fractal structures is noticeable for low speed basins (see Fig. 9b), but harder to appreciate in the case of high speed, e.g., Fig. 9d. Quantifying the different *degrees of fractality* can be done with the help of the basin entropy.

2.2 How to Compute Basin Entropy from Experimental Cold Atom Data

The scattering experiments with cold atoms and the procedure to calculate the basin entropy share significant similarities. In both cases we consider ensembles of trajectories instead of individual trajectories. In the experiments we have clouds of atoms with different values of position and velocity, and for the basin entropy calculation we must compute many trajectories with different initial conditions inside every box. Scattering experiments essentially study the output of the trajectories in order to gain knowledge about the system, just as the basin entropy does. We propose to use as the equivalent of boxes in the basin entropy scheme, wave packets of atoms which are fired towards the scattering region. Indeed, these wave packets correspond to a group of atoms distributed around a mean value of the velocity and the position following a Gaussian distribution. The experimental measurement through absorption pictures (see Fig. 8b) provides access to the population of different branches, and thus to the probabilities inside every box.

As described in Ref. [30], we must pay attention to some technical details concerning the basin entropy computation in scattering problems. Nonetheless, in the experiments we have described these issues can be easily tackled. First, the distributions of the clouds of atoms must be in a stationary regime before arriving to the scattering region. This can be achieved varying the launching distance x_0 appropriately. Second, the basin entropy is a statistical measure, so that we need a sufficient number of initial conditions per box to get accurate values for the probabilities of the different colors. But the number of trajectories in each box is directly related to the number of atoms in a wave packet, which in real experiments is in the thousands providing excellent statistics. In fact, it could be further increased by repeating the experiment for a wave packet with same initial mean values.

Remarkably, the Monte Carlo sampling of phase space can be done experimentally by selecting different sets of initial conditions with different mean velocity v_{y_0} and mean position y_0. In practice, small clouds of atoms will be successively delivered from a trap that accommodates a reservoir of atoms such as a Bose-Einstein condensate placed upstream. The transverse position for outcoupling the atoms can be tuned by modifying with optical means the reservoir trap geometry, while the mean transverse velocity can be transferred to the packet of atoms by applying a well-calibrated transverse magnetic gradient pulse. The successive repetition of such outcoupling procedures until the reservoir is empty permits to reduce drastically the number of experimental runs. We have checked numerically that for a realistic number of experimental runs ($N \sim 50$) the relative error in the basin entropy computation is below 10% [30].

Another important point is the size of the boxes used in the basin entropy computation, that is, the minimal resolution that can be reached in this experimental procedure. This corresponds to the size of the wave packet relative to the size of the range of phase space that we want to explore. Using experimental settings available at present time, a linear resolution of several tens can be obtained up to a maximum of one hundred [30]. This means that the experimental escape basins would have a resolution between 10×10 and 100×100 in practice.

In short, to compute the basin entropy S_b in the crossed beam configuration, one should perform a sufficient number of experiments. Each of these experiments consists in sending a wave packet with some mean transversal velocity and position. The experiments must be carried out for sufficiently long launching distances to assure the stationarity of the distributions. Then, the population escaping through each channel should be measured by absorption images. Each experimental run provides a value of the basin entropy in a box S_i. With an appropriate sampling of the region of phase space considered, the total basin entropy can be computed by averaging the basin entropy associated to each run.

2.3 *Detecting Fractal Structures in Experiments with Cold Atoms*

In this section, we investigate transient chaos and fractal structures appearing for low values of the horizontal velocity v_{x_0}. A low speed implies that particles spend more time in the scattering region, i.e., the crossing region of the two beams. Therefore, the exponential divergence of trajectories induced by the intricate shape of the potential at the crossing makes the system difficult to predict. The basin entropy captures this strong unpredictability due to the *highly fractalized* phase space. We have used a numerical Monte Carlo procedure to sample the basin entropy S_b for different shooting speeds v_{x_0} (see Fig. 10a). We can see that the basin entropy is lower for higher speeds, providing us a quantitative basis to our intuition: it is easier to predict the final destination of particles with high speed v_{x_0}. Indeed, the basin entropy seems to approach zero for very high launching speeds, where almost all the particles escape through the same exit.

In order to study the fractal nature of the basin boundaries, it is convenient to introduce the idea of *boundary basin entropy*, which can be defined as

$$S_{bb} = \frac{\sum_{i=1}^{N_b} S_i}{N_b}, \tag{13}$$

where $N_b < N$ is the number of boxes containing more than one color, that is, the number of boxes falling in the boundaries.

Using the boundary basin entropy S_{bb} we can derive a useful criterion to detect fractal boundaries in experiments. For the demonstration of this criterion we will

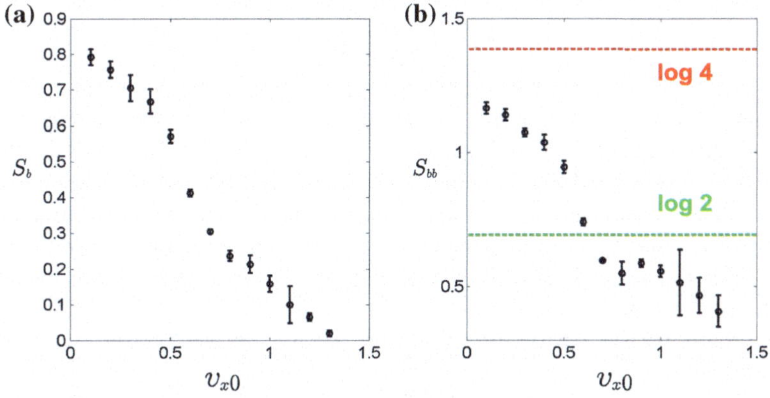

Fig. 10 **Basin entropy computation using different values of the launching speed.** The volume of the phase space investigated is $v_{y_0} \in [-1.5, 1.5]$, $y_0 \in [-1.5, 1.5]$, $x_0 \in [-760, -750]$. For each mean value of $\langle v_{x_0} \rangle$ we consider that $v_{x_0} \in [\langle v_{x_0} \rangle - \langle v_{x_0} \rangle /10, \langle v_{x_0} \rangle + \langle v_{x_0} \rangle /10]$. The basin entropy S_b is computed using 100 boxes (experimental runs) for each represented point, and this procedure is repeated three times so that we get the error bars displayed in the figures. **a** As we increase the horizontal speed v_{x_0} the basin entropy decreases. **b** The boundary basin entropy S_{bb} is above the log 2 threshold (dashed line) for low speeds v_{x_0}, and is below for high speeds

proceed by denying the premise, so that first we assume that our basins are separated by smooth basins. In this case, we have $\alpha = 1$, which means that the number of boxes lying in the boundary that separates two basins (boxes with two colors) grows as $N_2 = n_2 \varepsilon^{-(D-1)}$, where D is the dimension of the phase space. For $D = 2$, the boundary between two basins is a smooth line, for $D = 3$, the boundary separating two basins is a smooth surfaces and so forth.

However, if we have more than two basins, there might be some boxes N_k lying in the boundaries of $k > 2$ different basins. These boxes are in the intersection of at least two subspaces of dimension $D - 1$, that is, they are the intersection of two smooth boundaries. For instance, when $D = 2$, two or more smooth curves intersect in a point or collection of points, and when $D = 3$, two or more smooth surfaces intersect forming smooth curves. Thus, the dimension of the subspace separating more than two basins must be $D - 2$, and the boxes N_k belonging to this subspace must grow as $N_k = n_k \varepsilon^{-(D-2)}$.

Taking into account that the total number of boxes needed to cover the phase space grows as $N = \tilde{n}\varepsilon^{-D}$, we can express N_2 in terms of N as

$$N_2 = n_2 \left(\frac{N}{\tilde{n}} \right)^{\frac{D-1}{D}}, \tag{14}$$

and for the boundary boxes separating more than two basins N_k, we have

$$N_k = n_k \left(\frac{N}{\tilde{n}}\right)^{\frac{D-2}{D}}. \tag{15}$$

At this point, we recall that the maximum possible value of the entropy in a box S_i with m_i different colors is $S_i = \log m_i$, which is the Boltzmann expression for the entropy of m_i equiprobable microstates. Then, we can find that all the boxes in the boundary of two basins have $S_i \leq \log 2$, while for boxes in the boundary of k basins, $k > 2$, we have that $S_i \leq \log k$. Notice that the equality of the previous equations is only possible in pathological cases where all the boxes in the boundaries have equal proportions of the different colors.

Then, the boundary basin entropy S_{bb} for our hypothetical system with smooth boundaries is

$$S_{bb} \leq \frac{N_2 \log 2 + N_k \log k}{N_2 + N_k}. \tag{16}$$

After substituting N_2 and N_k using Eqs. 14 and 15 we get the following expression

$$S_{bb} \leq \frac{n_2 N \log 2 + n_k \tilde{n} \log k}{n_2 N + n_k \tilde{n}}, \tag{17}$$

where $\tilde{n}$, n_2, n_k are constants. Finally, we can take the limit of the previous inequality for a large number of boxes, that is when $N \to \infty$, leading to

$$\lim_{N\to\infty} S_{bb} \leq \log 2. \tag{18}$$

Therefore, we have proven that if the boundaries have uncertainty exponent $\alpha = 1$ (smooth boundaries), then $S_{bb} \leq \log 2$. This is equivalent to say that if $S_{bb} > \log 2$, then $\alpha < 1$, i.e., the boundaries are fractal. This is known as the log 2 criterion.

This criterion is especially useful for experimental situations where the resolution cannot be arbitrarily chosen. In these cases we have a fixed value $\varepsilon > 0$. Nevertheless, if we take a sufficient large number of boxes N, then the log 2 criterion holds. Moreover, the equality of Eq. 18 never takes place, so that there is some room for the possible deviations caused by the impossibility of making an infinite number of simulations or experiments.

The log 2 criterion is a sufficient but not necessary condition for fractality: some fractal basins do not pass this criterion, for instance those having only two outcomes. In the case of the double beam configuration for the scattering of cold atoms, the system presents four possible exits, and for low speeds the values of S_{bb} largely exceed the log 2 threshold, as shown in Fig. 10b. Furthermore, we can see that for very small values of the launching speed the S_{bb} approaches its maximum value for this system which is log 4. This can be seen as an asymptotic value similar to systems where exits get smaller [31], since for the limiting value $v_{x_0} = 0$ the particle would never escape. We have also checked in numerical experiments that the log 2 criterion can be fulfilled for all the angles θ (except the limit cases $\theta = 0°, 90°$) as shown in

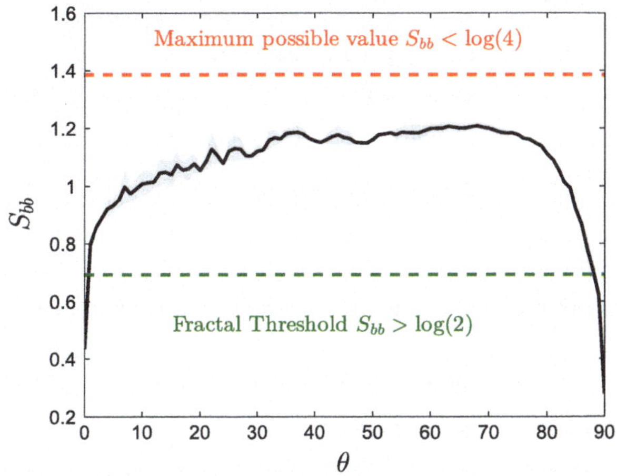

Fig. 11 Log 2 criterion and the angle of the beams. The boundary basin entropy S_{bb} is computed for different angles θ. The region of initial conditions sampled is $v_x \in [0.09, 0.11]$, $x \in [-250, -200]$, $v_y \in [-1.5, 1.5]$, $y \in [-1.5, 1.5]$. The black line is for a computation made with 100 boxes composed of 5^4 trajectories each one, and the shaded region is the absolute error with respect to an asymptotic value taken at 800 boxes. We can see that the log 2 criterion is fulfilled for all the angles except the limit cases $\theta = 0°, 90°$

Fig. 11. If such values were obtained in real experiments, it could be considered an experimental demonstration that the phase space is fractal.

Nevertheless, it is important to recall that the log 2 criterion detects fractals *at a given resolution*. Indeed, given a finite resolution it is impossible to distinguish a real fractal from something which is not a fractal, but that looks like it at that resolution. The log 2 criterion presents a major advantage compared to other techniques like implementing directly the box-counting algorithm: it avoids the use of different scales of velocity and position, which, in the context of experiments and in particular with cold atoms, is fundamental. The log 2 criterion is a strong argument to test fractal structures using minimal requirements. Of course, we will detect fractal structures at the resolution that can be achieved in the experiments, which depends on the size of the wave packet compared to the size of the region of phase space considered.

Finally, it is remarkable that some escape basins are not only fractal, but also may posses the stronger property of Wada [15, 16]. This means, that all the basins have a common boundary separating them. The experimental evidence of the Wada property would be that in this regime every time that more than one branch is populated, all the branches are populated. If the experiment is in the Wada regime, we will never detect atoms escaping through only two or three different branches.

3 Other Tools from Nonlinear Dynamics Applied to the Chaotic Scattering of Cold Atoms

In previous sections we have shown how the basin entropy can be very useful to characterize the unpredictability in experiments with cold atoms. Now we discuss how other methods like the basin stability allow to predict the efficiency of the switch and splitter regimes in a cross beam configuration. The escape time distribution can also be obtained and gives access to the dynamical evolution of the system. All these proposals can be implemented with current experimental techniques.

3.1 Splitter and Switch Regimes

Incident particles with high initial horizontal speed spend less time in the scattering region and most of them tend to escape through exits in the positive x direction. As a consequence, their asymptotic behavior is easier to predict, implying a decrease of the basin entropy for high $v_x(t = 0)$ (see Fig. 10a). Despite the fact that the phase space is still fractal, the $\log 2$ criterion is no longer fulfilled, as shown in Fig. 10b. This happens because there are dominant basins occupying most of the phase space, and the number of boxes lying in the boundaries decreases (see Fig. 10c).

Nevertheless, the appearance of a dominant basin is crucial for the efficiency of the switch regime, an experimental regime where we try to get as much atoms as possible escaping through the second beam. The basin entropy can give us a clue to find the parameters for this switch regime: if most particles escape through an exit, then the basin entropy must be low. Then, we can also apply the basin stability [9] to fully characterize the efficiency of the switch. The basin stability is simply the portion of phase space occupied by each basin, so $B_{Si} \in [0, 1]$ for $i = 1, \dots, 4$ and $\sum_{i=1}^{4} B_{Si} = 1$. Therefore, computing the basin stability for the exit basin 2 is equivalent to calculate its efficiency. In cold atom experiments the basin stability can be computed using the same Monte Carlo sampling method used for the basin entropy computation.

Some angles like $\theta = 33°$ display a large switch efficiency for high speeds, as shown in Fig. 12a. This prediction could be checked in real experiments. We have also tested the robustness of these results against small perturbations of the laser parameters $(\alpha_1, \alpha_2, \beta_1, \beta_2)$. Sometimes in chaotic dynamics small perturbations of the system parameters may lead to different dynamical behaviors [21]. However this is not the case here, and the switch regime turns out to be robust against perturbations of the wave guide parameters.

In the splitter regime, we try to get approximately half of the atoms escaping through the first beam and the other half through the second beam. Using the basin stability, we can define the efficiency of the switch regime as the correlation between basin stability of exits 1 and 2, which can be calculated as their normalized product $c = 4B_{S1}B_{S2}$, where the factor 4 is to normalize at the maximum correlation value

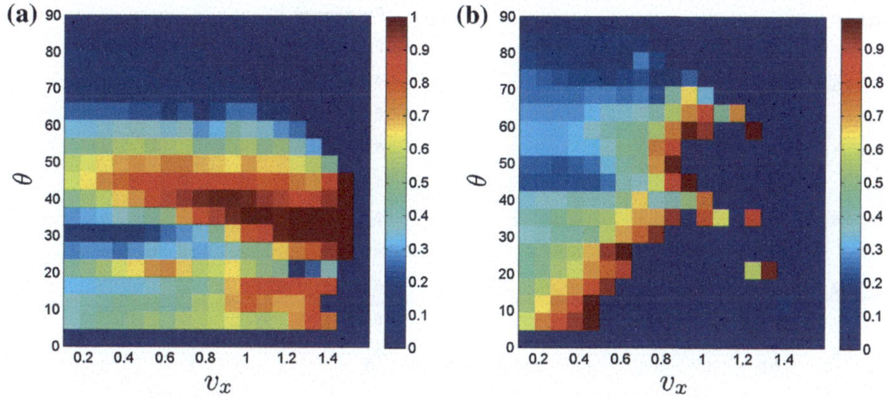

Fig. 12 Efficiency of the switch and splitter regimes. a Color map representing the fraction of trajectories escaping through exit 2, that is, the efficiency of the switch regime. For these computations we have used initial conditions in the region $v_y \in [-0.5, 0.5]$, $y \in [-0.5, 0.5]$ and $x_0 = -250$. **b** Color map for the correlation of the basin stability of exits 1 and 2, defined as the normalized product of their basin stability $c = 4B_{S1}B_{S2}$. For values close to 1 the system is close to a perfect 50-50 splitter regime. This takes place for larger angles as the speed increases

of $B_{S1} = B_{S2} = 0.5$. This efficiency of the splitter c is calculated for different v_x and θ and represented in Fig. 12b. We can see that as the horizontal speed v_x is increased, the splitter regime happens for larger angles. The splitter is more sensitive to perturbations of the parameters than the switch regime, as can be inferred from the non-trivial structure of Fig. 12b.

3.2 *Survival Probability*

The experimental setup described in Ref. [28] allows to measure not only the atom population of the branches, but also the population that lies in the crossing region for some time. Therefore, we can define the escape time as the time spent by atoms in a region of radius 3σ centered in $(0, 0)$, which we call the scattering region. We also define the survival probability as the probability P of finding an atom at a time t in the scattering region, which exactly corresponds with the measurements made in experiments.

Depending on the hyperbolic or non-hyperbolic nature of the system, the survival probability is expected to present exponential or algebraic decay for long times. In numerical simulations, we normalize time dividing by $t_0 = x/v_{x_0}$, which is the time that a particle would take to cross the scattering region if there were only one laser, and we find curves of probability versus time like the ones depicted in Fig. 13. The first plateau of this curve reflects that all the particles take at least $t = t_0$ to escape the scattering region. After the plateau, we can see an exponential decay for short times

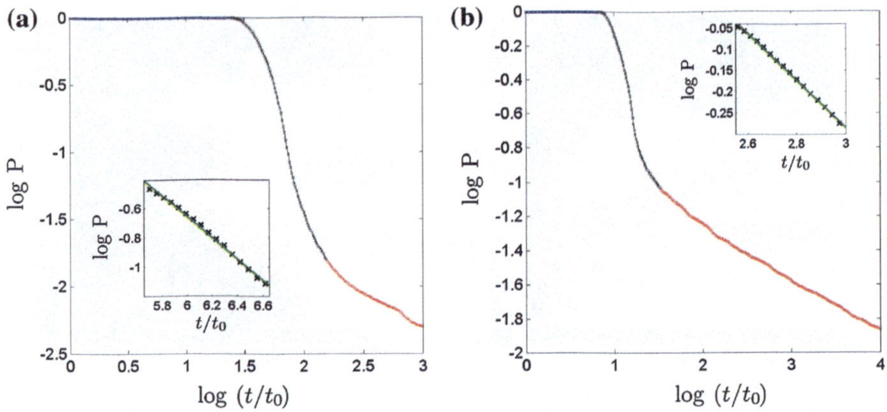

Fig. 13 **Survival probability as a function of time**. The survival probability P of the atoms in the scattering region as a function of time. After the initial plateau (in blue), there is an exponential decay (see inset) and for very long times an algebraic decay (in red). **a** $v_x = 0.3$. **b** $v_x = 0.9$. The rest of parameters are $\theta = 45°$, $v_y \in [-0.5, 0.5]$, $y \in [-0.5, 0.5]$ and $x_0 = -250$

(see insets of Fig. 13). For very long times the decay is algebraic, a typical behavior of non-hyperbolic systems [32]. However, in real experiments we expect to see only the exponential decay for two reasons. The first one is that non-hyperbolic systems are structurally unstable [33, 34]. This means that the slightest perturbation provokes the change from algebraic to exponential decay for long times. The second reason is that in real experiments, the long time behavior is hard to follow because small atom populations are difficult to detect. Moreover, when a non-hyperbolic system is weakly perturbed the curve of probability versus time behaves as the first part of the non-perturbed system, that is, it shows an exponential decay characterized by the same mean-life τ [35].

4 Conclusions

In nonlinear dynamics, different tools are commonly used to gain knowledge of a system. For instance, Lyapunov exponents are used to characterize its dynamics. On its behalf, basins of attraction contain much information about the asymptotic behavior of the system. Some efforts had already been made in the past to characterize the complex structure of basins of attraction, such as the uncertainty exponent [7] and the notion of basin stability [9]. The uncertainty exponent takes into account the nature of the boundary between two basins, and the basin stability informs about the percentage of phase space occupied by each basin. However, in many situations these concepts are insufficient to describe the complex structure of the basins of attraction [19].

The basin entropy integrates these concepts from the theoretical perspective of information entropy. It provides a quantitative measure of the uncertainty associated to the basins of attraction for a given scaling box size. This should become a very useful tool with a wide range of applications, as exemplified by the different systems that we have used to illustrate this concept.

In particular, here we have shown how the basin entropy can be applied to investigate the dynamics of propagating matter waves. We have focused on a double guide configuration, where the atoms can escape through four different exits. So far, nonlinear dynamics had only been used as an approximation to explain *a posteriori* some results concerning the chaotic dynamics of the atoms. But here we propose to go far beyond. In real experiments, we can measure the atom population escaping through each branch, that is, we can measure the probabilities of the atoms of escaping through the different exits. Gathering this information through an appropriate Monte Carlo sampling, we can measure the basin entropy for a given set of parameters. This enables the characterization of the final state unpredictability associated to different experimental parameters. Using the same data, we have shown how the presence of fractal structures in phase space could be detected for a certain ranges of parameters. An interesting modification of the experimental setting would be the inclusion of more guides. Indeed, with more exits, the log 2 criterion would be more easily fulfilled, facilitating the detection of fractality.

In terms of applicability, the tools developed here can be used more systematically to investigate the efficiency and robustness of the different experimental regimes of the crossed beam configuration in order to use it reliably as part of a matter wave circuit [36]. The protocols that we propose have been designed for a direct implementation with state of the art experimental techniques.

Acknowledgements This work was supported by the Spanish Ministry of Economy and Competitiveness under Project No. FIS2013-40653-P and by the Spanish State Research Agency (AEI) and the European Regional Development Fund (FEDER) under Project No. FIS2016-76883-P. MAFS acknowledges the jointly sponsored financial support by the Fulbright Program and the Spanish Ministry of Education (Program No. FMECD-ST-2016). Financial support from the Programme Investissements d'Avenir under the program ANR-11-IDEX-0002-02, reference ANR-10-LABX-0037-NEXT is also acknowledged.

References

1. Nusse, H.E., Yorke, J.A.: Science **271**, 1376 (1996)
2. Aguirre, J., Viana, R.L., Sanjuán, M.A.F.: Rev. Mod. Phys. **81**, 333 (2009)
3. Kolmogorov, A.N.: Doklady Russ. Acad. Sci. **119**, 861 (1959)
4. Sinai, Y.G.: Doklady Russ. Acad. Sci. **124**, 754 (1959)
5. Adler, R.L., Konheim, A.G., McAndrew, M.H.: Trans. Am. Math. Soc. **114**, 309 (1965)
6. Hunt, B.R., Ott, E.: Chaos **25**, 097618 (2015)
7. Grebogi, C., McDonald, S.W., Ott, E., Yorke, J.A.: Phys. Lett. A **99**, 415 (1983)
8. Grebogi, C., Kostelich, E., Ott, E., Yorke, J.A.: Phys. Lett. A **118**, 448 (1986)
9. Menck, P.J., Heitzig, J., Marwan, N., Kurths, J.: Nat. Phys. **9**, 89 (2013)
10. Alexander, J., Yorke, J.A., You, Z., Kan, I.: Int. J. Bifurcat. Chaos **02**, 795 (1992)

11. Ott, E., Sommerer, J.C., Alexander, J.C., Kan, I., Yorke, J.A.: Phys. Rev. Lett. **71**, 4134 (1993)
12. Lai, Y.-C., Winslow, R.L.: Phys. Rev. Lett. **74**, 5208 (1995)
13. Daza, A., Wagemakers, A., Georgeot, B., Guéry-Odelin, D., Sanjuán, M.A.F.: Sci. Rep. **6**, 31416 (2016)
14. Alligood, K.T., Sauer, T.D., Yorke, J.A.: Chaos: An Introduction to Dynamical Systems. Springer, New York (1996)
15. Kennedy, J., Yorke, J.A.: Phys. D **51**, 213 (1991)
16. Daza, A., Wagemakers, A., Sanjuán, M.A.F., Yorke, J.A.: Sci. Rep. **5**, 16579 (2015)
17. Aguirre, J., Vallejo, J.C., Sanjuán, M.A.F.: Phys. Rev. E **64**, 066208 (2001)
18. Hnon, M., Heiles, C.: Astron. J. **69**, 73 (1964)
19. Blesa, F., Seoane, J.M., Barrio, R., Sanjun, M.A.F.: Int. J. Bifurcat. Chaos **22**, 1230010 (2012)
20. Epureanu, B., Greenside, H.: SIAM Rev. **40**, 102 (1998)
21. Sanjun, M.A.F.: Phys. Rev. E **58**, 4377 (1998)
22. Cassettari, D., Hessmo, B., Folman, R., Maier, T., Schmiedmayer, J.: Phys. Rev. Lett. **85**, 5483 (2000)
23. Renn, M.J., Montgomery, D., Vdovin, O., Anderson, D.Z., Wieman, C.E., Cornell, E.A.: Phys. Rev. Lett. **75**, 3253 (1995)
24. Müller, D., Cornell, E.A., Prevedelli, M., Schwindt, P.D.D., Wang, Y.-J., Anderson, D.Z.: Phys. Rev. A **63**, 041602 (2001)
25. Houde, O., Kadio, D., Pruvost, L.: Phys. Rev. Lett. **85**, 5543 (2000)
26. Dumke, R., Volk, M., Müther, T., Buchkremer, F.B.J., Birkl, G., Ertmer, W.: Phys. Rev. Lett. **89**, 097903 (2002)
27. Gattobigio, G.L., Couvert, A., Georgeot, B., Guéry-Odelin, D.: Phys. Rev. Lett. **107**, 254104 (2011)
28. Gattobigio, G.L., Couvert, A., Reinaudi, G., Georgeot, B., Guéry-Odelin, D.: Phys. Rev. Lett. **109**, 030403 (2012)
29. Torrontegui, E., Echanobe, J., Ruschhaupt, A., Guéry-Odelin, D., Muga, J.G.: Phys. Rev. A **82**, 043420 (2010)
30. Daza, A., Georgeot, B., Guéry-Odelin, D., Wagemakers, A.A., Sanjuán, M.A.F.: Phys. Rev. A **95**, 013629 (2017)
31. Aguirre, J., Sanjuán, M.A.F.: Phys. Rev. E **67**, 056201 (2003)
32. Lai, Y.C., Tél, T.: Transient Chaos. Springer, New York (2011)
33. Seoane, J.M., Sanjuán, M.A.F., Lai, Y.C.: Phys. Rev. E **76**, 016208 (2007)
34. Seoane, J.M., Huang, L., Sanjuán, M.A.F., Lai, Y.C.: Phys. Rev. E **79**, 047202 (2009)
35. Motter, A.E., Lai, Y.C.: Phys. Rev. E **65**, 015205 (2001)
36. Seaman, B.T., Krämer, M., Anderson, D.Z., Holland, M.J.: Phys. Rev. A **75**, 023615 (2007); Ryu, C., Boshier, M.G.: New J. Phys. **17**, 092002 (2015); Caliga, S.C., Straatsma, C.J.E., Anderson, D.Z.: New J. Phys. **18**, 025010 (2016)

Fireflies: A Paradigm in Synchronization

G.M. Ramírez-Ávila, J. Kurths and J.L. Deneubourg

1 Introduction

All living beings exhibit oscillatory behavior manifested internally in metabolic, cellular, and molecular processes. Among those, glycolytic oscillations observed in muscles and yeast, oscillations of cyclic AMP found in *Dictyostelium* amoebae, mitotic oscillations leading to cell division cycle in eukaryotes, the pulsatile hormone signaling, the calcium oscillation observed at the level of internal part of cells, and circadian rhythms [37]. Nevertheless, the animated matter is also able to manifest oscillatory features that can be perceived by the senses of other living beings and especially by humans. Among the temporal patterns that humans can recognize in other species, we can mention the locust mass migration, and several synchronous behaviors such as in chewing, chirping, breeding, and flashing [15]. Synchronization is a widespread phenomenon both in nature and in artificial systems; it consists in

G.M. Ramírez-Ávila (✉)
Instituto de Investigaciones Físicas, Universidad Mayor de San Andrés, Casilla 8635, La Paz, Bolivia
e-mail: gramirez@ulb.ac.be

J. Kurths
Institut für Physik, Humboldt-Universität zu Berlin, Robert-Koch-Platz 4, 10115 Berlin, Germany
e-mail: Juergen.Kurths@pik-potsdam.de

J. Kurths
Potsdam Institut für Klimafolgenforschung, P.O. Box 60 12 03, 14412 Potsdam, Germany

J. Kurths
Institute for Complex Systems and Mathematical Biology, University of Aberdeen, Aberdeen AB24 3FX, UK

J. Kurths
Department of Control Theory, Nizhny Novgorod State University, 606950 Nizhny Novgorod, Russia

J.L. Deneubourg
Center for Nonlinear Phenomena and Complex Systems, CP231, Boulevard du Triomphe, 1050 Brussels, Belgium
e-mail: jldeneub@ulb.ac.be

M. Edelman et al. (eds.), *Chaotic, Fractional, and Complex Dynamics: New Insights and Perspectives*, Understanding Complex Systems,
https://doi.org/10.1007/978-3-319-68109-2_3

the adjustment of the time scales among oscillators due to a weak coupling [73] implying the emergence of structural order analogous to phase transitions [63]. Synchronization has deserved a lot of interest in the last two decades in which, thousands of papers, extensive reviews, and books treated this phenomenon related to chaotic aspects [5], and its applications to living [69], communication [47], networks [61], and mechanical systems [70]. There is also an excellent popularization book dealing with the most important features of synchronization and surely with fireflies [91]. Perhaps, the human light perception is the primary and most important connection with the environment. The sunlight, the moonlight, the brightness of different objects in the sky, the fire, and the light of certain animals, have undoubtedly triggered an intense curiosity in the human beings. The above-mentioned phenomena were not only observed by humans, but they intended to explain how and why those events occur, and they also tried to manage them finding numerous applications leading to a better lifestyle, establishing in this way, a qualitative difference with other species. Among the light emitting animals, fireflies possibly constitute the most charismatic and typical behavior because of their proximity, accessibility, and innocuity for humans. Several works have been carried out to describe, classify and study fireflies originating publications about taxonomy [50, 60], geographical distribution of certain species [32], diverse issues on the light organ [11], and recently genetic analyses associated to phylogenetics and systematics of some species [3] or in evolutionary problems linked to the flashing [64]. Throughout this chapter, we explain many biological, chemical, physical and mathematical aspects to unravel the firefly flashing synchronization, its consequences and potentials. In Sect. 2, we cover the bioluminescent features such as flashes colors and spectra of the fireflies [32], describing them under different perspectives going from their chemical aspects related to enzymes [66] or oxidation mechanisms [8] to their physical ones in connection with the influence of static magnetic fields [43]. In Sect. 3, we give the basis to understand why the evolution drives the fireflies to synchronize and the functional interpretation of this behavior [59] mainly implying communication aspects [16] in flash pattern recognition [14] as a fireflies' "language" related to courtship [13]. The heart of this chapter constitutes the physical-mathematical approach attempting to explain how and why diverse species of fireflies synchronize, the latter is considered with certain detail in Sect. 4 where we start with a toy model called the "solitary flash" [90]; then, we consider multi-agent systems (MASs) based models using well-known platforms and dealing with features related to firefly-inspired synchronization [7]. Thereafter, the details of phase and relaxation oscillators as the first approach for understanding fireflies synchronization [1, 23, 99] are given. Before to end Sect. 4, synchronization of pulse-coupled oscillators (PCOs) are described [62, 67, 76] mainly from a biological application perspective emphasizing the family of integrate-and-fire oscillators (IFOs) [42]. In Sect. 5, a consistent explanation of the phenomenon of response to synchronization [31] as a complete process of courtship is done supported by experimental results using artificial flashes [14, 68] and by a formal model [75, 78]. Finally, in Sect. 6, we explore the firefly-inspired synchronization and its applications including evolvable systems [92], wireless and other technological networks [7, 34, 54, 95], electronic and robotic devices [7, 21, 28, 76,

85] or even in improving light extraction efficiency [4]. All the aspects mentioned above give us a large bundle of concepts, models, and applications related to fireflies' collective behavior which constitutes a paradigmatic example of synchronization.

2 The Light of Fireflies

The first question that arises concerning the glowing insects is how and why fireflies emit light flashes. First of all, it is important to grasp the mechanisms leading to fireflies bioluminescence. We can approach this phenomenon considering three viewpoints: (i) Phenomenologically, in which, macroscopic aspects of the light are important such as the functionality of the emission comprising defense, offense, communication and propagation [98]. (ii) Chemically, as firstly stated by Dubois in 1887 [25], the processes involved in the production of light might be seen as a complex machinery, where two key substances luciferin (LH_2) and luciferase (Luc) allow the phenomenon to takes place in the presence of oxygen (O_2). This mechanism is common to several species exhibiting bioluminescence. (iii) Physically, the reactions in which single-electron-transfer seems to be essential for the mechanism leading to the production of the photon involved in bioluminescence [8].

The curiosity for the lightning innards of some living beings, in particular for fireflies that are the most accessible to the sense of sight triggered scientific work with the aim of deciphering the involved processes leading to firefly flashing. Detailed and systematic observations of flashing fireflies started perhaps with the experiences of Leconte in 1881, who described in detail the American Lampirydae [50]. Almost at the same time, in France, Dubois studied the bioluminescent beetles called *Pyrophorus* [24] and also the Pholadidae, a family of bivalve mollusks [25] trying to understand the "light production" in these animals through observing the mores, the morphological features, and the structure of the luminous organ. As above mentioned, Dubois identified the essential substances for the chemical reactions conducting to the light emission, and also generalized the idea that these chemical substances might be similar in different phosphorescent animals [26]. Several qualitative observations have been made in diverse fireflies species. Firstly, studying mating behaviors [65], then focusing on the control of flashing in fireflies [41], or on the effects of chemical compounds [36] and physical variables on the flashing features, such as temperature [27], light [20], static magnetic fields [44] or pulsed ones [43].

From a chemical point of view, the term bioluminescence introduced by Harvey in 1916 [40] is a fundamental concept to explain the cold light emission by living organisms, in particular by fireflies. The chemical aspects related to fireflies light emission had started to be developed since the origin of the identification of LH_2 and Luc. There were some works where these types of substances were extracted to show their phosphorescence properties [38]. After the discovery of these compounds in fireflies [39], scientists realized their importance and significant role in the production of light. The mechanisms of bioluminescence were unraveled both generically in all the living beings exhibiting this feature [88, 98], and particularly in fireflies [66]. It is possible to summarize the bioluminescent processes by simple chemical reactions as those shown below:

$$LH_2 + O_2 \xrightarrow{\text{Luc}} \text{Oxyluciferin} + CO_2 + h\nu \tag{1}$$

$$\text{Photoprotein} + Ca^{2+} \longrightarrow \text{Protein-coelenteramide} + CO_2 + h\nu, \tag{2}$$

where the first reaction depicts the oxidation of luciferin giving place to the protein-bound Oxyluciferin and a photon with frequency ν. On the other hand, in the second reaction, a conventional Photoprotein activated by calcium ions results into protein-bound coelenteramide. Specifically for fireflies, the reactions are

$$LH_2 + ATP + Luc \longrightarrow Luc \cdot LH_2 - AMP + PP_i \tag{3}$$

$$E \cdot LH_2 - AMP + O_2 \longrightarrow Luc + AMP + CO_2 + \text{Oxyluciferin} + h\nu, \tag{4}$$

with a wavelength photon $\lambda_{max} = 560\,\text{nm}$ (yellow-green).

Finally, from a physical viewpoint, apart from some variables susceptible to affect the firefly flashing (temperature, light, static and pulsed magnetic fields, etc.); there are microscopic aspects especially in relationship with the process of oxidation. More specifically, the oxygen supply mechanism [94] or the single-electron-transfer pathway for the critical oxidative process [8]. These microscopic studies are carried out using modern experimental techniques such as synchrotron phase-contrast microtomography and transmission x-ray microscopy.

Certainly, chemiluminescence and bioluminescence are closely related and although, in a first stage, the efforts were devoted to the explanation of the chemical processes giving rise to luminescence in some living beings. Nowadays, several works are contributing not only to unravel the mechanisms of bioluminescence but also to find applications in several fields including clinics, imagery, drug discovery, genetics, forensics, environmental monitoring, and conservation of cultural heritage among others. A detailed description of the above-mentioned applications is done in [81]. Recently, the relationship of firefly light-extraction efficiency [4], inspired similar mechanism for light-emitting diodes. The Nobel prize 2008 in chemistry awarded to Osamu Shimomura, Martin Chalfie, and Roger Y. Tsien for the discovery and development of the green fluorescent protein, GFP constitutes a milestone in bioluminescence and also a reward for the efforts performed in advance of the knowledge of the phenomenon firstly observed in fireflies.

3 Why Fireflies Synchronize?

The succession of observations made to conclude that the primary function of synchronization in fireflies is that of courtship has meant a long and tortuous road that has often led to incorrect interpretations. Fortunately, the scientific approach was imposed, and it is now possible to state that fireflies collective flashing is perhaps the most cited phenomenon as an example of synchronization.

3.1 First Encounters with Synchronous Fireflies

As stated by Roda [82], the oldest known written document describing qualitative and nonsystematic observations on flashing fireflies and glow-worms were made in China, dating roughly from 1500 to 1000 BCE. The knowledge and observation of fireflies were common in several cultures such as the Mayas where they played a role in religious practices and also in mythology associated with the cigar smoking. However, there is no doubt that the most interesting phenomenon when sighting ensembles of fireflies, is the ability that certain species exhibit to attain collective synchronous flashing. The first reported observation on firefly-synchronization is due to Engelbert Kaempfer a Dutch physician, naturalist, and explorer born in Westphalia, in the present German territory. As a result of his observations during his voyage to Japan and Siam (1690–1692) he wrote (quoted in [10]):

> The Glowworms (*Cicindelae*) represent another shew, which settle on some Trees, like a fiery cloud, with this surprising circumstance, that a whole swarm of these Insects, having taken possession of one Tree, and spread themselves over its branches, sometimes hide their Light all at once, and a moment after make it appear again with the utmost regularity and exactness, as if they were in perpetual Systole and Diastole.

The latter constitutes the first description of synchronization in a large population of coupled oscillators. After that, several observations of firefly synchronization were reported among the most interesting; we only mention the citation due to Theobald [93] who based in a comment about the unison light of fireflies, wrote:

> In Pegu, however, I have witnessed the exhibition in question; myriads of fireflies emitting their light, and again relapsing into darkness, in the most perfect unison .... The bushes overhanging the water were one mass of fireflies .... The light of this great body of insects was given out ...in rhythmic flashes, and for a second or two lighted up the bushes in a beautiful manner; heightened, no doubt, by the sudden relapse into darkness which followed each flash. There are the facts of the case (and I may add that it was towards the end of the year) and the only suggestion I would throw out, to account for the unusual method of luminous emanation, is that the close congregation of large numbers of insects, from the small space afforded them by the bushes in question, may have given rise to the synchronous emission of the flash by the force of imitation or sympathy.

Buck in 1938 cited more than 30 reports on synchronous fireflies [10] with observations carried out in different locations of the planet (e.g. Siam, Burma, Singapore, Borneo, Malaya, Philippines, New Guinea, Jamaica, Mexico, United States, and Brazil). In this historic article, he also pointed out some explanations for firefly synchronization, including the insubstantial ones (the wind and other environmental influences, twitching eyelids [49], the effect of the sap of the trees, accident, illusion, sense of rhythm or "sympathy", and leader or pacemaker). None of these explanations were adequate to resolve the phenomenon of synchronous flashing. It is important to call attention to the fact that the statements mentioned above do not consider any functionality in the achievement of firefly synchronization. From an energetic point of view, the lack of functionality of synchronization could be regarded as an inefficient, useless and unjustified process. Evidently, a more detailed study of firefly synchronization showed that this phenomenon is significant for the survival of many of these species since, as it will be seen later, synchronization allows the identification of particular species as well as an intersexual communication language.

3.2 *Synchronization for Courtship and Mating*

During the first two decades of the twentieth century, several observations done principally by Mc Dermott had shown that the light emission in some American Lampyridae species, evinced the flashing (photogenic function) importance as a mating adaptation [65]. However, there is not any mentioning to the synchronous flashing already reported by numerous scientists as stated above. The discovery of the synchronization functionality in fireflies is due to Buck in 1935 who established that firefly synchronous flashing is related to the mating and it is persistent until copulation is produced or when there are no more unfertilized females capable of responding to synchronized males [9]. After that, the idea that synchronization is associated with processes of conspecific recognition and mating evolved until it was strongly accepted in the phenomenology of fireflies. A considerable number of studies and publications testified to the acceptance that the synchronization phenomenon is closely related to courtship and mating in fireflies. Concerning the topics discussed considering the aforementioned association, we highlight those related to mating protocols of synchronously flashing fireflies [14, 57], and to different aspects of flash communication [18, 59]. The well established and cumulated knowledge on firefly synchronization allowed a deeper qualitative understanding of this phenomenon and the popularization of these facts in excellent books such as those written by Lewis [51], and Ortler [71]. The formulation of mathematical models to explain firefly synchronization also played a significant role in improving the insight of the phenomenon; details of the most well-known ones are presented in Sect. 4.

4 Models to Explain the Fireflies' Synchronous Behavior

Several models intend to reproduce and explain the fireflies synchronous behavior. The basis of each model has different motivations that can go from simple guidelines of a game [90], via simple interaction rules [97] and mathematical considerations [67, 99] to more sophisticated analysis, where phase response curves (PRC) and Arnold's tongues are useful tools for describing dynamical features and synchronization.

4.1 A Toy Model

The so-called "solitary flash" game is a friendly and straightforward model leading to an explanation of firefly synchronous flashing behavior. Originally, the game has been proposed by Stewart and Strogatz [90] and its simple rules as they were raised:

1. The game board consists of a polygon of n sides, each of them containing r boxes, i.e. $N = n \times r$ boxes on the board.
2. The first box plays the role of the flashing box, i.e. when a player (firefly) arrives at this box, it flashes.
3. Each firefly starts the game in any box (initial condition) except the flash one.
4. Each firefly advances clockwise one position per time step.
5. When a firefly flashes, it remains in the flash box one-time step, while the other fireflies go forward according to the place in the board in which they are. For instance, if the firefly is on a box of the first side, it continues to advance one position; on the other hand, if the firefly is on a box of the second side, it advances two spaces, and three if it is on a box of the third side and so on.
6. The goal of the game is that all fireflies flash synchronously in the shortest possible time.

The above-mentioned rules might allow or not the occurrence of synchronization. The dynamics of the game strongly depends on the initial conditions and also in rule 5 because it determines what happens to a firefly when it approaches, arrives or passes through the flash box. We consider the following four variants of rule 5:

(a) When a firefly is in a box located on the far side of the polygon, it could happen that it might overtake the flash box and consequently without flashing in its cycle. This fact imposes a difficulty in attaining of synchronization.
(b) When a firefly restarts a new cycle, it is mandatory that it arrives at the flash box and as a result, it flashes in each cycle. This rule facilitates the achievement of synchronization.
(c) When two or more fireflies are nearby the flash box, they wait until all of them are effective in this box; at this moment, all the fireflies advance one position. This modification respect to (b) makes it easier to attain synchronization.
(d) Finally, if we consider a similar situation than in (c) but with the modification that when the fireflies are forced to be in the flash box, the other fireflies advance according to rule 4 and not only one position as in the precedent case.

Some frames of the game evolution reflecting the rules (a)–(d), and some other possibilities for the cardboard are shown in Fig. 1.

Time series for all cases (a)–(d) are shown in Fig. 2, where we considered three fireflies and a specified set of initial conditions $(n_{01}, n_{02}, n_{03}) = (5, 10, 15)$. The basins of attraction for these cases are depicted in Fig. 2e–h where the condition is related to the lasted time to achieve synchronization with collective simultaneous flashing. It is interesting to observe in Fig. 2a that the fireflies do not reach the synchronization condition of simultaneous flashing of all the individuals of the system.

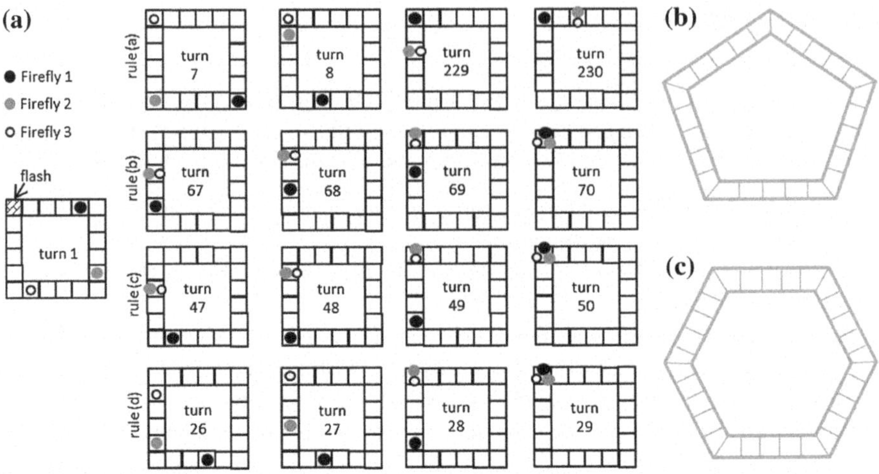

Fig. 1 **a** Frames of the game evolution for the rules (a)–(d) considering for all the four cases the same initial conditions $(n_{01}, n_{02}, n_{03}) = (5, 10, 15)$. Some other possibilities for the cardboard: **b** pentagon, and **c** hexagon

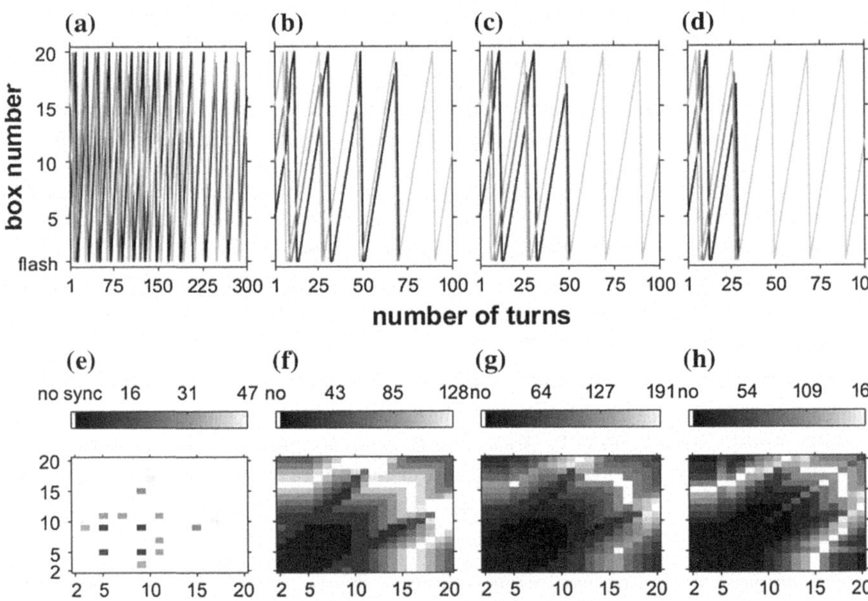

Fig. 2 First row: **a–d** Time series for the evolution of three fireflies following the rules (**a**)–(**d**) above mentioned. Second row: **e–h** Basins of attraction (n_{02} vs. n_{01}) for the rules (**a**)–(**d**) considering that the initial condition for the first firefly is the box 5. White boxes represent situations in which simultaneous, collective, and persistent flashing (in every cycle) are not achieved

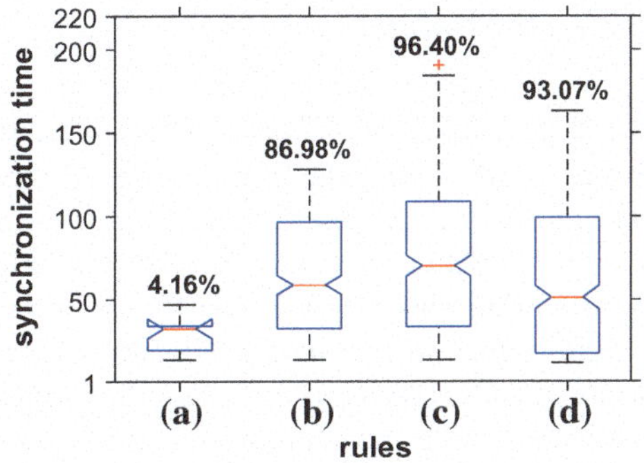

Fig. 3 Box plot of the statistical parameters (median and quartiles) related to Fig. 2 for the synchronization time associated with each of the rules. The percentage of the synchronous events are shown in the upper part, above the whisker. With all these information, it is possible to estimate which rule leads more often and quickly to synchronization

Nevertheless, considering the formal definition of synchronization, the phase difference among the fireflies remains constant from the 228th turn; as a consequence, exhibiting synchronization but fireflies 1 and 2 do not flash in every cycle. On the contrary, Fig. 2b–d show that simultaneous flashing is achieved respectively in 70, 50, and 29 turns (time steps). So that, in principle it indicates that in successive order the rules that facilitate synchronization are respectively (d), (c), (b), and (a). With a view to sustaining the recent affirmation, we compute statistical parameters related to the synchronization time when all fireflies flash simultaneously, as well as the percentage of this situation in each basin of attraction for the indicated cardboard related to Fig. 2 (four sides and five effective boxes per side). The results are shown through a box plot in Fig. 3 where its information supports our assumption related to facilitated synchronization. This simple game with its four basic rules (algorithms) allows to explain synchronization in some species of fireflies, and it has an intrinsic richness since these simple rules can be translated to a more technical language belonging to synchronization theory. Thus, it is possible to find a relation between the number of sides and boxes per side with the type of coupling: in the case explained above, we have an excitatory coupling between the fireflies (players). In some cases, as it is described in Sect. 4.5, for Light-controlled oscillators (LCOs) and for some species of fireflies, inhibitory coupling is also possible; this type of coupling might be incorporated in the game considering that for some boxes on the game board, the player must go back certain number of boxes according to its position. The high number of game choices (shape of the polygon, number of squares for each side of the polygon, the number of players, basic rules of the game, etc.) allow to extend the study of this model and analyze its isochronous dynamics as in [53].

4.2 Multi-agent Based Models

Conceptually, a MAS reflects one of the complex systems basis, namely the cooperativity due to the interactions among the components of the system giving rise to the accomplishment of a task or the emergence of a new property or functionality of the system. As stated in Sect. 1, several platforms are allowing to work with MASs. To explain the potentialities of MASs, we focus on the platform NetLogo [97] whose library contains a nice firefly model. This model is mainly based on the flashing behavior of some species: *Pteroptyx cribellata, Luciola pupilla*, and *Pteroptyx malaccae* described in [12], it has been built with simple interaction rules taking into account two main synchronization strategies: phase delay and phase advance (first and second columns of Fig. 4 respectively) [96]. The agents of the model are fireflies having as a main feature that they have their own period, and a cyclic behavior, i.e. each firefly has an initial period and position as well as its position in the cycle; in most cases, the fireflies period are initially considered as identical for the whole population. The interaction of fireflies is done by means of the flashes that they can emit (influencing the rest of the individuals) or perceive (being affected in their dynamics). Synchronization might be achieved according to the rules and the parameter values. SpikingLab is another interesting application done in NetLogo and related in some way to fireflies; actually to the integrate-and-fire oscillators, a popular model describing synchronization in fireflies (explained in more detail in Sect. 4.4). This NetLogo project introduces a Spiking Neural Network (SNN) phenomenological model mimicking the neural dynamics regardless of the biophysical processes [46]. As a consequence, all the neural features such as membrane and resting potential, spike threshold, inhibitory and excitatory postsynaptic response, exponential decay rate and refractory and absolute periods are embedded in two possible states: open and absolute refractory. The model is used for simulating a virtual insect able to process three types of information: visual and sensations related to pleasure and pain.

One of the most important applications of firefly synchronization is that related to communication networks and algorithms allowing a synchronous behavior on certain devices. In general, the above-mentioned applications use MAS concepts and programming. For instance, another way to solve the firefly synchronization task was carried out by Teuscher and Capcarrere, using two-dimensional (2-D) cellular automata (CA) and random boolean networks [92], programming in such a way; the performing computations are locally and based on co-evolution. They have also implemented employing an FPGA-Evolware. In other words, software and hardware implementation have been designed for solving the firefly synchronization task successfully. Other works dealing with synchronization in a framework of MAS have been mostly developed with PCOs. There are some other applications based on firefly synchronization and MAS that are pointed out in Sect. 6.

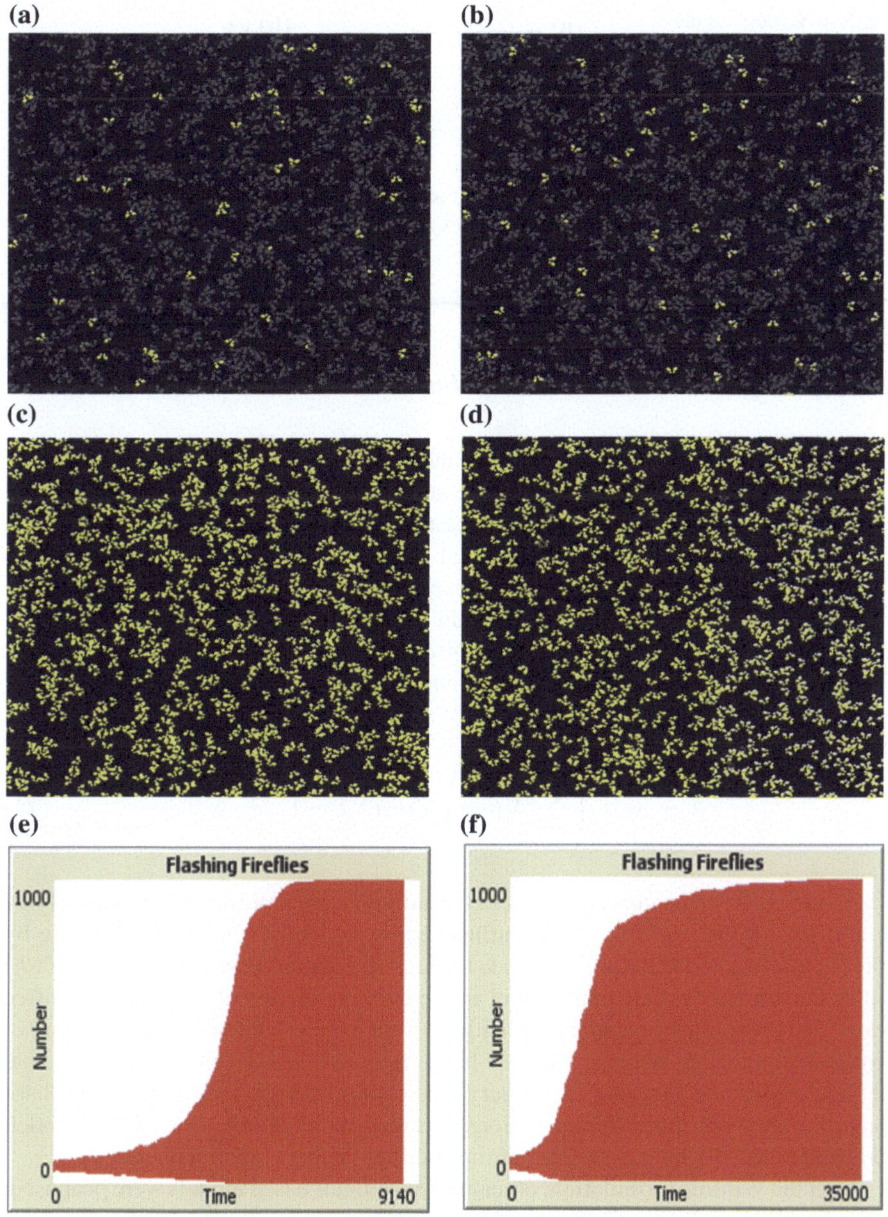

Fig. 4 Screenshots of NetLogo fireflies model interface using the strategy of phase delay (first column), and phase advance (second column). The general parameters are: number of fireflies = 1000, flash length = 2, flashes to reset = 2, and cycle-length = 35. **a**, **b** represent the arena showing the initial situation with few number of fireflies flashing simultaneously; **c**, **d** the final one where almost all fireflies are flashing in synchrony; and **e**, **f** the time series showing the number of fireflies versus the time steps. In both cases, the program runs until synchronization is fully established. For the delay strategy, the synchronization time is around 6200-time steps, and for the advance one, it is around 31000-time steps

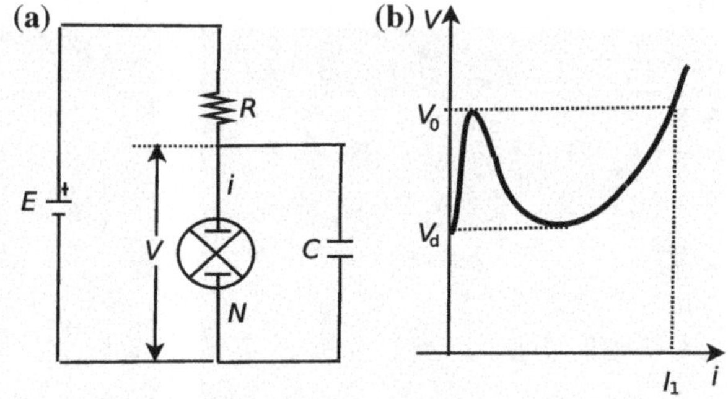

Fig. 5 **a** Scheme for the sawtooth oscillator, being the neon tube N the heart of the oscillator. **b** Current-voltage characteristic of N

4.3 Phase and Relaxation Oscillators

After a systematic study of the biological and chemical properties of flashing fireflies and their synchronous behavior, Winfree attempted to build a mathematical model capable of describing synchronization in large populations of *phase oscillators* and considering that each firefly is represented by an oscillator of this type [99]. Let us start by understanding the features of these oscillators. A phase oscillator might be considered as an oscillator whose periodic solution travels around a circular limit cycle with angular velocity $\Omega(r^*)$. Winfree proposed a model of coupled phase oscillators characterized by a sensitivity and an influence function, depending only on the phase. Mathematically, the Winfree's model might be written as:

$$\dot{\theta}_i = \omega_i + \left(\sum_{j=1}^{N} X(\theta_j) \right) Z(\theta_i), \quad i = 1, \ldots, N, \tag{5}$$

where the phase and natural frequency of oscillator i are represented respectively by θ_i and ω_i. All the oscillators i are influenced by oscillator j through the phase by $X(\theta_j)$; as a result, oscillator i responds through the phase-dependent function $Z(\theta_i)$ called the sensitivity. Using this model applied to a population of oscillators, Winfree found that there is a sort of phase transition towards the synchronization. Some other works dealing with Winfree's model have been developed and in particular that of Ariaratnam and Strogatz [2] is very interesting because they obtain the phase diagrams showing the different regions of the dynamical behavior of this model: total and partial locking, total and partial oscillation death, and incoherence.

After the Winfree formulation, other phase oscillator based models were proposed, one consisting of 25 sawtooth coupled oscillators with an experimental realization. Each one of these oscillators consists of a neon tube N connected to a battery E through a resistance R and shunted by a capacitance C as shown in Fig. 5a. The functioning of the oscillator is determined by the current-voltage characteristic (Fig. 5b). The oscillation is produced as follows: the condenser charges until it reaches the voltage V_0 and the oscillator fires generating a current I_1 during the discharge of C until V_d and then restarting the charging process; so that, giving rise to oscillation.

Following the intuitive model of Winfree, Kuramoto developed a more formal model carefully described in [48], where he used perturbation methods, weak coupling $K \geq 0$, almost identical oscillators, and the concept of mean field coupling to obtain its famous equation:

$$\dot{\theta}_i = \omega_i + \frac{K}{N} \sum_{j=1}^{N} \sin\left(\theta_j - \theta_i\right), \quad i = 1, \ldots, N. \tag{6}$$

The Kuramoto model with its purely sinusoidal coupling constitutes the simplest possible case of equally weighted, all-to-all coupled oscillators. The model can also be described in terms of the order parameters r and ψ, resulting:

$$re^{i\psi} = \frac{1}{N} \sum_{j=1}^{N} e^{\theta_j}, \tag{7}$$

where $r(t)$ is a normalized function that is a measure of the oscillators population coherence, and ψ is the average phase. Using Eq. (7), the original model states:

$$\dot{\theta}_i = \omega_i + Kr\sin(\psi - \theta_i), \quad i = 1, \ldots, N, \tag{8}$$

indicating that each oscillator is coupled to the common average phase with coupling strength given by Kr [1]. Under the consideration of certain assumptions, it is possible to find a critical value for the intensity of coupling K_c denoting a bifurcation point and also that for $K > K_c$, there is a dramatic increase in the coherence of the oscillators population when the bifurcation is supercritical. There are several reviews of the importance of Kuramoto model for synchronization and networks [1, 83] and also some recent extensions including adaptive frequencies [72].

Another firefly-inspired model is due to Ermentrout [29] who proposed a mechanism that allows the fireflies to synchronize at a nearly zero phase difference. The model has a PRC that it is the same as the determined for *Pteroptyx malaccae*, but the behavior under a train of periodic stimuli is different due to the adaptive character of the model. Firstly, he considered a single periodically forced oscillator stated by:

$$\dot{\theta} = \omega + P(t/\tau)\Delta(\theta) \quad \text{mod } (1), \tag{9}$$

where ω is the oscillator's natural frequency, $P(\phi)$ is the periodic forcing stimulus, and $\Delta(\theta)$ is the oscillator PRC. All functions are one-periodic. Under certain assumptions, it is possible to average Eq. (9) and obtain:

$$\dot{\theta} = \omega + H(t/\tau - \theta), \tag{10}$$

where H depends on the PRC as

$$H(\phi) = \int_0^1 P(s)\Delta(s - \phi)\mathrm{d}s. \tag{11}$$

The 1:1 phase-locked solutions of Eq. (9) are:

$$\theta_n \to \tilde{\theta},\ \Delta(\tilde{\theta} + \omega\tau) = 1 - \omega\tau. \tag{12}$$

The latter can be solved considering that $1 - \omega\tau$ does not exceed the maximum of Δ or fall below the minimum of Δ, giving:

$$\tilde{\theta} = 1 - \omega\tau + \Delta^{-1}(1 - \omega\tau). \tag{13}$$

If $\Delta(0) = 0$, then $\omega\tau \equiv 1$, means that the intrinsic frequency is the same than the forcing one. This model for N coupled oscillators might be written as:

$$\dot{\theta}_i = \omega_i + \sum_{j=1}^{N} H_{ij}(\theta_j - \theta_i). \tag{14}$$

The main result of the analysis of Eq. (14) states that there can be phase-locking but with phase differences not necessarily close to zero.

Relaxation oscillators are one of the most suitable models to study systems susceptible to synchronize: neurons, cardiac cells, and fireflies. An important feature of these oscillators is that within each cycle, there are two time scales: a slow one where an integration process takes place, and a fast one where a firing process occurs. Contrarily to the common phase oscillators, the relaxation oscillators waveform is very different from a sinusoidal wave; rather it looks like a sequence of pulses. There is no universal model for relaxation oscillators, having each of one their proper characteristics [73].

Numerous examples of relaxation oscillators may be found in literature, ranging from electronic devices generating relaxation oscillations [74] to those applied to biology, especially in neurons [45]. One of the most classical examples of a self-oscillating system is the van de Pol equation described by the equation of motion

$$\ddot{x} - \mu(1 - x^2)\dot{x} + x = 0, \tag{15}$$

where for large μ behaves as a relaxation oscillator (Fig. 6a). Equation (15) may be cast into a set of first-order differential equations:

$$\dot{x} = \mu\,[y - F(x)] \tag{16a}$$

$$\dot{y} = -\left(\frac{1}{\mu}\right)x, \tag{16b}$$

that allows us to observe the following: the x-nullcline given by the relation $y = F(x)$, has a cubic form and the y-nullcline, provided by the expression $x = 0$ is a vertical line. Both nullclines, as well as the corresponding limit cycle, are illustrated in Fig. 6c. This system has one fixed point, located at the origin, where the two nullclines cross one another. The motion along the limit cycle trajectory involves two time scales, a fast horizontal movement, and a slow vertical motion. When y is near the x-nullcline, both $\mathrm{d}x/\mathrm{d}t$ and $\mathrm{d}y/\mathrm{d}t$ vary gradually, and the movement is slow. When the trajectory departs from the cubic nullcline $\mathrm{d}y/\mathrm{d}t$ is large, and the horizontal movement is fast. Another well-known relaxation oscillator is the IFO that deserves our attention in Sect. 4.4 because it is one of the most used models to describe synchronization in fireflies.

4.4 Integrate and Fire Oscillators

These oscillators were extensively used to model a great variety of phenomena such as synchronization in fireflies [67] and several aspects of neuronal systems [30] among others. IFO models were also used to describe firing patterns [35] and critical phenomena [22] such as avalanches.

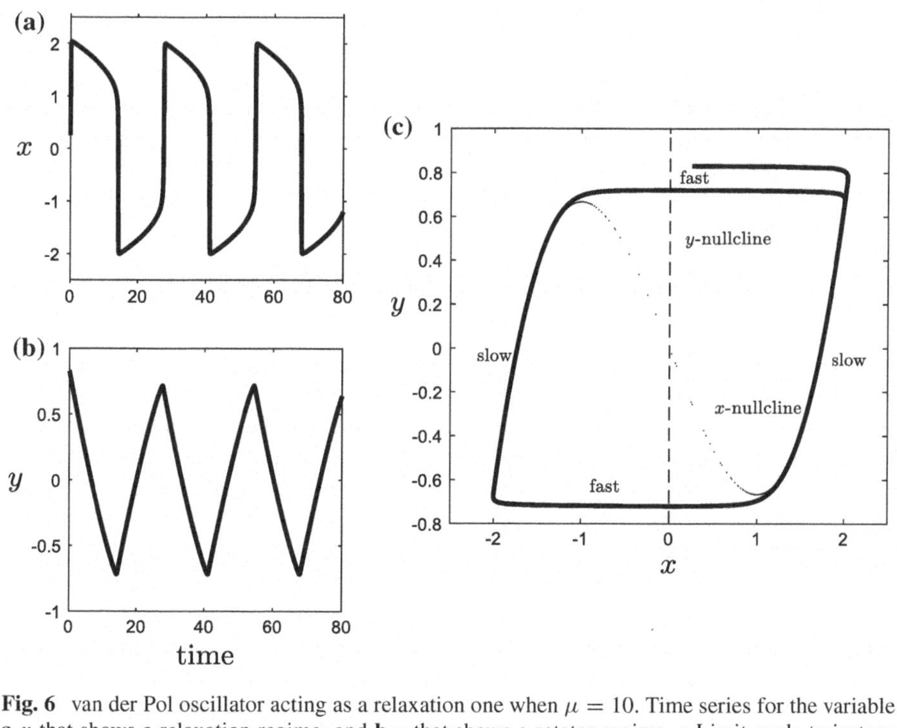

Fig. 6 van der Pol oscillator acting as a relaxation one when $\mu = 10$. Time series for the variable **a** x that shows a relaxation regime, and **b** y that shows a rotator regime. **c** Limit cycle trajectory and the accompanying fast and slow time scales

IFOs are principally used to describe collective behavior. To model self-synchronization of the cardiac pacemaker, Peskin considered a network of N IFOs, each characterized by a voltage-like state variable V_i, whose dynamics is:

$$\frac{\mathrm{d}V_i}{\mathrm{d}t} = I - \eta V_i, \qquad 0 \leq V_i \leq 1, \qquad i = 1, \ldots, N. \tag{17}$$

When the oscillator i reaches the threshold ($V_i = 1$), the oscillator "fires" and V_i is reset *instantaneously* to zero (Fig. 7a). The oscillators interact by a simple form of pulse coupling: when a given oscillator fires, all the other variables V_j, $j \neq i$ are increased by an amount β/N (the quotient by N is introduced in order to get reasonable behavior in the thermodynamic limit $N \to \infty$). That is,

$$\text{If} \quad V_i(t) = 1 \quad \Longrightarrow \quad V_j(t^+) = \min(1, V_j(t) + \beta/N), \quad \forall j \neq i. \tag{18}$$

Moreover, the oscillator at the state $V = 0$ (i.e. just after firing) cannot be affected by the others, so that the state $V = 0$ is absorbing. The latter ensures the possibility of perfect synchronization. To illustrate how this model works, we have numerically solved (18) for two mutually coupled IFOs and 500 globally coupled oscillators (see Fig. 7b), where we observe that at the beginning, each oscillator has its own natural frequency and as time goes by, groups of synchronous oscillators are formed, and finally, the entire population is synchronized, i.e. all the IFOs emitting their pulses simultaneously. Here, we have considered several phase oscillator models all of them related to the explanation of firefly synchronization in its simplest form. All these models could be applied to other oscillatory systems, but their original goal was to understand how and why fireflies synchronize.

4.5 *Light-Controlled Oscillators: Electronic Fireflies*

As it has been stated above, most of the analytical models of fireflies synchronization were based on mathematical equations driven the system of coupled oscillators to synchronize. Nevertheless, some experimental devices allow the study of synchronization by carrying out careful experiments. One of these devices is the LCO that also receives the appellation of *electronic firefly*. From a technical viewpoint, each LCO consists of an LM555 chip wired to function in its astable oscillating mode (Fig. 8a). The alternations of it are determined by a dual RC circuit in parallel with four photo-sensors that allow the LCO to interact with others by means of light pulses (Fig. 8b). Basically, an LCO is a relaxation oscillator in the sense that it has two time scales characterized by the binary variable $\varepsilon(t)$: within each cycle there are intervals of slow (charging stage, $\varepsilon(t) = 1$) and fast (discharging stage, $\varepsilon(t) = 0$) motion. The period is determined by the two external RC circuits and the output waveform takes the form of a pulse signal with minimum and maximum values set at $\frac{V_M}{3}$ and $\frac{2V_M}{3}$ respectively, V_M being the value of the supply voltage. These threshold

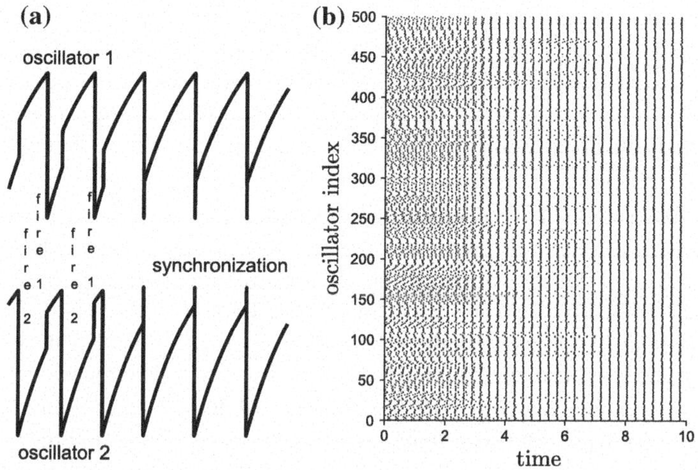

Fig. 7 **a** Synchronization mechanism in two coupled IFOs. **b** Points correspond to firing times to represent the dynamics of a population of 500 coupled IFOs showing the tendency towards complete synchronization. The parameter values used in both cases are $I = 2.5$, $\eta = 1$ and $\beta = 0.25$

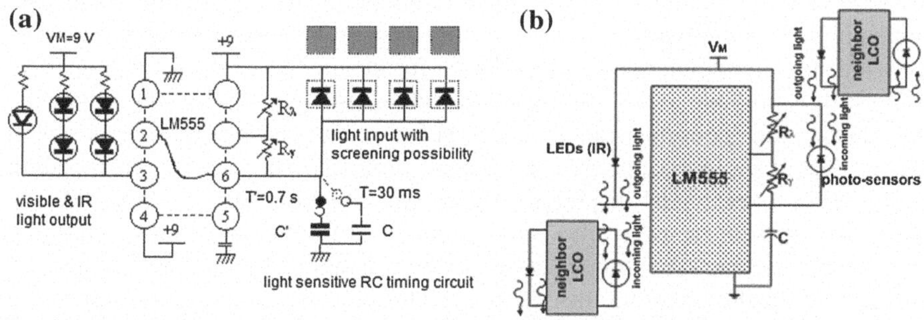

Fig. 8 **a** Block diagram of an LCO with the LM555 in its astable functioning mode. **b** Simplified diagram of the LCO and schematic view of the coupling between LCOs

voltages determine the value of $\varepsilon(t)$. If we consider a system composed of N LCOs, the equation we use to model the voltage evolution for the ith LCO is:

$$\frac{\mathrm{d}V_i(t)}{\mathrm{d}t} = \underbrace{\lambda_i(V_{Mi} - V_i(t))\varepsilon_i(t)}_{\text{charging term}} - \underbrace{\gamma_i V_i(t)[1 - \varepsilon_i(t)]}_{\text{discharging term}} + \underbrace{\sum_{j=1}^{N} \beta_{ij}\delta_{ij}[1 - \varepsilon_j(t)]}_{\text{coupling term}}, \quad (19)$$

where

$$\delta_{ij} = \begin{cases} 1, \text{ if } i \neq j \text{ and they may interact} \\ 0, \text{ otherwise} \end{cases}$$

indicates whether or not LCOs i and j interact. Note that the interaction term is active only when at least one of the other LCOs is discharging. In this model, we consider symmetric coupling, such that $\beta_{ij} = \beta_{ji}$. Another important experimental fact is the coupling distance-dependent, i.e. the coupling strength β is almost quadratically inverse with the distance, being the measured dependence: $\beta_{ij} \propto \frac{1}{r_{ij}^{\alpha}}$, being the power $\alpha = 2.11$.

Several experiments have been carried out with LCOs such as measurements of phase-locking and phase differences in various sets of locally coupled LCOs [76], transients [85] and determination of synchronous regions [84, 86]; in all of them, the model described by Eq. (19) has been validated. On the other hand, locally and globally coupled LCOs were studied comparing their dynamical features, finding astonishing changes in their dynamics, despite very small differences between the oscillators [80]. Studies dealing with the influence of noise on LCOs have also been done using uniform [77] and Gaussian [79] distributions, resulting in some impressive results. Such as those showing that white noise can enhance synchronization on a set of two LCOs under the condition that the noise acting on each of one has different variances, i.e. various noise intensities acting on each LCO.

From a biological point of view, the use of LCOs as a fireflies model, allows to describe realistic situations in which, there is a population of 10 fireflies distributed randomly both spatially and in what concerns their initial condition V_{i0} and considering a distance-dependent coupling (Fig. 9a for identical, and (b) for nonidentical fireflies). It is possible to describe synchronization employing polar plots where the radial and angular coordinate are related to period and to phase difference measured with respect to a reference firefly (in this case, firefly 2). The frames corresponding to Fig. 9a1–a4, and b1–b4 display the dynamical situations after 2, 250, 500, and 1000 flashing events respectively. Thus, for identical fireflies and after 250 flashing events, there are two synchronous clusters, one constituted by fireflies 7, 9, and 10, and the other by the rest of fireflies; these two clusters are maintained over time until 1000 flashing events. According to the fireflies positions into the arena, it does not surprise the emergence of the cluster formed by fireflies 7, 9, and 10, although it could also be expected that firefly 3 be part of it; but undoubtedly, the initial conditions made it possible to arrive at the situation shown in Fig. 9a4. On the other hand, for the configuration of nonidentical fireflies of Fig. 9b, it is observed that after 250 firing events, there are four groups of fireflies with the following distribution: (i) 1 and 8, (ii) 5 and 9, (iii) 10 and (iv) 2, 3, 4, 6 and 7 (Fig. 9b2). Synchronous cluster formation is interesting because it reflects in some way the position of the fireflies in the arena. As time lasts, groups (i) and (ii) are "absorbed" by (iv) (Fig. 9b3), situation that persists up to 1000 firing events (Fig. 9b4). Even tough, Firefly 10 has a very similar period to that of the large synchronous bunch, its phase difference is different. Again, the position of the fireflies seems to play an important role in the emergence of the synchronization.

Throughout this section, we have considered all the models are attempting to explain synchronization in fireflies as a self-organization process but without specific details on the features of each firefly species (except for the Ermentrout model). In

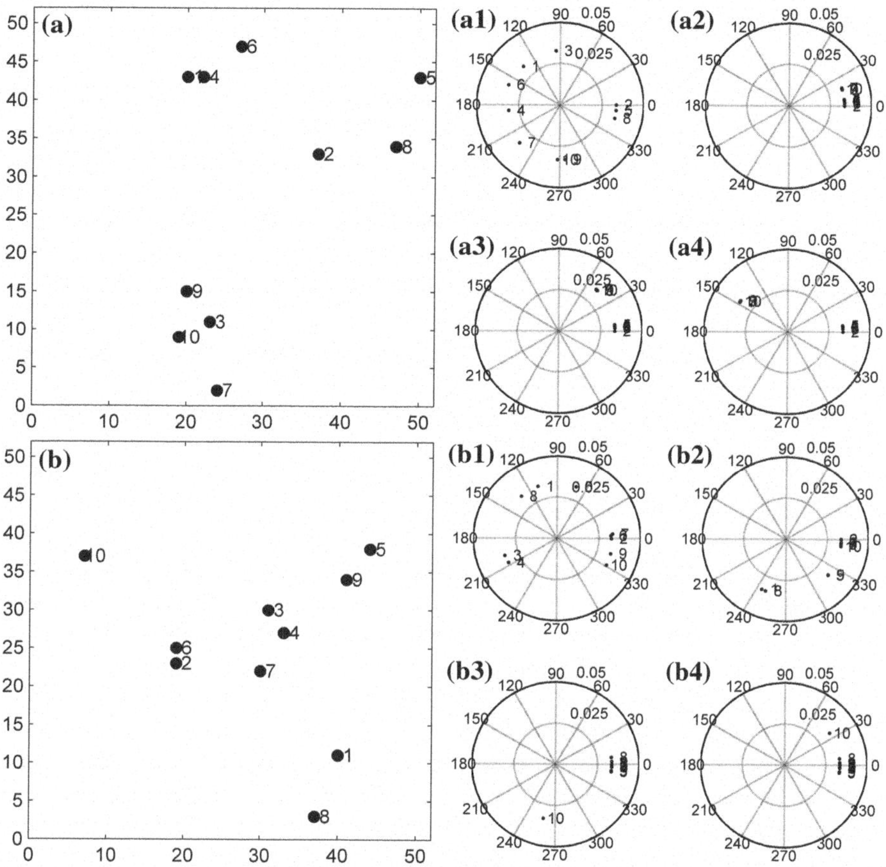

Fig. 9 Ensembles of 10 **a** identical and **b** nonidentical globally coupled fireflies with a distance-dependent coupling and randomly distributed in an arena consisting of 50 × 50 cells. Polar representation where the radial and angular coordinates are respectively related to the fireflies' period and phase difference. **a1–a4** Identical and **b1–b4** nonidentical LCOs. Frames showing the dynamical polar distribution after (**a1**) and (**b1**) 2, (**a2**) and (**b2**) 250, (**a3**) and (**b3**) 500, and (**a4**) and (**b4**) 1000 flashing events

general, most of the models consider only the oscillators ability to synchronize and that all of them have similar features (associated with the oscillators identity, e.g. males). Nevertheless, if we desire to understand the firefly courtship as a whole, it is necessary to take into account not only the emergence of males' synchronization but also the females' response, considering that the oscillators associated to each of the sexes are dissimilar to the other sex. Precisely in the next section we address this problem.

5 Response to Synchronization

Already in the beginnings of the century, the courting behavior of *Photinus pyralis* has been described and even some simple experiments to reinforce the males' behavior have been carried out. After that, the female response was observed and described in detail for *Photinus* fireflies [56], introducing the mechanism "flash-answer" that is part of a sexual function of bioluminescence and allows the conspecific identification, avoiding the interspecific one. Some exceptions must be pointed out that are related to the aggressive mimicry where females of the genus *Photuris* attract and devour males of the genus *Photinus* by mimicking the flash responses of *Photinus* females [55, 58]. The observation on *Photinus* flash-patterns describing their importance in fireflies communication, as well as their importance in courtship, deserved the attention and numerous publications came to light allowing a better understanding of the mating process as a whole [17, 19, 52, 59, 89].

In 2010, Moiseff and Copeland reported a surprising finding issued from experiments carried out with virtual males and a real *Photinus carolinus* female; they showed that males' synchronization is associated with the female's response [68]. The latter improved the knowledge concerning the synchronous behavior of fireflies, enhancing the fact that both males and females participate actively in the courtship. In other words, when referring to fireflies courtship, we must consider the females' response to males' synchronization. A first attempt to explain the response to synchronization [75] has been made using a modified LCOs model, in which, males and females exhibit dissimilar features when they are interacting. The model described in [75] not only reproduces the experimental results shown in [68], but it is also capable of predicting more complex and realistic situations; in particular, the behavior of other species of the genus *Photinus*. As stated in [52], there are species differences in the courtship flash signals of *Photinus* fireflies; this special feature permits even to identify *Photinus* species just by recognizing their courtship flashing patterns. Some examples for these flashing patterns in the case of males are [52]: two flashes per phrase for *Photinus consanguineus* and *Photinus greeni*, six flashes per phrase for *Photinus carolinus* (the species considered in [68, 75], eight flashes per phrase for *Photinus consimilis*.

In order to model the response to synchronization, it is considered the dissimilarity in their oscillatory features between males and females. In Fig. 10 is explained the terminology used in the description of a male and a female. The first type (Fig. 10a) fires a burst of n_f spikes (flashes) during the active phase, followed by a quiescent or silent time T_s, a parameter that remains constant even when the oscillators are coupled. The female has one flash in its fast discharging process T_d which is preceded by a long-lasting charging process T_c and followed by a silent time T_s (Fig. 10b). We define the interburst period or the duration of a phrase T_p as the complete cycle comprising the active phase and the silent time. Consequently, the active phase takes $n_f(T_c + T_d) = Tp - Ts$. Males and females are individually considered as relaxation oscillators because they have two different time scales, i.e. within each cycle, there is a slow process followed by a firing process. Each process ends at its

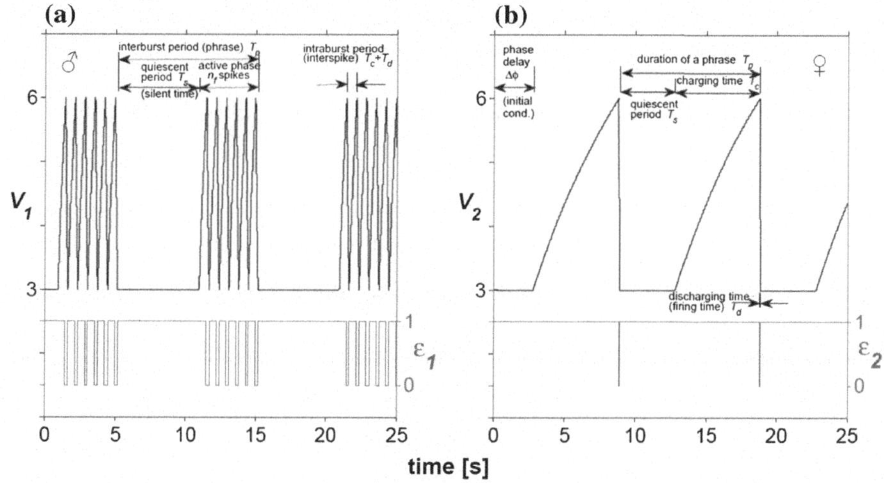

Fig. 10 Signals of the dynamic V and the binary ε variables for the two types of relaxation oscillators used in this work. They are characterized by the quiescent period T_s, the active phase with n_f spikes per burst, the interburst period or silent time T_s, the charging and the discharging times T_c and T_d respectively, the intraburst or interspike period $T_c + T_d$, the interburst period or duration of a phrase T_p, and the phase delay $\Delta\phi$ that plays the role of initial condition. **a** Male oscillator that in this case has the following parameter values: $T_s = 10.000$ s, $n_f = 6$ (*Photinus carolinus*), $T_c = 0.500$ s, $T_d = 0.200$ s, $T_s = 5.800$ s and $\Delta\phi = 0.603$ rad $\equiv 0.960$ s. **b** Female oscillator having in this particular case the parameter values: $T_s = 10.000$ s, $n_f = 1$, $T_c = 6.000$ s, $T_d = 0.100$ s, $T_s = 3.900$ s and $\Delta\phi = 1.750$ rad $\equiv 2.785$ s

own threshold, being the lower and the upper thresholds at $V^{\text{lower}} = V_M/3 = 3$ and $V^{\text{upper}} = 2V_M/3 = 6$ respectively. We take these threshold values in connection with the experimental aspects related to the LCO, namely, the oscillator serving as the basis of the model stated in Eq. (20a) and (20b). Note that we take $V_M = 9$ which is the considered value from an experimental point of view and related to the value of a voltage source. It is also important to note that in Fig. 10a, the parameter values and $n_f = 6$ which corresponds in biological terms to the *Photinus carolinus* flashing pattern. It is clear that the model could be adapted to other species just in changing the relevant parameters.

The equations describing the dynamical variable V_i of each oscillator i are given by:

$$\frac{dV_i(t)}{dt} = \frac{\ln 2}{T_{ci}} \left(V_{Mi} - V_i(t)\right) \varepsilon_i(t) - \frac{\ln 2}{T_{di}} V_i(t) \left(1 - \varepsilon_i(t)\right), \tag{20a}$$

$$V_i(t) = \left(V_i(t) - V_i^{\text{lower}}\right) \varepsilon_i(t) + V_i^{\text{lower}}. \tag{20b}$$

As stated above, V_M is a constant that determines the lower and upper thresholds and $\varepsilon_i(t)$ is a binary variable describing the state of the ith oscillator by:

$$\begin{array}{ll} \varepsilon_i(t) = 1: \text{extinguished oscillator} & \text{(charging and silent stage)} \\ \varepsilon_i(t) = 0: \text{fired oscillator} & \text{(discharging stage).} \end{array}$$

The transition between the states determined by ε is described by the following relation:

$$\textbf{If} \quad V_i(t) = V_i^{\text{lower}} \quad \textbf{and} \quad \varepsilon_i(t) = 0 \quad \textbf{then} \quad \varepsilon_i(t_+) = 1; \tag{21a}$$

$$\textbf{If} \quad V_i(t) = V_i^{\text{upper}} \quad \textbf{and} \quad \varepsilon_i(t) = 1 \quad \textbf{then} \quad \varepsilon_i(t_+) = 0; \tag{21b}$$

$$\textbf{If} \quad V_i(t) = V_i^{\text{lower}} \quad \textbf{and} \quad \varepsilon_i(t) = 1 \quad \textbf{then} \quad \varepsilon_i(t_+) = 1, \tag{21c}$$

where t_+ in the condition given by Eq. (21c) is defined in the interval

$$t = [t_+ \;\; (k-1)(T_p + n_f(T_c + T_d)) + \Delta\phi]$$

for every k interburst period or phrase, i.e. for every complete cycle comprising the active phase and the silent time.

The main feature of the considered oscillators dwells on its firing process which allows a pulsatile coupling with other oscillators that can receive these pulses or spikes leading to a modification in their oscillatory dynamics. The dynamical equations describing a generic group of N coupled oscillators are:

$$\frac{\mathrm{d}V_i(t)}{\mathrm{d}t} = \frac{\ln 2}{T_{c0i}}(V_{Mi} - V_i(t))\,\varepsilon_i(t) - \frac{\ln 2}{T_{d0i}}V_i(t)\,(1 - \varepsilon_i(t)) + \theta_i \sum_{i,j=1}^{N} \beta_{ij}(1 - \varepsilon_j(t)), \tag{22}$$

where $i, j = 1, \ldots, N$. Conditions that are given by Eqs. (20b) and (21), which take into account the existence of a silent time, must also be followed by Eq. (22). The quantities T_{c0i} and T_{d0i} are, respectively, the lasting time of the charge and the discharge when there is no action on the oscillator i by other oscillators. Furthermore, we consider that oscillators are mutually coupled with a coupling strength β_{ij} that represents the pulsatile action of the oscillator j spike during its discharge upon the oscillator i. Concurrently, β_{ij} are the elements of the weighted adjacency matrix of the set. A simple inspection of Eq. (22) shows that both charging and discharging stages might be modified by the effect of the coupling with other oscillator(s). The charging and the discharging times might be shortened or lengthened respectively when the pulsatile action due to the firing of other oscillator(s) takes place. The latter is determined by the value of θ that takes the values:

$$\theta = \begin{cases} 1, & \text{Males} \\ -1, & \text{Females} \end{cases}$$

This factor is significant because it determines the behavior of the oscillators when stimuli are applied to them.

Several studies have been carried out in [75] considering *Photinus carolinus* flash patterns. Here, we show in Fig. 11 the interaction between a set of eight males and four females that according to their flash patterns correspond to *Photinus consanguineus*. Figure 11a shows how the males' flashes evolve until all the population is completely synchronized and how females trigger their responses, firstly sporadically and then permanently as shown in Fig. 11b. This phenomenon that appears in *Photinus* fireflies might be extended to other types of oscillators under the condition that they are dissimilar as stated in [78] where the mechanisms of response to synchronization were unraveled. The response to synchronization phenomenon is illustrated in Fig. 12, where in (a) it is shown a set of males that can synchronize and, in (b) a set of females that their interactions do not lead to synchronization. Finally, when both populations are mingled as in Fig. 12c, the males still synchronize, and as a result of this, females respond following different patterns being the most interesting the simultaneous and permanent responses, as they have acquired the property to synchronize. When extending the results to other types of oscillators and considering the role of network topologies, it is possible to deeply study the collective behavior of these interacting dissimilar oscillators both from a theoretical and experimental point of view, especially when the sets are composed of a considerable number of oscillators. The latter could contribute to a better understanding of systems that exhibit the phenomenon of response to synchronization, viz. fireflies, neurons, and possibly other animals and other types of cells.

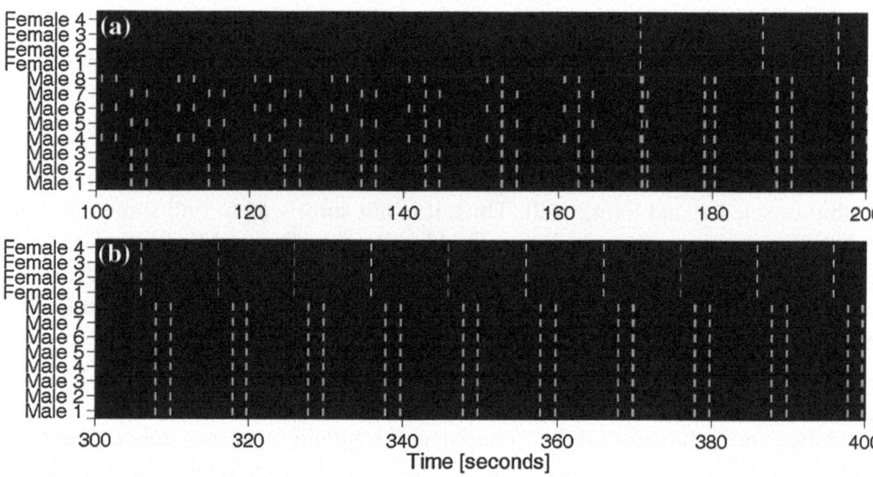

Fig. 11 Evolution of the flashes in a population of 12 slightly different males and four slightly different females considering similar parameters and features as in Fig. 10. **a** Sequence from 100 to 200 s in which, the males are not synchronized until around 180 s, and consequently, the females do not respond, or they do it sporadically as it appears for around 170 s. **b** Sequence from 300 to 400 s in which, the males are completely synchronized, resulting in permanent and simultaneous females' responses

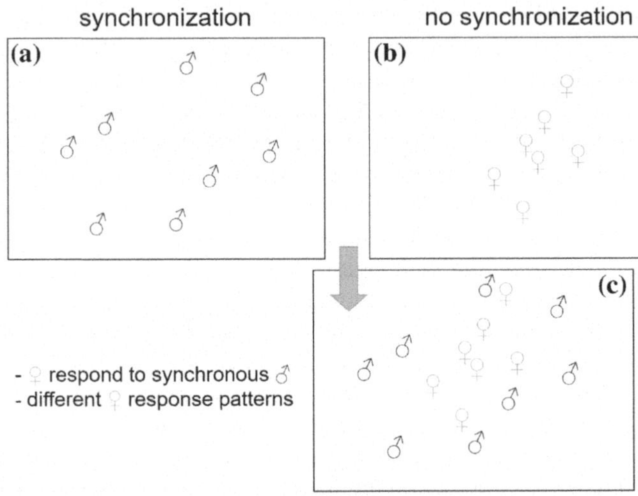

Fig. 12 Illustration of the response to synchronization event in *Photinus* fireflies. **a** A set of interacting males that can easily synchronize. **b** A set of interacting females that cannot synchronize. **c** Mingled set of males and females, where the males still synchronize and as a consequence, females respond exhibiting different response patterns according to the parameter values and initial conditions being one of the possible responses, the simultaneous and permanent females flashing in each cycle as shown in Fig. 11b

6 What Have We Learnt from Fireflies?

Sometimes talking about fireflies seems to be very romantic and without any relationship to science and living well. Thus, it is not surprising to find some opinions published in newspapers such as that cited by Strogatz [91] and Bojic et al. [6]: On May 18, 1993, an article entitled "Govt. Blows Your Tax $$ to Study Fireflies in Borneo-Not a Bright Idea! published in the tabloid *National Enquirer* recorded the opinion of the Representative Tom Petri who said:

> Spending taxpayers money studying fireflies doesn't sound like a very bright idea to me.

Certainly, is not a fortunate opinion and denotes a plain ignorance concerning how much the basic research can offer. As we have already mentioned, the knowledge acquired from fireflies is enormous and only referring to the most recent ones, we can mention the use of the firefly synchronization in the wireless network's world. Under the inspiration of the fireflies synchronous behavior, it is possible to argue that in communication processes, it is better to consider cooperation concepts rather than think in a system driven by a master. Hence, if all nodes cooperate, synchrony can be reached within few periods. Once nodes have agreed on a common time scale, they are then able to communicate in a synchronous manner using a slotted medium access protocol, benefiting from fewer collisions and higher throughput. When working with Wireless Sensor Networks (WSN), firefly synchronization is also exploited, and algorithms based on this concept are tested with the aim of giving greater robustness to the WSN. Other concepts such as the Meshed Emergent Firefly Synchronization (MEMFIS) that mitigates the acquisition phase by integrating synchronization into the communication phase have also been developed to enhance the robustness of the network [95]. Nowadays, the formulation of universal algorithms for WSNs is one of the main tasks. Other networks such as the cognitive radio ad hoc networks (CRAHNs) can also use a synchronization protocol based on fireflies and as a result, the convergence time to synchronization is shorter than convergence time using other protocols [54]. The use of fireflies as role models seems to have the best characteristics for synchronization mechanism related to Machine-to-Machine (M2M) systems (in general composed of networked computational devices) [6]. From the recent review of applications in networks, it is clear that firefly synchronization concept plays an important role and further development is also based on this fact.

Mobile networks or devices have also found significant developments with ideas issued from firefly synchronization. Thus, applications to the optimization of mobile networks have also been developed using firefly-synchronized agents [7]. On the other hand, the so-called firefly algorithm [100] is a swarm intelligence (collective behavior and decentralized systems) is a kind of stochastic, nature-inspired, meta-heuristic algorithm that can be applied to solving the hardest optimization problems. It solves problems of continuous optimization, combinatorial optimization, constraint optimization, multi-objective optimization, as well as dynamic and noisy environments, and even classification. It is widely applied in engineering problems [33]. Algorithms based on firefly synchronization have also been used for swarms of robots as a first approach for understanding real task-execution scenarios [21]. In the same line, possible interactions between real and electronic fireflies (mixed populations) were proposed and even with the possibility of considering males and females differently (mingled populations) that could help to a deeper understanding of the underlying behavior of fireflies and also find potential applications based on the response to synchronization. A picturesque view of this kind of mixed and mingled sets of electronic and real fireflies is shown in Fig. 13. Finally, we could mention some new impact applications such as those leading to improve the efficiency of LEDs [4], or those to conduct a new method of medical diagnostics based on the way in which fireflies produce cold light [87]. In summary, fireflies not only delight us with their beautiful, hypnotic flashes but also they allow us to better understand nature in their many facets from basic phenomena such as self-organization to all the

Fig. 13 A picturesque view of the interaction between LCOs (electronic fireflies) and real fireflies. Electronic fireflies would be designed either to simulate the behavior of males or females, and real males and females would be part of the system as well (Use of fireflies images with permission of Terry Priest)

applications in a wide variety of fields. The emergence of functional synchronization due to the collective behavior without any leader, the response to synchronization, the biochemical mechanism of luminescence, are only a few of the issues that fireflies "illuminated" us.

Acknowledgements J.K. acknowledges IRTG 1740 (DFG). J.-L.D. is senior research associate from the Belgian National Fund for Scientific Research (FNRS). We are grateful to S. Depickère for the careful reading of the manuscript.

References

1. Acebron, J., Bonilla, L., Vicente, C., Ritort, F., Spigler, R.: The Kuramoto model: a simple paradigm for synchronization phenomena. Rev. Mod. Phys. **77**(1), 137–49 (2005)
2. Ariaratnam, J., Strogatz, S.: Phase diagram for the Winfree model of coupled nonlinear oscillators. Phys. Rev. Lett. **86**(19), 4278–4281 (2001)
3. Ballantyne, L., Lambkin, C.: Systematics and phylogenetics of Indo-Pacific luciolinae fireflies (coleoptera: Lampyridae) and the description of new genera. Zootaxa **3653**, 1–162 (2013)
4. Bay, A., Cloetens, P., Suhonen, H., Vigneron, J.: Improved light extraction in the bioluminescent lantern of a Photuris firefly (lampyridae). Opt. Express **21**(1), 764–780 (2013)
5. Boccaletti, S., Kurths, J., Osipov, G., Valladares, D., Zhou, C.: The synchronization of chaotic systems. Phys. Rep. **366**(1–2), 1–101 (2002)

6. Bojic, I., Nymoen, K.: Survey on synchronization mechanisms in machine-to-machine systems. Eng. Appl. Artif. Intell. **45**, 361–375 (2015)
7. Bojic, I., Podobnik, V., Ljubi, I., Jezic, G., Kusek, M.: A self-optimizing mobile network: auto-tuning the network with firefly-synchronized agents. Inf. Sci. **182**(1), 77–92 (2012)
8. Branchini, B., Behney, C., Southworth, T., Fontaine, D., Gulick, A., Vinyard, D., Brudvig, G.: Experimental support for a single electron-transfer oxidation mechanism in firefly bioluminescence. J. Am. Chem. Soc. **137**(24), 7592–7595 (2015)
9. Buck, J.: Synchronous flashing of fireflies experimentally induced. Science **81**(2101), 339–340 (1935)
10. Buck, J.: Synchronous rhythmic flashing of fireflies. Q. Rev. Biol. **13**(3), 301–314 (1938)
11. Buck, J.: The anatomy and physiology of the light organ in fireflies. Ann. NY Acad. Sci. **49**(3), 397–485 (1948)
12. Buck, J.: Synchronous rhythmic flashing of fireflies. II. Q. Rev. Biol. **63**(3), 265–289 (1988)
13. Buck, J., Buck, E.: Flash synchronization as tool and as enabler in firefly courtship competition. Am. Nat. **116**(4), 591–593 (1980)
14. Buck, J., Case, J.: Physiological links in firefly flash code evolution. J. Insect. Behav. **15**(1), 51–68 (2002)
15. Camazine, S., Deneubourg, J.L., Franks, N., Sneyd, J., Theraulaz, G., Bonabeau, E.: Self-Organization in Biological Systems. Princeton University Press, Princeton (2001)
16. Carlson, A., Copeland, J.: Flash communication in fireflies. Q. Rev. Biol. **60**(4), 415–436 (1985)
17. Carlson, A., Copeland, J., Raderman, R., Bulloch, A.: Response patterns of female Photinus macdermotti firefly to artificial flashes. Anim. Behav. **25**, 407–413 (1977)
18. Carlson, A., Copeland, J., Shaskan, R.: Flash communication between the sexes of the firefly Photuris lucicrescens. Physiol. Entomol. **7**(2), 127–132 (1982)
19. Case, J.: Flight studies on photic communication by the firefly Photinus pyralis. Integr. Comp. Biol. **44**(3), 250–258 (2004)
20. Case, J., Trinkle, M.: Light-inhibition of flashing in the firefly Photuris missouriensis. Biol. Bull. **135**(3), 476–485 (1968)
21. Christensen, A., O'Grady, R., Dorigo, M.: From fireflies to fault-tolerant swarms of robots. IEEE Trans. Evol. Comput. **13**(4), 754–766 (2009)
22. Corral, A., Pérez, C., Díaz-Guilera, A., Arenas, A.: Self-organized criticality and synchronization in a lattice model of integrate-and-fire oscillators. Phys. Rev. Lett. **74**(1), 118–121 (1995)
23. Dörfler, F., Bullo, F.: Synchronization in complex networks of phase oscillators: a survey. Automatica **50**(6), 1539–1564 (2014)
24. Dubois, M.: Note sur la physiologie des pyrophores. CR Soc. Biol. **36**, 661–664 (1884)
25. Dubois, M.: Les elaterides lumineaux. CR Soc. Biol. **39**, 564–566 (1887)
26. Dubois, M.: Nouvelles recherches sur la phosphorescence animale. CR Soc. Biol. **41**, 611–614 (1889)
27. Edmunds, J.: The relation between temperature and flashing intervals in adult male fireflies, photinus pyralis1. Ann. Entomol. Soc. Am. **56**(5), 716–718 (1963)
28. Ercsey-Ravasz, M., Sárkzi, Z., Néda, Z., Tunyagi, A., Burda, I.: Collective behavior of electronic fireflies. Eur. Phys. J B **65**(2), 271–277 (2008)
29. Ermentrout, B.: An adaptive model for synchrony in the firefly Pteroptyx malaccae. J. Math. Biol. **29**(6), 571–585 (1991)
30. Ermentrout, G., Chow, C.: Modeling neural oscillations. Physiol. Behav. **77**(4–5), 629–633 (2002)
31. Faust, L.: Natural history and flash repertoire of the synchronous firefly photinus carolinus (coleoptera: Lampyridae) in the Great Smoky Mountains National Park. Fla. Entomol. **93**(2), 208–217 (2010)
32. Faust, L.: Fireflies, Glow-worms, and Lightning Bugs: Identification and Natural History of the Fireflies of the Eastern and Central United States and Canada. University of Georgia Press, Athens, Georgia (2017)

33. Fister, I., Fister Jr., I., Yang, X.S., Brest, J.: A comprehensive review of firefly algorithms. Swarm Evol. Comput. **13**, 34–46 (2013)
34. Gielow, F., Jakllari, G., Nogueira, M., Santos, A.: Data similarity aware dynamic node clustering in wireless sensor networks. Ad Hoc Netw. **24, Part A**, 29–45 (2015)
35. Goel, P., Ermentrout, B.: Synchrony, stability, and firing patterns in pulse-coupled oscillators. Phys. D **163**(3–4), 191–216 (2002)
36. Gohain Barua, A., Rajbongshi, S.: The light of the firefly under the influence of ethyl acetate. J. Biosci. **35**, 183–186 (2010)
37. Goldbeter, A.: Biochemical Oscillations and Cellular Rhythms. The Molecular Bases of Periodic and Chaotic Behaviour. Cambridge University Press, Cambridge (1996)
38. Harvey, E.: On the chemical nature of the luminous material of the firefly. Science **40**(1018), 33 (1914)
39. Harvey, E.: The light-producing substances, photogenin and photophelein, of luminous animals. Science **44**(1140), 652 (1916)
40. Harvey, E.: The mechanism of light production in animals. Science **44**(1128), 208 (1916)
41. Hastings, J., Buck, J.: The firefly pseudoflash in relation to photogenic control. Biol. Bull. **111**(1), 101–113 (1956)
42. Hopfield, J., Herz, A.: Rapid local synchronization of action potentials: toward computation with coupled integrate-and-fire neurons. Proc. Natl. Acad. Sci. USA **92**(15), 6655–6662 (1995)
43. Iwasaka, M., Miyashita, Y., Barua, A., Kurita, S., Owada, N.: Changes in the bioluminescence of firefly under pulsed and static magnetic fields. J. Appl. Phys. **109**(7), 07B303 (2011)
44. Iwasaka, M., Ueno, S.: Bioluminescence under static magnetic fields. J. Appl. Phys. **83**(11), 6456–6458 (1998)
45. Izhikevich, E.: Resonate-and-fire neurons. Neural Netw. **14**(6–7), 883–894 (2001)
46. Jimenez-Romero, C., Johnson, J.: Spikinglab: modelling agents controlled by spiking neural networks in netlogo. Neural Comput. Appl. 1–10 (2016)
47. Jovic, B.: Synchronization Techniques for Chaotic Communication Systems. Springer, Berlin (2012)
48. Kuramoto, Y.: Chemical Oscillations, Waves, and Turbulence. Springer, Berlin (1984)
49. Laurent, P.: The supposed synchronal flashing of fireflies. Science **45**(1150), 44 (1917)
50. Leconte, J.: Synopsis of the Lampyridæof the United States. Trans. Am.Entomol Soc. **9**(1), 15–72 (1881)
51. Lewis, S.: Silent Sparks: The Wondrous World of Fireflies. Princeton University Press (2016)
52. Lewis, S., Cratsley, C.: Flash signal evolution, mate choice, and predation in fireflies. Annu. Rev. Entomol. **53**(1), 293–321 (2008)
53. Li, P., Lin, W., Efstathiou, K.: Isochronous dynamics in pulse coupled oscillator networks with delay. Chaos **27**(5), 053,103 (2017)
54. Lipa, N., Mannes, E., Santos, A., Nogueira, M.: Firefly-inspired and robust time synchronization for cognitive radio ad hoc networks. Comput. Commun. **66**, 36–44 (2015)
55. Lloyd, J.: Aggressive mimicry in photuris: firefly femmes fatales. Science **149**(3684), 653 (1965)
56. Lloyd, J.: A new Photinus firefly, with notes on mating behavior and a possible case of character displacement (coleoptera: Lampyridae). Coleopts Bull. **22**(1), 1–10 (1968)
57. Lloyd, J.: Model for the mating protocol of synchronously flashing fireflies. Nature **245**, 268–270 (1973)
58. Lloyd, J.: Mimicry in the sexual signals of fireflies. Sci. Am. **245**, 138–145 (1981)
59. Lloyd, J.: Evolution of a firefly flash code. Fla. Entomol. **67**(2), 228–239 (1984)
60. Lloyd, J.: On research and entomological education vi: firefly species and lists, old and now. Fla. Entomol. **86**(2), 99–113 (2003)
61. Lu, X., Qin, B.: Synchronization in Complex Networks. Nova Science Pub Inc, New York (2011)
62. Lyu, H.: Synchronization of finite-state pulse-coupled oscillators. Phys. D **303**, 28–38 (2015)

63. Manrubia, S., Mikhailov, A., Zanette, D.: Emergence of Dynamical Order. World Scientific Publishing Co. Pte. Ltd., Singapore (2004)
64. Martin, G., Branham, M., Whiting, M., Bybee, S.: Total evidence phylogeny and the evolution of adult bioluminescence in fireflies (coleoptera: Lampyridae). Mol. Phylogenet. Evol. **107**, 564–575 (2017)
65. McDermott, F.: Observations on the light-emission of American lampyridæ: the photogenic function as a mating adaptation; 5th paper. Can. Entomol. **49**(2), 53–61 (1917)
66. McElroy, W., Seliger, H., White, E.: Mechanism of bioluminescence, chemi-luminescence and enzyme function in the oxidation of firefly luciferin*. Photochem. Photobiol. **10**(3), 153–170 (1969)
67. Mirollo, R., Strogatz, S.: Synchronization of pulse-coupled biological oscillators. SIAM J. Appl. Math. **50**(6), 1645–1662 (1990)
68. Moiseff, A., Copeland, J.: Firefly synchrony: a behavioral strategy to minimize visual clutter. Science **329**(5988), 181 (2010)
69. Mosekilde, E., Maistrenko, Y., Postnov, D.: Chaotic Synchronization: Applications to Living Systems. World Scientific, Singapore (2002)
70. Nijmeije, H., Rodriguez-Angeles, A.: Synchronization of Mechanical Systems. World Scientific Publishing, Singapore (2003)
71. Ortler, B.: The Fireflies Book: Fun Facts About the Fireflies You Loved as a Kid. Adventure Publications Inc, Cambridge (2014)
72. Ott, E., Antonsen, T.: Frequency and phase synchronization in large groups: Low dimensional description of synchronized clapping, firefly flashing, and cricket chirping. Chaos **27**(5), 051,101 (2017)
73. Pikovsky, A., Rosenblum, M., Kurths, J.: Synchronization: A Universal Concept in Nonlinear Sciences. Cambridge University Press, New York (2001)
74. van der Pol, B.: On relaxation oscillation. Philos. Mag. **2**, 978–992 (1926)
75. Ramírez-Ávila, G., Deneubourg, J., Guisset, J., Wessel, N., Kurths, J.: Firefly courtship as the basis of the synchronization-response principle. Europhys. Lett. **94**(6), 60,007 (2011)
76. Ramírez-Ávila, G., Guisset, J., Deneubourg, J.: Synchronization in light-controlled oscillators. Phys. D **182**(3–4), 254–273 (2003)
77. Ramírez-Ávila, G., Guisset, J., Deneubourg, J.: Influence of uniform noise on two light-controlled oscillators. Int. J. Bifurcat. Chaos **17**(12), 4453–4462 (2007)
78. Ramírez-Ávila, G., Kurths, J.: Unraveling the primary mechanisms leading to synchronization response in dissimilar oscillators. Eur. Phys. J. Spec. Top. **225**(13), 2487–2506 (2016)
79. Ramírez-Ávila, G., Kurths, J., Guisset, J., Deneubourg, J.: When does noise destroy or enhance synchronous behavior in two mutually coupled light-controlled oscillators? Phys. Rev. E **82**(5), 056,207 (2010)
80. Ramírez-Ávila, G., Kurths, J., Guisset, J., Deneubourg, J.: How do small differences in non-identical pulse-coupled oscillators induce great changes in their synchronous behavior? Eur. Phys. J. Spec. Top. **223**(13), 2759–2773 (2014)
81. Roda, A.: Chemiluminescence and Bioluminescence: Past, Present and Future. Royal Society of Chemistry, Cambridge (2011)
82. Roda, A.: A History of Bioluminescence and Chemiluminescence from Ancient Times to the Present, pp. 1–50. The Royal Society of Chemistry (2011)
83. Rodrigues, F., Peron, T., Ji, P., Kurths, J.: The Kuramoto model in complex networks. Phys. Rep. **610**, 1–98 (2016)
84. Rubido, N., Cabeza, C., Kahan, S., Ramírez-Ávila, G., Martí, A.: Synchronization regions of two pulse-coupled electronic piecewise linear oscillators. Eur. Phys. J. D **62**(1), 51–56 (2011)
85. Rubido, N., Cabeza, C., Martí, A., Ramírez-Ávila, G.: Experimental results on synchronization times and stable states in locally coupled light-controlled oscillators. Philos. Trans. R. Soc. A **367**, 32673280 (2009)
86. Santillán, M.: Periodic forcing of a 555-ic based electronic oscillator in the strong coupling limit. Int. J. Bifurcat. Chaos **26**(03), 1630,007 (2016)

87. Schena, A., Griss, R., Johnsson, K.: Modulating protein activity using tethered ligands with mutually exclusive binding sites. Nat. Commun. **6**, 7830 (2015)
88. Shimomura, O.: Bioluminescence: Chemical Principles and Methods. World Scientific (2006)
89. Stanger-Hall, K., Lloyd, J.: Flash signal evolution in Photinus fireflies: character displacement and signal exploitation in a visual communication system. Evolution **69**(3), 666–682 (2015)
90. Stewart, I., Strogatz, S.: Synchronisation, rythmes et allures. Pour la science (dossier hors-série) **44**(July/September), 90–93 (2004)
91. Strogatz, S.: Sync: The Emerging Science of Spontaneous Order. Hyperion Press, New York (2003)
92. Teuscher, C., Capcarrere, M.: On Fireflies, Cellular Systems, and Evolware, pp. 1–12. Springer (2003)
93. Theobald, W.: Notes and queries. J. Asiatic Soc. Bengal **35**(2), 73–76 (1866)
94. Tsai, Y.L., Li, C.W., Hong, T.M., Ho, J.Z., Yang, E.C., Wu, W.Y., Margaritondo, G., Hsu, S.T., Ong, E., Hwu, Y.: Firefly light flashing: Oxygen supply mechanism. Phys. Rev. Lett. **113**(25), 258,103 (2014)
95. Tyrrell, A., Auer, G., Bettstetter, C.: Emergent slot synchronization in wireless networks. IEEE T Mobile Comput. **9**(5), 719–732 (2010)
96. Wilensky, U.: NetLogo Fireflies model. Center for Connected Learning and Computer-Based Modeling, Northwestern University, Evanston, IL (1997). http://ccl.northwestern.edu/netlogo/models/Fireflies
97. Wilensky, U.: NetLogo. Center for Connected Learning and Computer-Based Modeling, Evanston, IL (1999). http://ccl.northwestern.edu/netlogo/
98. Wilson, T.: Bioluminescence: Living Lights, Lights for Living. Harvard University Press (2013)
99. Winfree, A.: Biological rhythms and the behavior of populations of coupled oscillators. J. Theoret. Biol. **16**, 15–42 (1967)
100. Yang, X.S.: Firefly Algorithm, pp. 221–230. Wiley (2010)

Mixed Synchronization in the Presence of Cyclic Chaos

Gokul PM, Tomasz Kapitaniak and Anandamohan Ghosh

1 Introduction

Low-dimensional dynamical systems exhibit a variety of complex behaviors including chaos and the study of the collective dynamics of many such coupled systems has drawn attention over the past few decades [1, 2]. It has been experimentally observed and theoretically explored that the collective behavior is often manifested in a synchronized response [3–5]. Examples can be drawn from a wide variety of systems ranging from the neural dynamics to the population models [6–8]. In many of these systems one can observe different types of synchronization like complete synchronization, anti synchronization, lag synchronization, mixed synchronization, etc. [9–22]. Recent studies have demonstrated the importance of mixed synchronization in devising effective control strategies and in secure communication [23]. The above study also experimentally demonstrates mixed synchronization in electronic circuits, where the in-phase or the anti synchronized states can be obtained by an appropriate tuning of the control parameters [23]. Studies of synchronization in a system of coupled networks have also drawn much attention in recent time [24, 25]. One can not only observe "inner synchronization" but also "outer synchronization" of two coupled networks which can also exhibit mixed synchronization states [26, 27]. In this work we demonstrate that in symmetrically coupled dynamical units it is possible to observe mixed synchronization and it is solely dictated by the initial conditions. An important finding of our work is that mixed synchronized states can occur intermittently owing to symmetry properties of the coupling.

The symmetries in dynamical systems play a crucial role in the collective behavior of coupled systems. The coupled system can be such that it visits a sequence of equilibrium states in a cyclic manner leading to a trajectory called the heteroclinic

Gokul PM · A. Ghosh
Indian Institute of Science Education and Research—Kolkata,
Mohanpur 741246, India

Gokul PM · T. Kapitaniak (✉)
Division of Dynamics, Lodz University of Technology,
Stefanowskiego 1/15, 90-457 Lodz, Poland
e-mail: tomasz.kapitaniak@p.lodz.pl; tomaszka@p.lodz.pl

M. Edelman et al. (eds.), *Chaotic, Fractional, and Complex Dynamics: New Insights and Perspectives*, Understanding Complex Systems,
https://doi.org/10.1007/978-3-319-68109-2_4

cycle [28–30]. The symmetry of the global coupling ensures that the trajectories are stable with respect to the perturbations resulting in a stable heteroclinic cycle (SHC from henceforth) and have been analyzed in detail in the past [31, 32]. Stable heteroclinic cycles also exist in discrete dynamical systems, where the local dynamics is governed by a nonlinear map and three such units are globally coupled by *wreath product coupling* [33, 34].

Moreover, depending on the strength of the nonlinearity, the dynamics can be such that it cycles between chaotic states as in *cycling chaos* [34–36]. There exists a large volume of work on SHC in continuous systems and their synchronization properties when subjected to time-dependent drives [37, 38]. In this paper we are interested in studying master-slave synchronization of discrete dynamical units. We consider discrete models, with cubic nonlinearity and exhibiting SHC as well as cycling chaos, as a simple prototype system to demonstrate the synchronization properties. First we report our results for coupling scheme based on contraction mapping, where the system can either synchronize or anti synchronize depending on suitable choice of coupling. Next we consider the response system with wreath product coupling and demonstrate that there exists mixed synchronization. Importantly, the emergence of mixed synchronized states does not require any parametric tuning. We also study the mixed synchronization in coupled phase oscillators with SHC and conclude by discussing the importance of the mixed synchronized states in the context of neural dynamics.

2 SHC in Discrete Systems

We study the coupled dynamics of discrete maps, where the *ith* component of the variable X evolves as

$$X^i_{n+1} = f(X^i_n) + \gamma g(X^i_n, X^j_n). \tag{1}$$

The *local* dynamics is given by the function $f(X^i_n)$ while the *global* dynamics $g(X^i_n, X^j_n)$ couples the *ith* and the *jth* components with coupling strength γ. This model was introduced earlier [34] as a generalization of a system of differential equations exhibiting heteroclinic cycles cycling between chaotic attractors [35]. More specifically, for the local dynamics we also consider a cubic nonlinear function

$$f(X_n) = \lambda X_n - X_n^3 \tag{2}$$

where $\lambda > 0$ is the parameter of the map. As λ is increased from unity the system exhibits a series of period doubling bifurcation to chaos. It is easy to check that if $\mu \in \mathcal{L}$ is an element in the group of local symmetries $\mathcal{L}$

$$f(\mu X^i) = \mu f(X^i). \tag{3}$$

The coupling function is chosen such that it follows the global symmetry. If $\sigma \in \mathcal{G}$ is an element in the group of global symmetries $\mathcal{G}$

$$g(\sigma X) = \sigma g(X). \tag{4}$$

The specific form of the coupling chosen is called *wreath product coupling*

$$g(x^i, x^j) = |x^j|^m x^i \tag{5}$$

where $0 < m < 1$ and determines the rate at which the global dynamics cycles between different states. The local symmetries also satisfy global symmetries [33]

$$g(X^i, \mu X^j) = \mu g(X^i, X^j) \tag{6}$$
$$g(\mu X^i, X^j) = g(X^i, X^j) \tag{7}$$

For example, if we consider μ as an element in the group of cyclic permutations, the above equalities are satisfied. Thus combining the local dynamics, Eq. (2), and the global coupling, Eq. (5), in Eq. (1), the three-dimensional discrete dynamical system assumes the form

$$\begin{aligned} x_{n+1} &= \lambda x_n - x_n^3 - \gamma |y_n|^m x_n \\ y_{n+1} &= \lambda y_n - y_n^3 - \gamma |z_n|^m y_n \\ z_{n+1} &= \lambda z_n - z_n^3 - \gamma |x_n|^m z_n. \end{aligned} \tag{8}$$

This is our Master system. In our simulation we take $\gamma = 3.0$ and $m = 1/2$, and we consider the positive values of the coupling term. Typical trajectories are shown in Fig. 1. It is clearly seen that when any one of the variables is *active* other two variables are *silent* and the dynamics is cycling in the sense that the variables become active in a certain order, $x \rightarrow y \rightarrow z \rightarrow x$. The cycling behavior corresponds to the heteroclinic cycles for weak coupling, and their existence has been rigorously established by linear stability analysis [39]. It has been numerically established that the system undergoes a sequence of period-doubling bifurcation leading to a pair of chaotic attractors [39]. In this work we will consider strong coupling, $\gamma = 3.0$, resulting in cycling behavior which is again weakly or strongly chaotic depending on the parameter, $\lambda = 1.4$ and $\lambda = 2.5$, respectively. For the lower value of $\lambda (= 1.4)$ the dynamics of a cubic map in Eq. (2) is periodic. However, when the cubic maps are coupled as in Eq. (8) the nonlinearity in the global dynamics renders the coupled system weakly chaotic : any active variable can be in the neighborhood of either of its two fixed points (positive/negative) and this switch is irregular and unpredictable. Typical Lyapunov exponents are $(0.001, 0.0005, -0.8)$ where the Maximal Lyapunov Exponent (MLE) is small but positive. While for the choice of $\lambda = 2.5$ the local dynamics is chaotic and the dynamics of Eq. (8) is strongly chaotic as confirmed by the values of the Lyapunov exponents $(0.25, 0.23, -0.09)$. In fact, two positive Lyapunov exponents indicate that the dynamics is hyperchaotic [40–43].

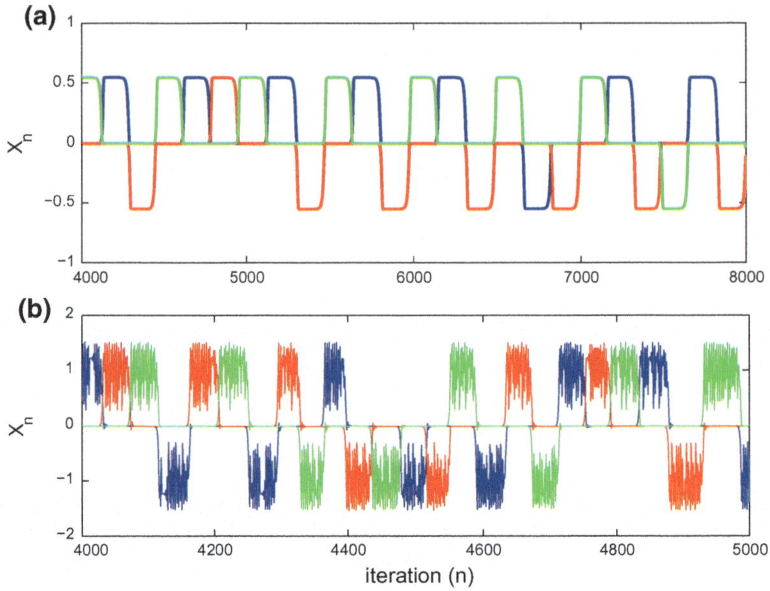

Fig. 1 Time series data for the three variables, $X_n = \{x_n, y_n, z_n\}$ in Eq. (8), are shown in red, green, blue respectively. We have set $\gamma = 3.0$, $m = 0.5$ and initial conditions $X_0 = \{0.11, 0.3, 0.6\}$. The value of the nonlinearity λ determines (**a**) weakly chaotic dynamics, $\lambda = 1.4$; (**b**) chaotic dynamics, $\lambda = 2.5$

3 Synchronization in Discrete Cycling Systems

Synchronization of a chaotic system can be of different kinds and depends on the coupling function. In many physical systems, it is commonly observed that two coupled systems are either synchronized or *anti* synchronized. We would now demonstrate that it is possible to design a coupling strategy that shows mixed synchronization in a system that has cycling chaotic behavior.

The driving (master) system is given by Eq. (8) as before while the driven (slave) system is now the following:

$$\begin{aligned} p_{n+1} &= \lambda p_n - p_n^3 - \gamma |y_n|^m p_n \\ q_{n+1} &= \lambda q_n - q_n^3 - \gamma |z_n|^m q_n \\ r_{n+1} &= \lambda r_n - r_n^3 - \gamma |x_n|^m r_n. \end{aligned} \tag{9}$$

Thus the driving variables enter the driven system through the wreath product coupling term. A representative dynamics is given in Fig. 2. In this example, the driving variable, x_n, and the driven variable, p_n, show in-phase synchronization while the other pairs of variables $\{y_n, q_n\}$ and $\{z_n, r_n\}$ show anti synchronization. This is a clear indication of the existence of mixed synchronization. The interesting feature

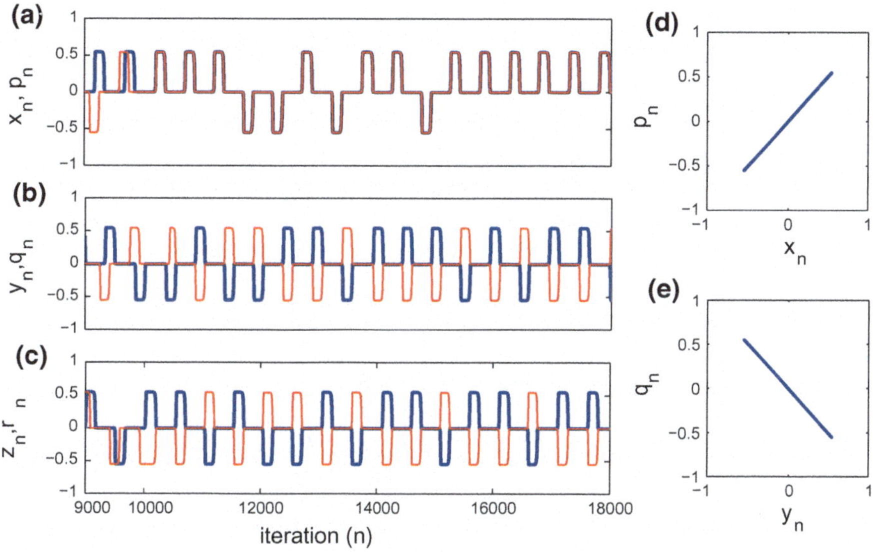

Fig. 2 Time series data for the driving and the driven system variables: (**a**) x_n, p_n (**b**) y_n, q_n (**c**) z_n, r_n. The coupling is set on at iteration number, $n = 10000$. The pair of variables $\{x_n, p_n\}$ is synchronized (**d**) and the pair of variables $\{y_n, q_n\}$ is anti-synchronized (**e**). The pair $\{z_n, r_n\}$ has also anti-synchronized but not shown here. The parameters are $\gamma = 3.0$, $m = 0.5$, $\lambda = 1.4$

of the dynamics is that the choice of different sets of initial conditions are sufficient to get different synchronized/anti synchronized states. Since each pair can either synchronize or anti synchronize all such $2^3 = 8$ possibilities coexist.

Let us now introduce the following variables $U = x - p$, $\tilde{U} = x + p$, $V = y - q$, $\tilde{V} = y + q$, $W = z - r$, $\tilde{W} = z + r$ and the Eqs. (8) and (9) can now be expressed as

$$U_{n+1} = \lambda U_n - \frac{U_n^3}{4} - \frac{3U_n\tilde{U}_n^2}{4} - \gamma U_n \left|\frac{V_n + \tilde{V}_n}{2}\right|^m$$

$$\tilde{U}_{n+1} = \lambda \tilde{U}_n - \frac{\tilde{U}_n^3}{4} - \frac{3U_n^2\tilde{U}_n}{4} - \gamma \tilde{U}_n \left|\frac{V_n + \tilde{V}_n}{2}\right|^m$$

$$V_{n+1} = \lambda V_n - \frac{V_n^3}{4} - \frac{3V_n\tilde{V}_n^2}{4} - \gamma V_n \left|\frac{W_n + \tilde{W}_n}{2}\right|^m$$

$$\tilde{V}_{n+1} = \lambda \tilde{V}_n - \frac{\tilde{V}_n^3}{4} - \frac{3V_n^2\tilde{V}_n}{4} - \gamma \tilde{V}_n \left|\frac{W_n + \tilde{W}_n}{2}\right|^m$$

$$W_{n+1} = \lambda W_n - \frac{W_n^3}{4} - \frac{3W_n\tilde{W}_n^2}{4} - \gamma W_n \left|\frac{U_n + \tilde{U}_n}{2}\right|^m$$

$$\tilde{W}_{n+1} = \lambda \tilde{W}_n - \frac{\tilde{W}_n^3}{4} - \frac{3W_n^2\tilde{W}_n}{4} - \gamma \tilde{W}_n \left| \frac{U_n + \tilde{U}_n}{2} \right|^m \tag{10}$$

As shown in the example of Fig. 2 the synchronization of $\{x_n, p_n\}$ implies $U_n = 0$, while anti synchronizations of $\{y_n, q_n\}$ and $\{z_n, r_n\}$ imply $\tilde{V}_n = 0$ and $\tilde{W}_n = 0$, respectively. Plugging $U_n = 0$, $\tilde{V}_n = 0$ and $\tilde{W}_n = 0$ in the above set of equations we obtain a simplified three-dimensional system

$$\begin{aligned}
\tilde{U}_{n+1} &= \lambda \tilde{U}_n - \frac{\tilde{U}_n^3}{4} - \gamma \tilde{U}_n \left| \frac{V_n}{2} \right|^m \\
V_{n+1} &= \lambda V_n - \frac{V_n^3}{4} - \gamma V_n \left| \frac{W_n}{2} \right|^m \\
W_{n+1} &= \lambda W_n - \frac{W_n^3}{4} - \gamma W_n \left| \frac{\tilde{U}_n}{2} \right|^m .
\end{aligned} \tag{11}$$

It turns out that $\{\tilde{U}, V, W\}$ form a dynamical system satisfying all the symmetry conditions in the local and the global dynamics. This can be easily seen if we divide both sides of Eq. (11) by 2 and identify $\{\tilde{U}/2, V/2, W/2\} \equiv \{x, y, z\}$ we recover Eq. (8). Stability of the cycling behavior governed by Eq. (8) has been rigorously established in Theorem 3.1 in [39] and is also true for our reduced system Eq. (11). The important finding is that this is true for any three variables chosen from the set of six variables $\{U, \tilde{U}, V, \tilde{V}, W, \tilde{W}\}$. Thus eight possible types of mixed synchronization can be observed. The system can settle for any one of the eight cycling states depending only on the initial conditions. It is trivial to show that if we assume only two variables, from the set of six elements, vanish the resultant four-dimensional system of difference equations do not satisfy the symmetry conditions, hence, can not form a stable cycle. The stability of the synchronized dynamics with respect to transverse perturbations can be quantified by computing the transverse Lyapunov exponents for a trajectory running in the synchronized subspace along a cycling trajectory [5]. The largest transverse Lyapunov exponent (Λ_T) stays negative for varying coupling strength γ, as seen in Fig. 3, establishing the stability of the mixed synchronized states. It is to be noted that the heteroclinic cycle crucially depends on γ and m and the cycle is destroyed if $\gamma \gtrsim 3$.

The dynamics of the complete system exhibits cycling chaos, typically, when any of the variables is locally active all other variables are ≈ 0. Thus to qualitatively understand the basin of attraction dynamics can be approximated by the local dynamics of the coupled system Eq. (11), i.e.,

$$\begin{aligned}
U_{n+1} &= \lambda U_n - \frac{U_n^3}{4} - \frac{3U_n\tilde{U}_n^2}{4} \\
\tilde{U}_{n+1} &= \lambda \tilde{U}_n - \frac{\tilde{U}_n^3}{4} - \frac{3U_n^2\tilde{U}_n}{4}
\end{aligned} \tag{12}$$

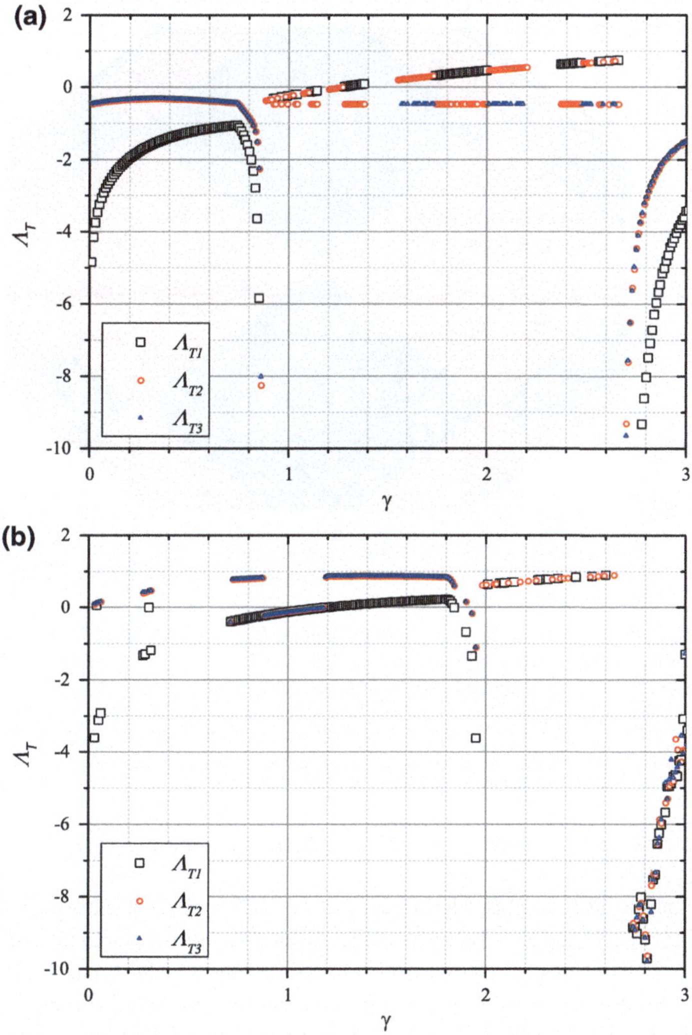

Fig. 3 Three Transverse Lyapunov exponent ($\Lambda_{T_{1,2,3}}$) is shown as a function of coupling strength γ. $m = 0.5$ for **a** $\lambda = 1.4$ and **b** $\lambda = 2.5$. **a** We see a region of stability ($\gamma = (0, 1.1)$) followed by a region of instability (till $\gamma = (1.1, 2.7)$ followed again by another region of stability ($\gamma = (2.7, 3)$) after which the system is then unstable ($\gamma > 3.0$). **b** We see an unstable region in the beginning ($\gamma = (0, 0.9)$) after which there is a small region of stability ($\gamma = (0.9, 1.2)$), which is followed by another unstable region ($\gamma = (1.1, 1.98)$) followed again by a small stable region ($\gamma = (1.98, 2)$) after which another unstable region exists ($\gamma = (2, 2.7)$)) which is then followed by a region of stability ($\gamma = (2.7, 3)$) beyond which the system is unstable ($\gamma > 3$)

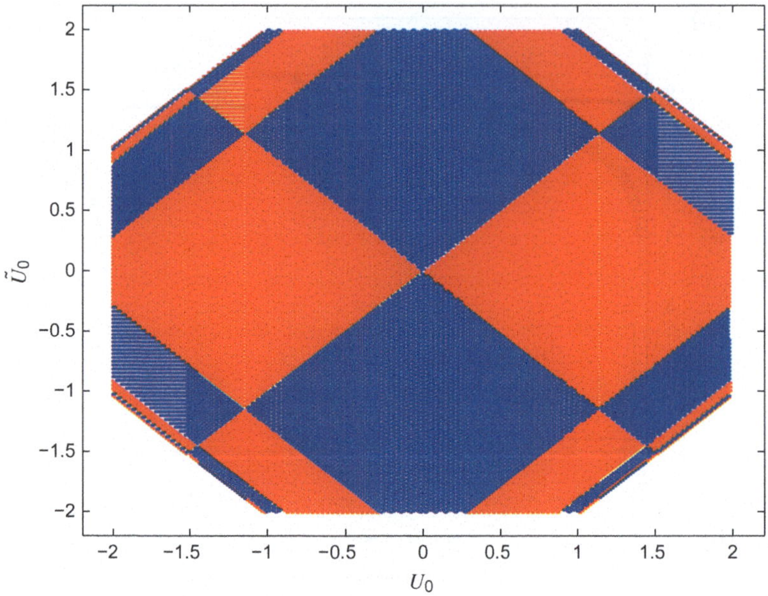

Fig. 4 The basin of attraction is shown as color map on the U_0-$\tilde{U}_0$ plane for the local dynamics as in Eq. (12). The different regions correspond to stable states: $(\pm 2\sqrt{\lambda - 1}, 0)$ in red (anti synchronized), $(0, \pm 2\sqrt{\lambda - 1})$ in blue (synchronized), $(0, 0)$ in white. The parameter $\lambda = 1.3$

The above two-dimensional map has the stable equilibria

$$(U^*, \tilde{U}^*) = \{(0, 0), (\pm 2\sqrt{\lambda - 1}, 0), (0, \pm 2\sqrt{\lambda - 1})\}. \tag{13}$$

If we consider weakly chaotic dynamics i.e. $\lambda = 1.3 < 2$, depending on the initial conditions $(U_{n=0}, \tilde{U}_{n=0})$ the dynamics asymptotically converges to one of the equilibrium states. The basin of attraction is shown in Fig. 4. which can be interpreted as the basins of attraction for the synchronized and the anti synchronized state. (It is obvious that for $(V, \tilde{V})$, $(W, \tilde{W})$ the basins of attraction will be identical.) For example, if the coupled system is iterated with $x_{n=0} = p_{n=0} = \beta$ (say), the corresponding $U = 0$ and $\tilde{U} = 2\beta$, then from Fig. 4, the asymptotic state is the synchronized state [region blue]. Similarly, knowing the initial states of all the driving and the driven variables it is possible to predict the emergent mixed synchronized states.

Now we study the coupled system when the local dynamics is chaotic. If we consider the local dynamics of Eq. (11), we have three identical maps of the form $V_{n+1} = \lambda V_n - V_n^3/4$. The fixed points of the maps $(\pm 2\sqrt{\lambda - 1})$ are stable for $1 < \lambda < 2$. We chose $\lambda = 2.5$, and the resultant chaotic dynamics is shown in Fig. 5. It is observed that the cycles in the driving and the driven system i.e., the global dynamics can synchronize or anti synchronize but the local dynamics do not synchronize. In the globally synchronized part the difference variable U_n shows fluctuations due to mismatch in local dynamics and the fluctuation is greater during anti synchronization

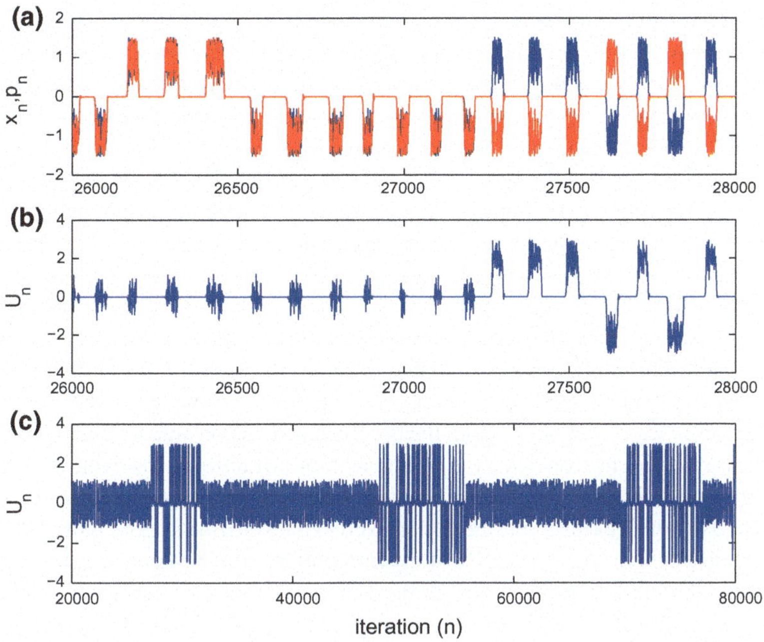

Fig. 5 **a** Time series data for x_n (blue) and p_n (red) for $\gamma = 3.0$, $m = 0.5$, $\lambda = 2.5$ **b** U_n **c** U_n is shown for large number of iterations

Fig. 5b. The transverse Lyapunov exponents for $\lambda = 2.5$ are still negative, Fig. 3, though complete synchronization is not achieved. This only indicates that the negative transverse Lyapunov exponents is a necessary condition and not a sufficent condition for synchronization. In the non-chaotic regime different initial conditions lead to different mixed synchronized states while in the chaotic regime starting with any initial condition it is possible to obtain both the synchronized and anti synchronized states intermittently. Since the intrinsic dynamics is strongly chaotic the transition from synchronized to anti synchronized states remains unpredictable. This is an interesting feature, the transverse error dynamics is intermittently synchronized or anti synchronized, in terms of the global heteroclinic cycles but the local dynamics clearly is not. However, this hyperchaotic parameter regime is not fully understood and rigorous studies needs to be undertaken.

4 Discussion

Our study indicates that symmetrically coupled oscillating units on master-slave coupling can exhibit mixed synchronization. The mixed synchronization is a property of systems exhibiting cycling chaos and their dependence on the initial conditions is quite interesting. Our observation is reminiscent of multivalued synchronization observed in forced systems where multistability leads to interesting dynamical basins of attraction [44]. Our study is also relevant in the context of neural dynamics. As we have shown in our model systems, the neuronal response can be such that the neurons form synchronous and asynchronous clusters. In a typical neural network, the number of neurons being large, the number of initial states is also large resulting in many possibilities of such mixed cluster states. A particular initial state will lead to one of the many possible mixed synchronized states. The existence of many mixed states results in a greater variability in the dynamics of neural networks, a subject of great interest [45, 46]. It will be interesting to explore the role of noise on the extent of synchronization. The noise can be introduced as an additive term or by making the intrinsic parameters randomly distributed. Sensitivity to noise strength can destroy or even stabilise a cycle leading to interesting synchronization behavior. Moreover, if the dynamics is weakly chaotic, from the output state it is possible to read out the input state, which is evident from the computed basin of attraction. However, the read-out will not be possible if the local dynamics is chaotic, but an important finding is that the the global dynamics exhibit intermittent mixed synchronization. The global coupling maintains the synchrony in the global cycling pattern reminiscent of central pattern generators. A detailed study, with more realistic model of neurons with synaptic coupling is required to understand the implications of mixed synchronization in neural systems and studies in that direction are in progress.

Acknowledgements One of us (G.P M) has been supported by the Polish National Centre, Maestro program—Project No. 2013/08/ST8/00/780. We also acknowledge Soumitro Banerjee and Subhasis Sinha for their input and useful discussions in relation to this work.

References

1. Boccaletti, S., Kurths, J., Osipov, G., Valladares, D.L., Zhou, C.S.: The synchronization of chaotic systems. Phys. Rep. **366**(1), 1 (2002)
2. Pikovsky, A.S., Rosenblum, M.G., Kurths, J.: Synchronization: A Universal Concept in Nonlinear Sciences. Cambridge University Press, Cambridge, England (2002)
3. Kocarev, L., Parlitz, U.: Generalized synchronization, predictability, and equivalence of unidirectionally coupled dynamical systems. Phys. Rev. Lett. **76**(11), 1816 (1996)
4. Kuntsevich, B.F., Pisarchik, A.N.: Synchronization effects in a dual-wavelength class-B laser with modulated losses. Phys. Rev. E **64**, 046221 (2001)
5. Pecora, L.M., Carroll, T.L.: Synchronization in chaotic systems. Phys. Rev. Lett. **64**(8), 821 (1990)
6. Blasius, B., Huppert, A., Stone, L.: Complex dynamics and phase synchronization in spatially extended ecological systems. Nature **399**, 354 (1999)
7. Rabinovich, M.I., Varona, P., Selverston, A.I., Abarbanel, H.D.: Dynamical principles in neuroscience. Rev. Mod. Phys. **78**, 1213 (2006)
8. Watts, D.J., Strogatz, S.H.: Collective dynamics of small-worldnetworks. Nature **393**, 440 (1998)
9. Chen, Y., Xiao, J., Liu, W., Li, L., Yang, Y.: Dynamics of chaotic systems with attractive and repulsive couplings. Phys. Rev. E **80**, 046206 (2009)

10. Kim, C.M., Rim, S., Kye, W.H., Ryu, J.W., Park, Y.J.: Anti-synchronization of chaotic oscillators. Phys. Lett. A **320**(1), 39 (2003)
11. Liu, W., Xiao, J., Qian, X., Yang, J.: Antiphase synchronization in coupled chaotic oscillators. Phys. Rev. E **73**, 057203 (2006)
12. Prasad, A.: Universal occurrence of mixed-synchronization in counter-rotating nonlinear coupled oscillators. Chaos, Solitons and Fractals **43**(1), 42 (2010)
13. Rosenblum, M.G., Pikovsky, A.S., Kurths, J.: From phase to lag synchronization in coupled chaotic oscillators. Phys. Rev. Lett. **78**(22), 4193 (1997)
14. Stefanski, A., Perlikowski, P., Kapitaniak, T.: Ragged synchronization of coupled oscillators. Phys. Rev. E **75**, 016210 (2007)
15. Maistrenko, Y., Kapitaniak, T.: Different types of synchronization in two coupled piecewise linear maps. Phys. Rev. E **54**, 3285 (1996)
16. Maistrenko, Y., Kapitaniak, T., Szuminski, P.: Locally and globally riddled basins in two coupled piecewise linear maps. Phys. Rev. E **56**, 6393 (1997)
17. Stefanski, A., Kapitaniak, T.: Using chaos synchronization to estimate the largest lyapunov exponent of nonsmooth systems. Discret. Dyn. Nat. Soc. **4**, 207 (2000)
18. Perlikowski, P., Yanchul, S., Wolfrom, M.: Routes to complex dynamics in ring of unidirectionally coupled systems. Chaos **20**, 013111 (2010)
19. Stefanski, A., Dabrowski, A., Kapitaniak, T.: Evaluation of the largest Lyapunov exponent in dynamical systems with time delay. Chaos Solutions and fractals **23**, 1651 (2005)
20. Chudzik, A., Perlikowski, P., Stefanski, A., et al.: Multistability and rare attractors in van der pol—duffing oscillator. IJBC **21**, 1907 (2011)
21. Czolczynski, K., Perlikowski, P., Stefanski, A., et al.: Why two clocks synchronize: energy balance of the synchronized clocks. Chaos **21**, 023129 (2011)
22. Brezetskyi, S., Dudkowski, D., Kapitaniak, T.: Rare and hidden attractors in Van der Pol-Duffing oscillators. Eur. Phys. J. **224**, 14591467 (2015)
23. Bhowmick, S.K., Hens, C., Ghosh, D., Dana, S.K.: Mixed synchronization in chaotic oscillators using scalar coupling. Phys. Lett. A **376**, 2490 (2012)
24. Li, C., Sun, W., Kurths, J.: Synchronization between two coupled complex networks. Phys. Rev. E **76**, 046204 (2007)
25. Sun, Y., Shi, H., Bakare, E.A., Meng, Q.: Noise-induced outer synchronization between two different complex dynamical networks. Nonlinear Dyn. **76**, 519 (2014)
26. Li, Z., Xue, X.: Outer synchronization of coupled networks using arbitrary coupling strength. Chaos **20**, 023106 (2010)
27. Wang, J.W., Ma, Q., Abd-Elouahab, M.S.: Mixed outer synchronization of coupled complex networks with time-varying coupling delay. Chaos **21**, 013121 (2011)
28. Armbruster, D., Guckenheimer, J., Holmes, P.: Heteroclinic cycles and modulated travelling waves in systems with O(2) symmetry. Physica D **29**, 257 (1988)
29. Guckenheimer, J., Holmes, P.: Structurally stable heteroclinic cycles. Math. Proc. Camb. Phil. Soc. **103**, 189 (1988)
30. Melbourne, I., Chossat, P., Golubitsky, M.: Heteroclinic cycles involving periodic solutions in mode interactions with O (2) symmetry. Proc. Roy. Soc. Edinburgh. **113A**, 315 (1989)
31. Buono, P.L., Golubitsky, M., Palacios, A.: Heteroclinic cycles in rings of coupled cells. Physica D **143**, 74 (2000)
32. Krupa, M., Melbourne, I.: Asymptotic stability of heteroclinic cycles in systems with symmetry. Ergod. Th. Dynam. Sys. **15**, 121 (1995)
33. Dionne, B., Golubitsky, M., Stewart, I.: Coupled cells with internal symmetry: I. Wreath products. Nonlinearity **9**(2), 559 (1996)
34. Palacios, A.: Cycling chaos in one-dimensional coupled iterated maps. Int. J. Bifur. Chaos **12**(8), 1859 (2002)
35. Dellnitz, M., Field, M., Golubitsky, M., Ma, J., Hohmann, A.: Cycling chaos. Int. J. Bif. Chaos **5**, 1243 (1995)
36. Palacios, A., Juarez, H.: Cryptography with cycling chaos. Phys. Lett. A **303**(5), 345 (2002)

37. Afraimovich, V.S., Rabinovich, M.I., Varona, P.: Heteroclinic contours in neural ensembles and the winnerless competition principle. Int. J. Bifur. Chaos Appl. Sci. Engrg. **14**, 1195 (2004)
38. Ashwin, P., Burylko, O., Maistrenko, Y., Popovych, O.: Extreme sensitivity to detuning for globally coupled phase oscillators. Phys. Rev. Lett. **96**(5), 054102 (2006)
39. Palacios, A.: Heteroclinic cycles in coupled systems of difference equations. Special Issue J. Differ. Equ. Appl. **9**(7), 671 (2003)
40. Baier, G., Klein, M.: Maximum hyperchaos in generalized Henon map. Phys. Lett. A **151**, 281 (1990)
41. Rossler, O.E.: An equation for hyperchaos. Phys. Lett. A **71**(2), 155 (1979)
42. Yanchuk, S., Kapitaniak, T.: Symmetry-increasing bifurcation as a pedictor of a chaos-hyperchaos transition in coupled systems. Phys. Rev. E **64**, 056235 (2001)
43. Yanchuk, S., Kapitaniak, T.: Chaos-hyperchaos transition in coupled Rossler systems. Phys. Lett. A **290**, 139 (2001)
44. Ontanon-Garcia, L.J., Campos-Canton, E.: Preservation of a two-wing Lorenz-like attractor with stable equilibria. J. Franklin Inst. **350**, 2867 (2013)
45. Arieli, A., Sterkin, A., Grinvald, A., Aertsen, A.D.: Dynamics of ongoing activity: explanation of the large variability in evoked cortical responses. Science **273**, 5283 (1996)
46. Ding, M., Glanzman, D. (Eds.): The Dynamic Brain: An Exploration of Neuronal Variability and Its Functional Significance. Oxford University Press, USA (2011)

Time-Delay Effects on Periodic Motions in a Duffing Oscillator

Albert C.J. Luo and Siyuan Xing

In this chapter, time-delay effects on periodic motions in a periodically forced, time-delayed, hardening Duffing oscillator are reviewed and further discussed. Bifurcation trees of periodic motions to chaos varying with time-delay are presented for such a time-delayed, Duffing oscillator. From the analytical prediction, periodic motions in the time-delayed, hardening Duffing oscillator are simulated numerically. Through numerical illustrations, time-delay effects on period-1 motions to chaos in nonlinear dynamical systems are strongly related to the distributions and quantity levels of harmonic amplitudes.

1 Introduction

To study time-delay systems, one can approximate the time-delay term through

$$x^{\tau} = x(t - \tau) \approx x(t) - \dot{x}(t)\tau \tag{1}$$

For a very small time-delay, such an approximation in dime-delay systems may be acceptable. If the time-delay becomes large, such an approximation cannot be used to study the time-delay systems.

Recently, periodic motions in time-delayed dynamical systems were of great interest. The perturbation method has been adopted for approximate solutions of periodic motions in the time-delayed nonlinear oscillators (e.g., Hu et al. [1], Wang and Hu [2]). The traditional harmonic balance method was also employed for approximate solutions of periodic motions in time-delayed nonlinear oscillators (e.g., MacDonald

A.C.J. Luo (✉) · S. Xing
Department of Mechanical and Industrial Engineering, Southern Illinois University Edwardsville, Edwardsville, IL 62026-1805, USA
e-mail: aluo@siue.edu

S. Xing
e-mail: sxing@siue.edu

M. Edelman et al. (eds.), *Chaotic, Fractional, and Complex Dynamics: New Insights and Perspectives*, Understanding Complex Systems,
https://doi.org/10.1007/978-3-319-68109-2_5

[3], Liu and Kalmar-Nagy [4], Lueng and Guo [5]). However, such approximate solutions of periodic motions in the time-delayed oscillators were based on one or two harmonic terms without enough accuracy. To improve accuracy of approximate solutions of periodic motions, in 2013, Luo [6] presented the generalized harmonic balance method for periodic motions in time-delayed, nonlinear dynamical systems. Luo and Jin [7] applied such a generalized harmonic balance method for the time-delayed, quadratic nonlinear oscillator, and the analytical bifurcation trees of period-1 motions to chaos were obtained. Luo and Jin [8] determined complex period-1 motions in the periodically forced Duffing oscillator with a time-delayed displacement. Further, Luo and Jin [9] analytically determined the bifurcation trees of the period-1 motions to chaos in the time-delayed Duffing oscillator, and complex period-m motions were studied in such a time-delayed Duffing oscillator. The generalized harmonic balance method cannot be used the nonlinear dynamical systems with non-polynomial nonlinear systems. In 2015, Luo [10] developed a semi-analytical method to determine periodic motions in nonlinear dynamical systems through discrete implicit maps. From a specific mapping structure, periodic motions in the nonlinear dynamical systems can be determined. From such solutions of discrete nodes on the periodic motions, the corresponding frequency-amplitudes can be determined. Luo and Guo [11] applied such a semi-analytical method for bifurcation trees of periodic motions to chaos in the Duffing oscillator. The semi-analytical method in Luo [10] was also developed for time-delay nonlinear systems. Luo and Xing [12] used such a method for complicated period-1 motions and the corresponding bifurcation trees to chaos in the time-delayed Duffing oscillator. Luo and Xing [13] analytically predicted the bifurcation trees of period-1 motions to chaos. For such studies, the time-delay is specifically chosen. In 2017, Luo and Xing [14] discussed the time-delay effects on periodic motions in the time-delayed Duffing oscillator. In this chapter, the time-delay effects on periodic motions in the time-delayed dynamical systems will be reviewed and further discussed.

2 A Semi-analytical Method

From Luo [10], a period-m flow in a time-delayed, nonlinear dynamical system can be described through discrete nodes for period-mT. To determine period-m motion in time-delay dynamical systems, the following theorem is presented herein.

Theorem 1 *Consider a time-delay nonlinear dynamical system*

$$\dot{\mathbf{x}} = \mathbf{f}(\mathbf{x}, \mathbf{x}^{\tau}, t, \mathbf{p}) \in R^n,$$
$$\textit{with } \mathbf{x}(t_0) = \mathbf{x}_0, \mathbf{x}(t) = \Phi(\mathbf{x}_0, t - t_0, \mathbf{p}) \textit{ for } t \in [t_0 - \tau, \infty). \tag{2}$$

If such a time-delay dynamical system has a period-m flow $\mathbf{x}^{(m)}(t)$ *with finite norm* $||\mathbf{x}^{(m)}||$ *and period* $mT (T = 2\pi/\Omega)$*, there is a set of discrete time* $t_k (k = 0, 1, \ldots, mN)$ *with* $(N \to \infty)$ *during m-periods* (mT)*, and the corresponding solu-*

tion $\mathbf{x}^{(m)}(t_k)$ *and vector field* $\mathbf{f}(\mathbf{x}^{(m)}(t_k), \mathbf{x}^{\tau(m)}(t_k), t_k, \mathbf{p})$ *are exact. Suppose discrete nodes* $\mathbf{x}_k^{(m)}$ *and* $\mathbf{x}_k^{\tau(m)}$ *are on the approximate solution of the periodic flow under* $||\mathbf{x}^{(m)}(t_k) - \mathbf{x}_k^{(m)}|| \leq \varepsilon_k$ *and* $||\mathbf{x}^{\tau(m)}(t_k) - \mathbf{x}_k^{\tau(m)}|| \leq \varepsilon_k^{\tau}$ *with small* $\varepsilon_k, \varepsilon_k^{\tau} \geq 0$ *and*

$$||\mathbf{f}(\mathbf{x}^{(m)}(t_k), \mathbf{x}^{\tau(m)}(t_k), t_k, \mathbf{p}) - \mathbf{f}(\mathbf{x}_k^{(m)}, \mathbf{x}_k^{\tau(m)} t_k, \mathbf{p})|| \leq \delta_k \tag{3}$$

with a small $\delta_k \geq 0$. *During a time interval* $t \in [t_{k-1}, t_k]$, *there is a mapping* $P_k : (\mathbf{x}_{k-1}^{(m)}, \mathbf{x}_{k-1}^{\tau(m)}) \to (\mathbf{x}_k^{(m)}, \mathbf{x}_k^{\tau(m)})(k = 1, 2, \ldots, mN)$ *as*

$$\begin{aligned}
&(\mathbf{x}_k^{(m)}, \mathbf{x}_k^{\tau(m)}) = P_k(\mathbf{x}_{k-1}^{(m)}, \mathbf{x}_{k-1}^{\tau(m)}) \text{ with } \mathbf{g}_k(\mathbf{x}_{k-1}^{(m)}, \mathbf{x}_k^{(m)}; \mathbf{x}_{k-1}^{\tau(m)}, \mathbf{x}_k^{\tau(m)}, \mathbf{p}) = \mathbf{0},\\
&\mathbf{x}_j^{\tau(m)} = \mathbf{h}_j(\mathbf{x}_{r_j-1}^{(m)}, \mathbf{x}_{r_j}^{(m)}, \theta_{r_j}), \ j = k, k-1; r_j = j - l_j, k = 1, 2, \ldots, mN;\\
&(e.g., \mathbf{x}_r^{\tau(m)} = \mathbf{x}_{s_r}^{(m)} + \theta_r(\mathbf{x}_{r_r-1}^{(m)} - \mathbf{x}_{r_r}^{(m)}), \theta_r = \frac{1}{h_{r_j}}[\tau - \sum_{i=1}^{l_{r_j}} h_{r_j+i}]).
\end{aligned} \tag{4}$$

where $\mathbf{g}_k$ *is an implicit vector function and* $\mathbf{h}_j$ *is an interpolation vector function. Consider a mapping structure as*

$$\begin{aligned}
&P = P_{mN} \circ P_{mN-1} \circ \ldots \circ P_2 \circ P_1 : \mathbf{x}_0^{(m)} \to \mathbf{x}_{mN}^{(m)};\\
&\text{with } P_k : (\mathbf{x}_{k-1}^{(m)}, \mathbf{x}_{k-1}^{\tau(m)}) \to (\mathbf{x}_k^{(m)}, \mathbf{x}_k^{\tau(m)})(k = 1, 2, \ldots, mN).
\end{aligned} \tag{5}$$

For $\mathbf{x}_{mN}^{(m)} = P(\mathbf{x}_0^{(m)}, \mathbf{x}_0^{\tau(m)})$, *if there is a set of points* $(\mathbf{x}_k^{(m)*}, \mathbf{x}_k^{\tau(m)*})(k = 0, 1, \ldots, mN)$ *computed by*

$$\begin{aligned}
&\left.\begin{aligned}
&\mathbf{g}_k(\mathbf{x}_{k-1}^{(m)*}, \mathbf{x}_k^{(m)*}; \mathbf{x}_{k-1}^{\tau(m)*}, \mathbf{x}_k^{\tau(m)*}, \mathbf{p}) = \mathbf{0},\\
&\mathbf{x}_j^{\tau(m)*} = \mathbf{h}_j(\mathbf{x}_{r_j-1}^{(m)*}, \mathbf{x}_{r_j}^{(m)*}, \theta_{r_j}), \ j = k, k-1
\end{aligned}\right\} (k = 1, 2, \ldots, mN)\\
&\mathbf{x}_{r_j-1}^{(m)*} = \mathbf{x}_{\text{mod }(r_j-1+mN, mN)}^{(m)*}, \mathbf{x}_{r_j}^{(m)*} = \mathbf{x}_{\text{mod }(r_j+mN, mN)}^{(m)*};\\
&\mathbf{x}_0^{(m)*} = \mathbf{x}_{mN}^{(m)*} \text{ and } \mathbf{x}_0^{\tau(m)*} = \mathbf{x}_{mN}^{\tau(m)*}
\end{aligned} \tag{6}$$

Then the points $\mathbf{x}_k^{(m)*}$ and $\mathbf{x}_k^{\tau(m)*}(k = 0, 1, \ldots, mN)$ *are the approximation of points* $\mathbf{x}^{(m)}(t_k)$ *and* $\mathbf{x}^{\tau(m)}(t_k)$ *of periodic solutions. In the neighborhoods of* $\mathbf{x}_k^{(m)*}$ *and* $\mathbf{x}_k^{\tau(m)*}$, $\mathbf{x}_k^{(m)} = \mathbf{x}_k^{(m)*} + \Delta\mathbf{x}_k^{(m)}$ *and* $\mathbf{x}_k^{\tau(m)} = \mathbf{x}_k^{\tau(m)*} + \Delta\mathbf{x}_k^{\tau(m)}$, *the linearized equation is given by*

$$\begin{aligned}
&\sum_{j=k-1}^{k} \frac{\partial \mathbf{g}_k}{\partial \mathbf{x}_j^{(m)}} \Delta\mathbf{x}_j^{(m)} + \frac{\partial \mathbf{g}_k}{\partial \mathbf{x}_j^{\tau(m)}} \left(\frac{\partial \mathbf{x}_j^{\tau(m)}}{\partial \mathbf{x}_{r_j}^{(m)}} \Delta\mathbf{x}_{r_j}^{(m)} + \frac{\partial \mathbf{x}_j^{\tau(m)}}{\partial \mathbf{x}_{r_j-1}^{(m)}} \Delta\mathbf{x}_{r_j-1}^{(m)}\right) = \mathbf{0}\\
&\text{with } r_j = j - l_j, \ j = k-1, k; \ (k = 1, 2, \ldots, mN).
\end{aligned} \tag{7}$$

The resultant Jacobian matrices of the periodic flow are

$$DP_{k(k-1)\dots 1} = \left[\frac{\partial \mathbf{y}_k^{(m)}}{\partial \mathbf{y}_0^{(m)}}\right]_{(\mathbf{y}_0^{(m)*},\dots,\mathbf{y}_k^{(m)*})} = \mathbf{A}_k\mathbf{A}_{k-1}\dots\mathbf{A}_1 (k = 1, 2, \dots, mN),$$

$$\text{and } DP = DP_{mN(mN-1)\dots 1} = \left[\frac{\partial \mathbf{y}_{mN}^{(m)}}{\partial \mathbf{y}_0^{(m)}}\right]_{(\mathbf{y}_0^{(m)*},\dots,\mathbf{y}_{mN}^{(m)*})} = \mathbf{A}_{mN}\mathbf{A}_{mN-1}\dots\mathbf{A}_1 \quad (8)$$

where

$$\Delta\mathbf{y}_k^{(m)} = \mathbf{A}_k^{(m)}\Delta\mathbf{y}_{k-1}^{(m)},\ \mathbf{A}_k^{(m)} \equiv \left[\frac{\partial \mathbf{y}_k^{(m)}}{\partial \mathbf{y}_{k-1}^{(m)}}\right]_{(\mathbf{y}_{k-1}^{(m)*},\mathbf{y}_k^{(m)*})} \quad (9)$$

and

$$\begin{aligned}
\mathbf{A}_k^{(m)} &= \begin{bmatrix} \mathbf{B}_k^{(m)} & (\mathbf{a}_{k(r_{k-1}-1)}^{(m)})_{n\times n} \\ \mathbf{I}_k^{(m)} & \mathbf{0}_k^{(m)} \end{bmatrix}_{n(s+1)\times n(s+1)}, s = 1 + l_{k-1} \\
\mathbf{B}_k^{(m)} &= [(\mathbf{a}_{k(k-1)}^{(m)})_{n\times n}, \mathbf{0}_{n\times n}, \dots, (\mathbf{a}_{k(r_k-1)}^{(m)})_{n\times n}], \\
\mathbf{I}_k^{(m)} &= diag(\mathbf{I}_{n\times n}, \mathbf{I}_{n\times n}, \dots, \mathbf{I}_{n\times n})_{ns\times ns}, \\
\mathbf{0}_k^{(m)} &= (\underbrace{\mathbf{0}_{n\times n}, \mathbf{0}_{n\times n} \dots, \mathbf{0}_{n\times n}}_{s})^T;
\end{aligned} \quad (10)$$

$$\begin{aligned}
\mathbf{y}_k^{(m)} &= (\mathbf{x}_k^{(m)}, \mathbf{x}_{k-1}^{(m)}, \dots, \mathbf{x}_{r_{k-1}}^{(m)})^T, \\
\mathbf{y}_{k-1}^{(m)} &= (\mathbf{x}_{k-1}^{(m)}, \mathbf{x}_{k-2}^{(m)}, \dots, \mathbf{x}_{r_{k-1}-1}^{(m)})^T, \\
\Delta\mathbf{y}_k^{(m)} &= (\Delta\mathbf{x}_k^{(m)}, \Delta\mathbf{x}_{k-1}^{(m)}, \dots, \Delta\mathbf{x}_{r_{k-1}}^{(m)})^T, \\
\Delta\mathbf{y}_{k-1}^{(m)} &= (\Delta\mathbf{x}_{k-1}^{(m)}, \Delta\mathbf{x}_{k-2}^{(m)}, \dots, \Delta\mathbf{x}_{r_{k-1}-1}^{(m)})^T;
\end{aligned} \quad (11)$$

$$\begin{aligned}
\mathbf{a}_{kj}^{(m)} &= [\frac{\partial \mathbf{g}_k}{\partial \mathbf{x}_k^{(m)}}]^{-1}\frac{\partial \mathbf{g}_k}{\partial \mathbf{x}_j^{(m)}}, \\
\mathbf{a}_{kr_j}^{(m)} &= [\frac{\partial \mathbf{g}_k}{\partial \mathbf{x}_k^{(m)}}]^{-1}\sum_{\alpha=j}^{j+1}\frac{\partial \mathbf{g}_k}{\partial \mathbf{x}_\alpha^{\tau(m)}}\frac{\partial \mathbf{x}_\alpha^{\tau(m)}}{\partial \mathbf{x}_{r_j}}, \\
\mathbf{a}_{k(r_j-1)}^{(m)} &= [\frac{\partial \mathbf{g}_k}{\partial \mathbf{x}_k^{(m)}}]^{-1}\sum_{\alpha=j-1}^{j}\frac{\partial \mathbf{g}_k}{\partial \mathbf{x}_\alpha^{\tau(m)}}\frac{\partial \mathbf{x}_\alpha^{\tau(m)}}{\partial \mathbf{x}_{r_j-1}} \\
&\text{with } r_j = j - l_j, j = k-1, k.
\end{aligned} \quad (12)$$

The properties of discrete points $\mathbf{x}_k^{(m)}(k = 1, 2, \dots, mN)$ *can be estimated by the eigenvalues of* $DP_{k(k-1)\dots 1}$ *as*

$$|DP_{k(k-1)\dots 1} - \bar{\lambda}\mathbf{I}_{n(s+1)\times n(s+1)}| = 0\ (k = 1, 2, \dots, mN). \quad (13)$$

The eigenvalues of DP for such a periodic flow are determined by

$$|DP - \lambda \mathbf{I}_{n(s+1)\times n(s+1)}| = 0, \tag{14}$$

and the stability and bifurcation of the periodic flow can be classified by the eigenvalues of $DP(\mathbf{y}_0^*)$ with

$$([n_1^m, n_1^o] : [n_2^m, n_2^o] : [n_3, \kappa_3] : [n_4, \kappa_4] | n_5 : n_6 : [n_7, l, \kappa_7]). \tag{15}$$

(i) *If the magnitudes of all eigenvalues of DP are less than one (i.e., $|\lambda_i| < 1, i = 1, 2, \ldots, n(s+1)$), the approximate periodic solution is stable.*
(ii) *If at least the magnitude of one eigenvalue of DP is greater than one (i.e., $|\lambda_i| > 1, i \in \{1, 2, \ldots, n(s+1)\}$), the approximate periodic solution is unstable.*
(iii) *The boundaries between stable and unstable periodic flow with higher order singularity give bifurcation and stability conditions with higher order singularity.*

Proof See Luo [10]. ∎

3 Discretization of Dynamical Systems

In Luo and Xing [12, 13], consider a time-delayed Duffing oscillator as

$$\ddot{x} + \delta\dot{x} + \alpha_1 x - \alpha_2 x^\tau + \beta x^3 = Q_0 \cos \Omega t \tag{16}$$

where $x = x(t)$ and $x^\tau = x(t - \tau)$. The parameters ($\delta, \alpha_1, \alpha_2$ and β) are damping coefficient, linear stiffness, linear displacement time-delay term coefficient, and nonlinear term coefficient, respectively. Ω and Q_0 are excitation frequency and amplitude, respectively. Let $\mathbf{x} = (x, y)^{\mathrm{T}}$ and $\mathbf{x}^\tau = (x^\tau, y^\tau)^{\mathrm{T}}$. For discrete time $t_k = kh$ ($k = 0, 1, 2, \ldots$), $\mathbf{x}_k = (x_k, y_k)^{\mathrm{T}}$ and $\mathbf{x}_k^\tau = (x_k^\tau, y_k^\tau)^{\mathrm{T}}$. Using a midpoint scheme for the time interval $t \in [t_{k-1}, t_k]$ ($k = 1, 2, \ldots$), the differential equation is discretized as

$$\begin{aligned} &P_k : (\mathbf{x}_{k-1}, \mathbf{x}_{k-1}^\tau) \to (\mathbf{x}_k, \mathbf{x}_k^\tau) \\ &\Rightarrow (\mathbf{x}_k, \mathbf{x}_k^\tau) = P_k(\mathbf{x}_{k-1}, \mathbf{x}_{k-1}^\tau). \end{aligned} \tag{17}$$

The implicit map can be expressed as

$$\begin{aligned} x_k &= x_{k-1} + \frac{1}{2}h(y_k + y_{k-1}), \\ y_k &= y_{k-1} + h[Q_0 \cos \Omega(t_{k-1} + \frac{h}{2}) - \frac{1}{2}\delta(y_k + y_{k-1}) \\ &\quad - \frac{1}{2}\alpha_1(x_k + x_{k-1}) + \frac{1}{2}\alpha_2(x_k^\tau + x_{k-1}^\tau) - \frac{1}{8}\beta(x_k + x_{k-1})^3]. \end{aligned} \tag{18}$$

The time-delay node $\mathbf{x}_k^\tau \approx \mathbf{x}(t_{k-\tau})$ of $\mathbf{x}_k \approx \mathbf{x}(t_k)$ lies between $\mathbf{x}_{k-l_k}$ and $\mathbf{x}_{k-l_k-1}$ (l_k =int(τ/h)). The time-delay node is determined by an interpolation of two points $\mathbf{x}_{k-l_k}$ and $\mathbf{x}_{k-l_k-1}$. For a time-delay node $\mathbf{x}_j^\tau$ $(j = k-1, k)$,

$$\mathbf{x}_j^\tau = \mathbf{h}_j(\mathbf{x}_{r_j-1}, \mathbf{x}_{r_j}, \theta_{r_j}) \text{ for } r_j = j - l_j. \tag{19}$$

The simple Lagrange interpolation gives $\mathbf{x}_j^\tau = \mathbf{h}_j(\mathbf{x}_{r_j-1}, \mathbf{x}_{r_j}, \theta_{r_j})$ $(j = k, k-1)$ as

$$\begin{aligned} x_j^\tau &= x_{j-l_j-1} + (1 - \frac{\tau}{h} + l_j)(x_{j-l_j} - x_{j-l_j-1}), \\ y_j^\tau &= y_{j-l_j-1} + (1 - \frac{\tau}{h} + l_j)(y_{j-l_j} - y_{j-l_j-1}). \end{aligned} \tag{20}$$

Thus, the time-delay nodes are expressed by non-time-delay nodes. The discretization of differential equation for the time-delayed, hardening Duffing oscillator is completed.

4 Period-m Motions

A period-m motion in the time-delayed, hardening Duffing oscillator is described by a discrete mapping structure:

$$\begin{aligned} &P = \underbrace{P_{mN} \circ P_{mN-1} \circ \ldots \circ P_2 \circ P_1}_{mN-\text{actions}} : (\mathbf{x}_0^{(m)}, \mathbf{x}_0^{\tau(m)}) \to (\mathbf{x}_{mN}^{(m)}, \mathbf{x}_{mN}^{\tau(m)}) \\ &(\mathbf{x}_{mN}^{(m)}, \mathbf{x}_{mN}^{\tau(m)}) = P(\mathbf{x}_0^{(m)}, \mathbf{x}_0^{\tau(m)}) \end{aligned} \tag{21}$$

with

$$\begin{aligned} &P_k : (\mathbf{x}_{k-1}^{(m)}, \mathbf{x}_{k-1}^{\tau(m)}) \to (\mathbf{x}_k^{(m)}, \mathbf{x}_k^{\tau(m)}) \\ &(k = 1, 2, \ldots, mN). \end{aligned} \tag{22}$$

The points $\mathbf{x}_k^{(m)*}$ $(k = 1, 2, \ldots, mN)$ on the period-m motion for the time-delayed Duffing oscillator are determined by

$$\begin{aligned} &\left.\begin{aligned} &\mathbf{g}_k(\mathbf{x}_{k-1}^{(m)*}, \mathbf{x}_k^{(m)*}; \mathbf{x}_{k-1}^{\tau(m)*}, \mathbf{x}_k^{\tau(m)*}, \mathbf{p}) = \mathbf{0} \\ &\mathbf{x}_j^{\tau(m)*} = \mathbf{h}_j(\mathbf{x}_{r_j-1}^{(m)*}, \mathbf{x}_{r_j}^{(m)*}, \theta_{rj}), \ j = k, k-1 \end{aligned}\right\} (k = 1, 2, \ldots, mN) \\ &\mathbf{x}_0^{(m)*} = \mathbf{x}_{mN}^{(m)*} \ and \ \mathbf{x}_0^{\tau(m)*} = \mathbf{x}_{mN}^{\tau(m)*}. \end{aligned} \tag{23}$$

In Eq. (23), with $\mathbf{g}_k = (g_{k1}, g_{k2})^{\mathrm{T}}$, the algebraic equations for period-m motion are

$$
\begin{aligned}
g_{k1} &= x_k^{(m)} - [x_{k-1}^{(m)} + \frac{1}{2}h(y_k^{(m)} + y_{k-1}^{(m)})] = 0, \\
g_{k2} &= y_k^{(m)} - \{y_{k-1}^{(m)} + h[Q_0 \cos\Omega(t_{k-1} + \frac{1}{2}h) - \frac{1}{2}\delta(y_k^{(m)} + y_{k-1}^{(m)}) \\
&\quad - \frac{1}{2}\alpha_1(x_k^{(m)} + x_{k-1}^{(m)}) + \frac{1}{2}\alpha_2(x_k^{\tau(m)} + x_{k-1}^{\tau(m)}) - \frac{1}{8}\beta(x_k^{(m)} + x_{k-1}^{(m)})^3]\} \\
&= 0 \\
&(k = 1, 2, \ldots, mN)
\end{aligned} \tag{24}
$$

In Eq. (23), time-delay node $\mathbf{x}_j^{\tau(m)} (j = k, k-1)$ are from Eq. (20)

$$
\begin{aligned}
x_j^{\tau(m)} &= x_{k-l_j-1}^{(m)} + (1 - \frac{\tau}{h} + l_j)(x_{k-l_j}^{(m)} - x_{k-l_j-1}^{(m)}), \\
y_j^{\tau(m)} &= y_{k-l_j-1}^{(m)} + (1 - \frac{\tau}{h} + l_j)(y_{k-l_j}^{(m)} - y_{k-l_j-1}^{(m)}).
\end{aligned} \tag{25}
$$

From Eqs. (23)–(25), discrete nodes of periodic motions in the time-delayed Duffing nonlinear oscillator are obtained by the $2(mN+1)$ equations. If the discrete nodes $\mathbf{x}_k^{(m)*}$ $(k = 1, 2, \ldots, mN)$ of the period-m motion is determined, the corresponding stability of the period-m motion can be discussed by the eigenvalue analysis of the Jacobian matrix of the mapping structure based on the discrete nodes. For $\mathbf{x}_k^{(m)*}$ and $\mathbf{x}_k^{\tau(m)*}$, $\mathbf{x}_k^{(m)} = \mathbf{x}_k^{(m)*} + \Delta\mathbf{x}_k^{(m)}$ and $\mathbf{x}_k^{\tau(m)} = \mathbf{x}_k^{\tau(m)*} + \Delta\mathbf{x}_k^{\tau(m)}$. The linearized equation of implicit mapping is

$$
\sum_{j=k-1}^{k} \frac{\partial \mathbf{g}_k}{\partial \mathbf{x}_j^{(m)}} \Delta\mathbf{x}_j^{(m)} + \frac{\partial \mathbf{g}_k}{\partial \mathbf{x}_j^{\tau(m)}} \left(\frac{\partial \mathbf{x}_j^{\tau(m)}}{\partial \mathbf{x}_{r_j}^{(m)}} \Delta\mathbf{x}_{r_j}^{\tau(m)} + \frac{\partial \mathbf{x}_j^{\tau(m)}}{\partial \mathbf{x}_{r_j-1}^{(m)}} \Delta\mathbf{x}_{r_j-1}^{(m)} \right) = \mathbf{0} \tag{26}
$$

with $r_j = j - l_j,\ j = k-1, k;\ (k = 1, 2, \ldots, mN)$.

Define

$$
\begin{aligned}
\mathbf{y}_k^{(m)} &= (\mathbf{x}_k^{(m)}, \mathbf{x}_{k-1}^{(m)}, \ldots, \mathbf{x}_{r_{k-1}}^{(m)})^{\mathrm{T}}, \\
\mathbf{y}_{k-1}^{(m)} &= (\mathbf{x}_{k-1}^{(m)}, \mathbf{x}_{k-2}^{(m)}, \ldots, \mathbf{x}_{r_{k-1}-1}^{(m)})^{\mathrm{T}}, \\
\Delta\mathbf{y}_k^{(m)} &= (\Delta\mathbf{x}_k^{(m)}, \Delta\mathbf{x}_{k-1}^{(m)}, \ldots, \Delta\mathbf{x}_{r_{k-1}}^{(m)})^{\mathrm{T}}, \\
\Delta\mathbf{y}_{k-1}^{(m)} &= (\Delta\mathbf{x}_{k-1}^{(m)}, \Delta\mathbf{x}_{k-2}^{(m)}, \ldots, \Delta\mathbf{x}_{r_{k-1}-1}^{(m)})^{\mathrm{T}}.
\end{aligned} \tag{27}
$$

The resultant Jacobian matrix of the period-m motion is

$$
\begin{aligned}
DP = DP_{mN(mN-1)\ldots 1} &= \left[\frac{\partial \mathbf{y}_{mN}^{(m)}}{\partial \mathbf{y}_0^{(m)}} \right]_{(\mathbf{y}_0^{(m)*}, \mathbf{y}_1^{(m)*}, \ldots, \mathbf{y}_N^{(m)*})} \\
&= \mathbf{A}_{mN}^{(m)} \mathbf{A}_{mN-1}^{(m)} \cdots \mathbf{A}_1^{(m)} = \mathbf{A}^{(m)}
\end{aligned} \tag{28}
$$

where

$$\Delta \mathbf{y}_k^{(m)} = \mathbf{A}_k^{(m)} \Delta \mathbf{y}_{k-1}^{(m)}, \mathbf{A}_k^{(m)} \equiv \left[\frac{\partial \mathbf{y}_k^{(m)}}{\partial \mathbf{y}_{k-1}^{(m)}} \right]_{(\mathbf{y}_{k-1}^{(m)*}, \mathbf{y}_k^{(m)*})} \tag{29}$$

and

$$\begin{aligned}
\mathbf{a}_{kj}^{(m)} &= [\frac{\partial \mathbf{g}_k}{\partial \mathbf{x}_k^{(m)}}]^{-1} \frac{\partial \mathbf{g}_k}{\partial \mathbf{x}_j^{(m)}}, \\
\mathbf{a}_{kr_j}^{(m)} &= [\frac{\partial \mathbf{g}_k}{\partial \mathbf{x}_k^{(m)}}]^{-1} \sum_{\alpha=j}^{j+1} \frac{\partial \mathbf{g}_k}{\partial \mathbf{x}_\alpha^{\tau(m)}} \frac{\partial \mathbf{x}_\alpha^{\tau(m)}}{\partial \mathbf{x}_{r_j}^{(m)}}, \\
\mathbf{a}_{k(r_j-1)}^{(m)} &= [\frac{\partial \mathbf{g}_k}{\partial \mathbf{x}_k^{(m)}}]^{-1} \sum_{\alpha=j-1}^{j} \frac{\partial \mathbf{g}_k}{\partial \mathbf{x}_\alpha^{\tau(m)}} \frac{\partial \mathbf{x}_\alpha^{\tau(m)}}{\partial \mathbf{x}_{r_j-1}^{(m)}} \\
&\text{with } r_j = j - l_j,\ j = k-1, k; \\
\mathbf{A}_k^{(m)} &= \begin{bmatrix} \mathbf{B}_k^{(m)} & (\mathbf{a}_{k(r_{k-1}-1)}^{(m)})_{2\times 2} \\ \mathbf{I}_k^{(m)} & \mathbf{0}_k^{(m)} \end{bmatrix}_{2(s+1)\times 2(s+1)}, s = 1 + l_{k-1} \\
\mathbf{B}_k^{(m)} &= [(\mathbf{a}_{k(k-1)}^{(m)})_{2\times 2}, \mathbf{0}_{2\times 2}, \ldots, (\mathbf{a}_{k(r_k-1)}^{(m)})_{2\times 2}], \\
\mathbf{I}_k^{(m)} &= diag(\mathbf{I}_{2\times 2}, \mathbf{I}_{2\times 2}, \ldots, \mathbf{I}_{2\times 2})_{2s\times 2s}, \\
\mathbf{0}_k^{(m)} &= (\underbrace{\mathbf{0}_{2\times 2}, \mathbf{0}_{2\times 2}, \ldots, \mathbf{0}_{2\times 2}}_{s})^{\mathrm{T}}
\end{aligned} \tag{30}$$

where

$$\begin{aligned}
\frac{\partial \mathbf{g}_k}{\partial \mathbf{x}_{k-1}^{(m)}} &= \begin{bmatrix} -1 & -\frac{1}{2}h \\ \Delta & \frac{1}{2}h\delta - 1 \end{bmatrix}, \frac{\partial \mathbf{g}_k}{\partial \mathbf{x}_k^{(m)}} = \begin{bmatrix} 1 & -\frac{1}{2}h \\ \Delta & \frac{1}{2}h\delta + 1 \end{bmatrix}, \\
\frac{\partial \mathbf{x}_j^{\tau(m)}}{\partial \mathbf{x}_{r_j-1}^{(m)}} &= \begin{bmatrix} 0 & 0 \\ (\frac{\tau}{h} - l_j) & 0 \end{bmatrix}, \frac{\partial \mathbf{x}_j^{\tau(m)}}{\partial \mathbf{x}_{r_j}^{(m)}} = \begin{bmatrix} 0 & 0 \\ (1 - \frac{\tau}{h} + l_j) & 0 \end{bmatrix}, \\
\frac{\partial \mathbf{g}_k}{\partial \mathbf{x}_j^{(m)}} &= \begin{bmatrix} 0 & 0 \\ 0 & -\frac{1}{2}h\alpha_2 \end{bmatrix}, \\
\Delta &= \frac{1}{8}h[4\alpha_1 + 3\beta(x_k + x_{k-1})^2].
\end{aligned} \tag{31}$$

The eigenvalues of DP for the period-m motion in the time-delayed Duffing oscillator are computed by

$$|DP - \lambda \mathbf{I}_{2(s+1)\times 2(s+1)}| = 0, \tag{32}$$

(i) If the magnitudes of all eigenvalues of DP are less than one (i.e., $|\lambda_i| < 1$, $i = 1, 2, \ldots, 2(s+1)$), the approximate periodic solution is stable.
(ii) If at least the magnitude of one eigenvalue of DP is greater than one (i.e., $|\lambda_i| > 1$, $i \in \{1, 2, \ldots, n(s+1)\}$), the approximate periodic solution is unstable.
(iii) The boundaries between stable and unstable periodic flow with higher order singularity give bifurcation and stability conditions with higher order singularity.

The bifurcation conditions are given as follows.

(iv) If $\lambda_i = 1$ with $|\lambda_j| < 1 (i, j \in \{1, 2, \ldots, 2(s+1)\}$ and $i \neq j)$, the saddle-node bifurcation (SN) occurs.
(v) If $\lambda_i = -1$ with $|\lambda_j| < 1 (i, j \in \{1, 2, \ldots, 2(s+1)\}$ and $i \neq j)$, the period-doubling bifurcation (PD) occurs.
(vi) If $|\lambda_{i,j}| = 1$ with $|\lambda_l| < 1 (i, j, l \in \{1, 2, \ldots, 2(s+1)\}$ and $\lambda_i = \bar{\lambda}_j$ $l \neq i, j)$, Neimark bifurcation (NB) occurs.

5 Bifurcation Trees Varying with Time-Delay

The bifurcation trees of period-1 motion to chaos in the time-delayed Duffing oscillator will be presented through the analytical predictions of period-1 to period-4 motions. Illustration of periodic motions for such time-delayed system will be illustrated. As in Luo and Xing [14], consider a set of system parameters as

$$\delta = 0.5,\ \alpha_1 = 10.0, \alpha_2 = 5.0,\ \beta = 10, \Omega = 1.8, Q_0 = 200.0 \tag{33}$$

with $T = 2\pi/\Omega$.

The bifurcation trees of period-1 to period-4 motions varying with time-delay in the time-delayed Duffing oscillator are predicted analytically, and the discrete nodes are analytically determined by the implicit mapping. The bifurcation trees varying with time-delay are illustrated by displacement and velocity of the periodic nodes with $\text{mod}(k, N) = 0$, as shown in Fig. 1. The solid and dashed curves represent the stable and unstable motions, respectively. The acronyms 'SN', 'PD' and 'NB' represent the saddle-node, period-doubling, and Neimark bifurcations, respectively. The period-1, period-2, and period-4 motions are labeled by P-1, P-2, and P-4, respectively. The period-2 motions appear from the PD bifurcations of the period-1 motions, and the period-4 motions appear from the PD bifurcations of the period-2 motions. The global view of the bifurcation trees is presented in Fig. 1 for $\tau/T \in [0, 1]$ and the zoomed views of the bifurcation trees are presented for a specific frequency ranges in Fig. 1c–f. The ranges of time-delay for the stable and unstable periodic motions are listed in Table 1, and the bifurcation points are tabulated in Table 2.

In Fig. 1a and b, a global view of the bifurcation trees of period-1 to period-4 motion is presented for $\tau/T \in [0, 1]$. For $\tau = 0$, the time-delay Duffing oscillator becomes the non-time-delayed Duffing oscillator. For $\tau = T$, the time-delay is the

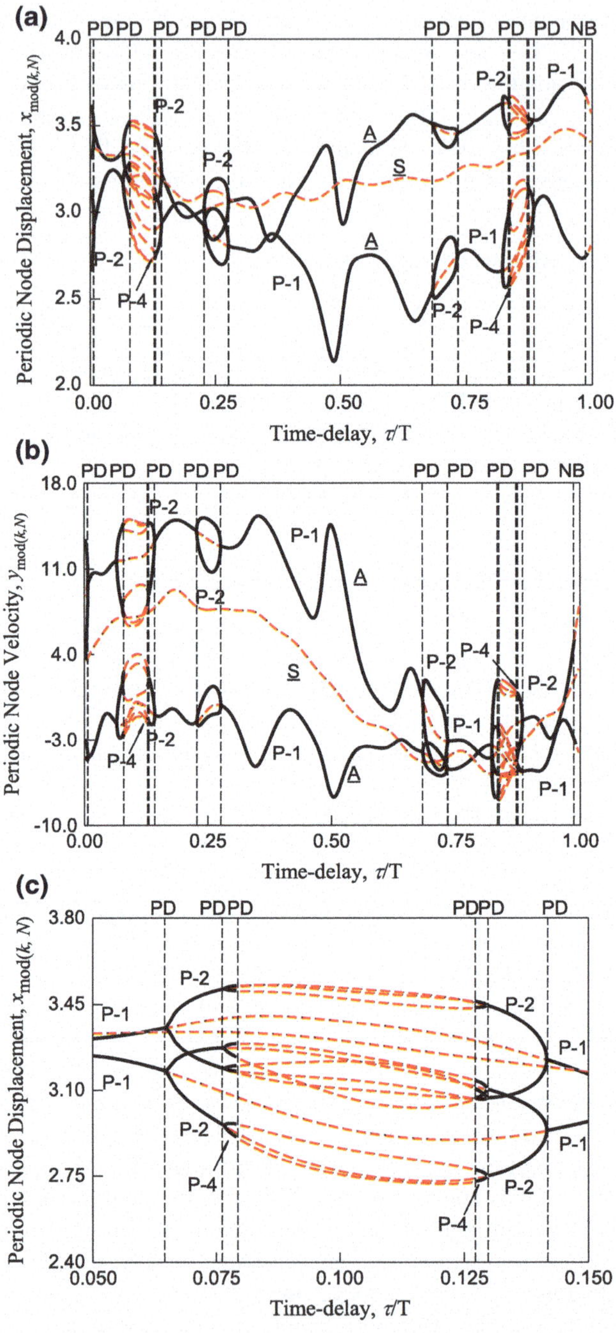

Fig. 1 Bifurcation trees of period-1 motions to chaos varying with time-delay: **a** periodic node displacement, **b** periodic node velocity. The first zoomed view: **c** periodic node displacement, **d** periodic node velocity, the second zoomed view: **e** periodic node displacement, **f** periodic node velocity ($\alpha_1 = 10.0, \alpha_2 = 5.0, \beta = 10.0, \delta = 0.5, Q_0 = 200, \Omega = 1.8$), $\mathrm{mod}(R, N) = 0$

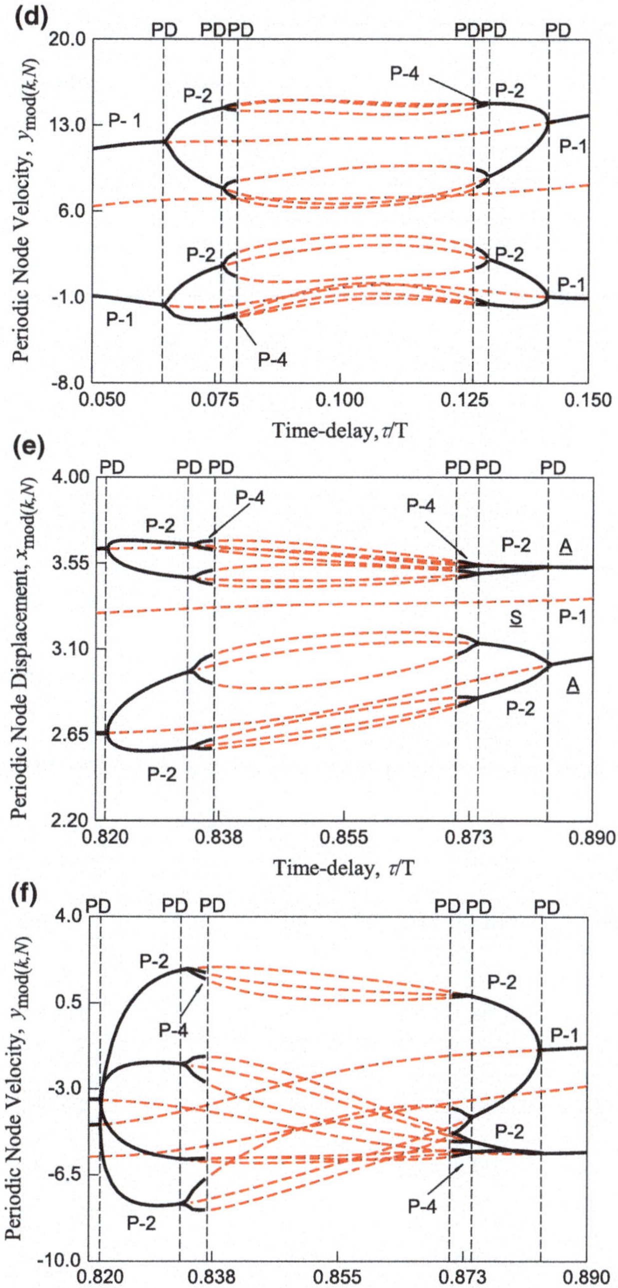

Fig. 1 (continued)

Table 1 Bifurcations for symmetric period-1 motions ($\alpha_1 = 10.0, \alpha_2 = 5.0, \ \beta = 10.0, \delta = 0.5, \ Q_0 = 200, \Omega = 1.8$)

	τ/T (stable)	τ/T (unstable)	Motion type
Over all range	(0.0074, 0.0645)	(0.0, 0.0074)	P-1
	(0.1418, 0.2278)	(0.0645, 0.1418)	
	(0.2765, 0.6284)	(0.2278, 0.2765)	
	(0.7334, 0.8213)	(0.6824, 0.7334)	
	(0.8853, 0.9869)	(0.8213, 0.8853)	
		(0.9869, 1.0)	
1st branch	(0.0, 0.0074)	–	P-2
2nd branch	(0.0645, 0.0762)	(0.0762, 0.1298)	P-2
	(0.1298, 0.1418)		
	(0.0762, 0.0794)	(0.0794, 0.1266)	P-4
	(0.1266, 0.1298)		
3rd branch	(0.2278, 0.2765)	–	P-2
4th branch	(0.6824, 0.7334)	–	P-2
5th branch	(0.8213, 0.8331)	(0.8331, 0.8738)	P-2
	(0.8738, 0.8853)		
	(0.8331, 0.8368)	(0.8368, 0.8706)	P-4
	(0.8706, 0.8738)		
6th branch	(0.9869, 1.0)	–	Quasiperiodic

Table 2 Bifurcations for periodic motions ($\alpha_1 = 10.0, \alpha_2 = 5.0, \beta = 10.0, \delta = 0.5, Q_0 = 200, \Omega = 1.8$)

	τ/T	Bifurcations	Motion type
1st branch (0.0, 0.0074)	0.0074	PD	For P-2
2nd branch (0.0645, 0.1418)	0.0645, 0.1418	PD	For P-2
	0.0762, 0.1298	PD	For P-4
	0.0794, 0.1266	PD	For P-8
3rd branch (0.2278, 0.2765)	0.2278, 0.2765	PD	For P-2
4th branch (0.6824, 0.7334)	0.6824, 0.7334	PD	For P-2
5th branch (0.8213, 0.8853)	0.8213, 0.8853	PD	For P-2
	0.8331, 0.8738	PD	For P-4
	0.8368, 0.8706	PD	For P-8
6th branch (0.9869, 1.0)	0.9869	NB	Quasiperiodic

Note NB-Neimark Bifurcation, PD-period-doubling bifurcation between stable and unstable symmetric periodic motions same as the excitation period. There is an unstable period-1 motion for the entire range of $\tau/T \in [0, 1]$. For $\tau/T \in [0, 0.0074)$, there is a period-2 motion in the first bifurcation tree. At $\tau/T \approx 0.0074$, the saddle-node bifurcation of the period-2 motion exists, and the period-2 motion vanishes, where the period-1 motion also appears. The period-doubling bifurcation for the period-1 motion occurs at $\tau/T \approx 0.0074$. The period-1 motion is stable for $\tau/T \in (0.0074, 0.0645)$. For $\tau/T \in (0.0645, 0.1418)$, there exists the second bifurcation tree from period-1 to period-4 motions, as shown in Fig. 1c and d. The period-doubling bifurcation of period-1 motions is at $\tau/T \approx 0.0645, 0.1418$. The period-2 motion exists in $\tau/T \in (0.0645, 0.1418)$. At $\tau/T \approx 0.0762, 0.1298$, the period-doubling bifurcation of period-2 motions occurs, and the onset of period-4 motions takes place. For $\tau/T \in (0.0762, 0.1298)$, the period-4 motion exists. The period-doubling bifurcations occur at $\tau/T \approx 0.0794, 0.1266$. The unstable period-4 motion lies in $\tau/T \in (0.0794, 0.1266)$ where the period-8 motion exists. For the third branch, the period-doubling bifurcations of the period-1 motion are at $\tau/T \approx 0.2278, 0.2765$. Only the stable period-2 motion exists at $\tau/T \in (0.2278, 0.2765)$. Similarly, for the fourth branch, the period-doubling bifurcations of the period-1 motion are at $\tau/T \approx 0.6824, 0.7334$. Only the stable period-2 motion exists at $\tau/T \in (0.6824, 0.7334)$. For $\tau/T \in (0.8213, 0.8853)$, the fifth bifurcation tree from period-1 to period-4 motions is shown in Fig. 1e and f. The period-doubling bifurcations of period-1 motions are at $\tau/T \approx 0.8213, 0.8853$. The period-2 motion exists in $\tau/T \in (0.8213, 0.8853)$. At $\tau/T \approx 0.8331, 0.8738$, the period-doubling bifurcations of period-2 motions occur for onsets of the period-4 motion. For $\tau/T \in (0.8331, 0.8738)$, the period-4 motion exists, and the period-doubling bifurcations occur at $\tau/T \approx 0.8368, 0.8706$. The unstable period-4 motion lies in $\tau/T \in (0.8368, 0.8706)$ where the period-8 motion also exists.

6 Discrete Fourier Series

As in Luo [12], consider the predicted nodes of period-m motions as $\mathbf{x}_k^{(m)} = (x_k^{(m)}, y_k^{(m)})^{\mathrm{T}}$ for $k = 0, 1, 2, \ldots, mN$ in the time-delayed Duffing oscillator. The period-m motion is determined by the finite Fourier series as

$$\mathbf{x}^{(m)}(t) \approx \mathbf{a}_0^{(m)} + \sum_{j=1}^{M} \mathbf{b}_{j/m} \cos(\frac{j}{m}\Omega t) + \mathbf{c}_{j/m} \sin(\frac{j}{m}\Omega t). \tag{34}$$

There are $(2M + 1)$ unknown vector coefficients of $\mathbf{a}_0^{(m)}, \mathbf{b}_{j/m}, \mathbf{c}_{j/m}$ $(j = 1, 2, \ldots, M)$. From the given nodes $\mathbf{x}_k^{(m)}$ $(k = 0, 1, 2, \ldots, mN)$, such vector unknown coefficients $(2M + 1 \leq mN + 1)$ can be determined. The predicted nodes $\mathbf{x}_k^{(m)}$ on the period-m motion is expressed by the finite Fourier series as for $t_k \in [0, mT]$

$$\begin{aligned}\mathbf{x}^{(m)}(t_k) \equiv \mathbf{x}_k^{(m)} &= \mathbf{a}_0^{(m)} + \sum_{j=1}^{mN/2} \mathbf{b}_{j/m} \cos(\frac{j}{m}\Omega t_k) + \mathbf{c}_{j/m} \sin(\frac{j}{m}\Omega t_k) \\ &= \mathbf{a}_0^{(m)} + \sum_{j=1}^{mN/2} \mathbf{b}_{j/m} \cos(\frac{j}{m}\frac{2k\pi}{N}) + \mathbf{c}_{j/m} \sin(\frac{j}{m}\frac{2k\pi}{N}) \\ &(k = 0, 1, \ldots, mN-1)\end{aligned} \tag{35}$$

where

$$\begin{aligned}&T = \frac{2\pi}{\Omega} = N\Delta t;\ \Omega t_k = \Omega k \Delta t = \frac{2k\pi}{N}, \\ &\mathbf{a}_0^{(m)} = (a_{01}^{(m)}, a_{02}^{(m)})^{\mathrm{T}}, \mathbf{b}_{j/m} = (b_{j/m1}, b_{j/m2})^{\mathrm{T}}, \mathbf{c}_{j/m} = (c_{j/m1}, c_{j/m2})^{\mathrm{T}}.\end{aligned} \tag{36}$$

From discrete nodes on the period-m motion, Eq. (35) gives

$$\begin{aligned}&\mathbf{a}_0^{(m)} = \frac{1}{mN} \sum_{k=0}^{mN-1} \mathbf{x}_k^{(m)}, \\ &\left.\begin{aligned}\mathbf{b}_{j/m} &= \tfrac{2}{mN} \textstyle\sum_{k=1}^{mN-1} \mathbf{x}_k^{(m)} \cos(k\tfrac{2j\pi}{mN}), \\ \mathbf{c}_{j/m} &= \tfrac{2}{mN} \textstyle\sum_{k=1}^{mN-1} \mathbf{x}_k^{(m)} \sin(k\tfrac{2j\pi}{mN})\end{aligned}\right\} (j = 1, 2, \ldots, mN/2)\end{aligned} \tag{37}$$

Thus, the approximate solution for period-m motion in Eq. (16) is

$$\mathbf{x}^{(m)}(t) \approx \mathbf{a}_0^{(m)} + \sum_{j=1}^{mN/2} \mathbf{b}_{j/m} \cos(\frac{j}{m}\Omega t) + \mathbf{c}_{j/m} \sin(\frac{j}{m}\Omega t). \tag{38}$$

The foregoing equation can be rewritten as

$$\begin{Bmatrix} x^{(m)}(t) \\ y^{(m)}(t) \end{Bmatrix} \equiv \begin{Bmatrix} x_1^{(m)}(t) \\ x_2^{(m)}(t) \end{Bmatrix} \approx \begin{Bmatrix} a_{01}^{(m)} \\ a_{02}^{(m)} \end{Bmatrix} + \sum_{j=1}^{mN/2} \begin{Bmatrix} A_{j/m1} \cos(\frac{j}{m}\Omega t - \phi_{j/m1}) \\ A_{j/m1} \cos(\frac{j}{m}\Omega t - \phi_{j/m2}) \end{Bmatrix} \tag{39}$$

where the harmonic amplitudes and harmonic phases for period-m motion are

$$\begin{aligned}A_{j/m1} &= \sqrt{b_{j/m1}^2 + c_{j/m1}^2}, \phi_{j/m1} = \arctan\frac{c_{j/m1}}{b_{j/m1}}, \\ A_{j/m2} &= \sqrt{b_{j/m2}^2 + c_{j/m2}^2}, \phi_{j/m2} = \arctan\frac{c_{j/m2}}{b_{j/m2}}.\end{aligned} \tag{40}$$

For simplicity, harmonic amplitudes of displacement $x^{(m)}(t)$ for period-m motions will be presented only. Thus the displacement is expressed as

$$x^{(m)}(t) \approx a_0^{(m)} + \sum_{j=1}^{mN/2} b_{j/m} \cos(\frac{j}{m}\Omega t) + c_{j/m} \sin(\frac{j}{m}\Omega t) \tag{41}$$

and

$$x^{(m)}(t) \approx a_0^{(m)} + \sum_{j=1}^{mN/2} A_{j/m} \cos(\frac{j}{m}\Omega t - \phi_{j/m}) \tag{42}$$

where

$$A_{j/m} = \sqrt{b_{j/m}^2 + c_{j/m}^2}, \phi_{j/m} = \arctan \frac{c_{j/m}}{b_{j/m}}. \tag{43}$$

7 Illustrations

To illustrate complexity of periodic motions in the time-delayed Duffing oscillator, initial conditions for numerical simulations are taken from the analytical prediction in the bifurcation trees of period-1 to period-4 motions, and the corresponding harmonic amplitudes of periodic motions are presented to show harmonic terms effects on periodic motions. The system parameters in Eq. (33) are used. Numerical and analytical results are presented by solid curves and symbols, respectively. The initial time-delay are presented through blue circular symbols. The delay-initial-starting and delay-initial finishing points are "D.I.S." and "D.I.F.", respectively.

From the fifth circular bifurcation tree of period-1 motion to chaos, consider an asymmetric period-1 motion with $\tau = 2.70$, and initial conditions for a pair of the two asymmetric period-1 motions are from the analytical prediction (i.e., $x_0 = 3.531702$, $y_0 = -3.246012$ (left); $x_0 = 2.734844$, $y_0 = -4.877491$ (right)). $\tau/T \approx 0.7735$ implies that the time-delay takes about 77% excitation period. The initial time-delay is presented by green symbols. The numerical solution of the period-1 motion is presented by solid curves and the analytical prediction is depicted by circular symbols. Displacements for the two asymmetric period-1 motion are presented in Fig. 2(i) and (ii). The two trajectories with time-delay for the two period-1 motions are shown in Fig. 2 (iii) and (iv). The trajectories have two small cycles on both sides plus a big circle. The corresponding harmonic amplitudes and phases are presented in Fig. 2(v) and (vi), respectively. $A_0 = a_0^R = -a_0^L \approx 0.2268$, $A_1 \approx 2.8346$, $A_2 \approx 0.2516$, $A_3 \approx 0.4459$, $A_4 \approx 0.3166$, $A_5 \approx 0.1482$, $A_6 \approx 0.3558$, $A_7 \approx 0.5709$, $A_8 \approx 0.2826$, $A_9 \approx 0.1807$, $A_{10} \approx 0.0199$, $A_{11} \approx 0.0244$, $A_{12} \approx 0.0142$, $A_{13} \approx 0.0180$, $A_{14} \approx 0.0316$, $A_{15} \approx 0.0274$, $A_{16} \approx 0.0189$. Other harmonic amplitudes lie in $A_k \in (10^{-15}, 10^{-3})$ $(k = 17, 19, \ldots, 100)$ and $A_{100} \approx 1.3004\text{e-}14$. With increasing harmonic orders, the harmonic amplitudes decrease. Thus, one can use 100 harmonic terms to approximate the two asymmetric period-1 motions. The harmonic phases changes with harmonic orders from 0 to 2π with $\varphi_k^L = \text{mod}(\varphi_k^R + (k+1)\pi, 2\pi)$.

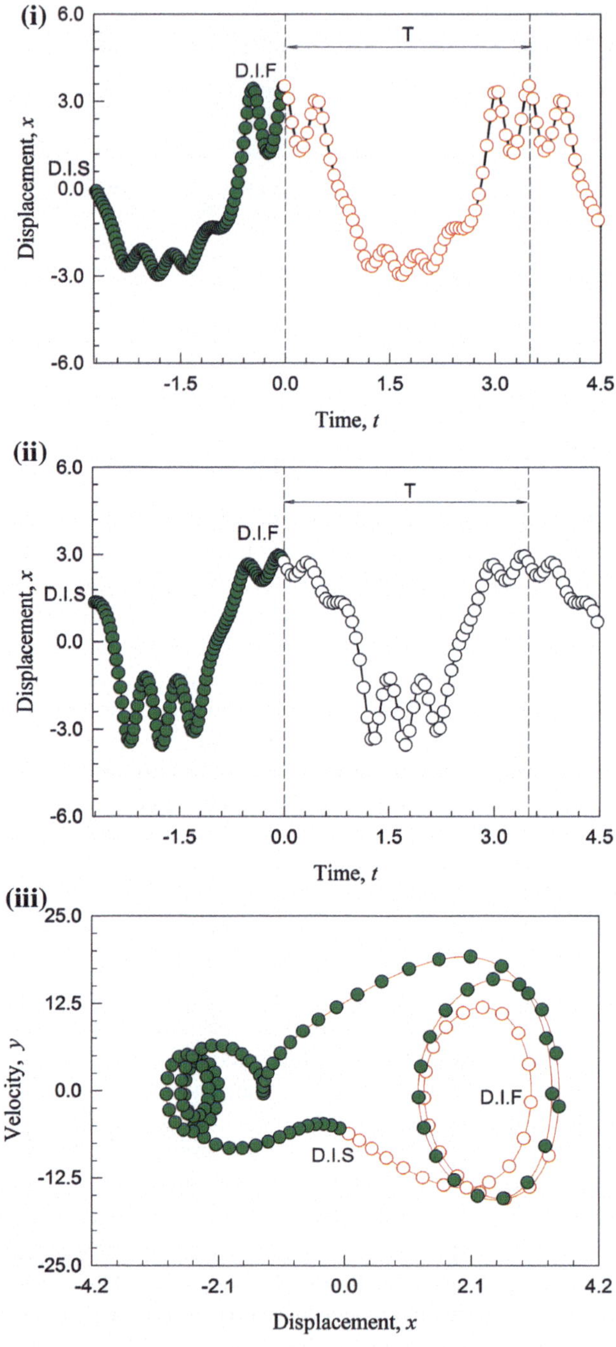

Fig. 2 Period-1 motion with large time-delay ($\tau = 2.7$): (i) displacement (left), (ii) displacement (right); (iii) trajectory (left), (iv) trajectory (right), (v) harmonic amplitudes, (vi) harmonic phases. (I.C.: $x_0 = 3.531702$, $y_0 = -3.246012$ (left); $x_0 = 2.734844$, $y_0 = -4.877491$ (right)) ($\alpha_1 = 10.0, \alpha_2 = 5.0, \beta = 10.0, \delta = 0.5, Q_0 = 200, \Omega = 1.8$)

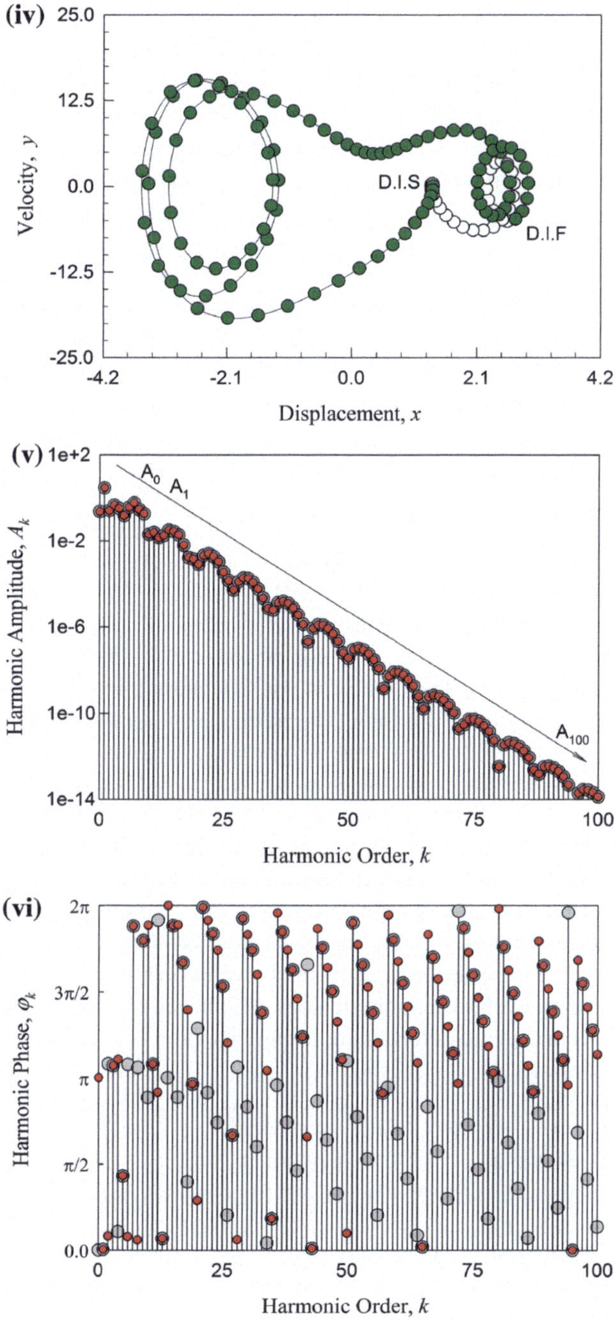

Fig. 2 (continued)

For a pair of asymmetric period-2 motions with $\tau = 2.90$, initial conditions are $x_0 = 3.626654$, $y_0 = 4.551817$ (left); and $x_0 = 2.590282$, $y_0 = -2.033225$ (right). $\tau/T \approx 0.8308$ and the time-delay is also very large. The initial time-delay is still presented by green symbols. Displacements for the two asymmetric period-2 motion are presented in Fig. 3(i) and (ii). The time for period-2 motion is doubled from the period-1 motion, and the period-2 motion is almost repeated the period-1 motion responses two times. The two trajectories with time-delay for the two period-2 motions are shown in Fig. 3(iii) and (iv). The trajectories have four small cycles on both sides plus two big circles to connect small cycles on the both sides. The harmonic amplitudes and phases for period-2 motions are presented in Fig. 3(v) and (vi), respectively. The magnitudes are $A_0 = a_0^{(2)R} = -a_0^{(2)L} \approx 0.2201$, $A_{1/2} \approx 0.0628$, $A_1 \approx 2.8550$, $A_{3/2} \approx 0.0141$, $A_2 \approx 0.2201$, $A_{5/2} \approx 0.0734$, $A_3 \approx 0.4541$, $A_{7/2} \approx 0.0270$, $A_4 \approx 0.2661$, $A_{9/2} \approx 0.1229$, $A_5 \approx 0.0706$, $A_{11/2} \approx 0.0385$, $A_6 \approx 0.3328$, $A_{13/2} \approx 0.2107$, $A_7 \approx 0.5403$, $A_{15/2} \approx 0.0191$, $A_8 \approx 0.2805$, $A_{17/2} \approx 0.1255$, $A_9 \approx 0.1849$, $A_{19/2} \approx 8.2525\text{e-}3$, $A_{10} \approx 0.0178$, $A_{21/2} \approx 5.7368\text{e-}3$, $A_{11} \approx 0.0180$, $A_{23/2} \approx 9.2916\text{e-}3$, $A_{12} \approx 6.5333\text{e-}3$, $A_{25/2} \approx 3.5730\text{e-}3$, $A_{13} \approx 0.0170$, $A_{27/2} \approx 0.0108$, $A_{14} \approx 0.0280$, $A_{29/2} \approx 0.0177$, $A_{15} \approx 0.0272$, $A_{31/2} \approx 0.0111$, $A_{16} \approx 0.0157$, $A_{33/2} \approx 7.1351\text{e-}3$, and $A_{17} \approx 6.6818\text{e-}3$. Other harmonic amplitudes lie in $A_{k/2} \in (10^{-15}, 10^{-3})$ $(k = 35, 36, \ldots, 200)$ and $A_{100} \approx 2.0283\text{e-}14$. With increasing harmonic orders, the harmonic amplitudes also decrease. Harmonic terms $A_{k/2}$ $(\text{mod}(k, 2) \neq 0)$ for period-2 motion only is much smaller than $A_{k/2}$ $(\text{mod}(k, 2) = 0)$. Thus, one can use 200 harmonic terms to approximate the two asymmetric period-2 motions. The harmonic phases from 0 to 2π are with $\varphi^L_{k/2} = \text{mod}(\varphi^R_{k/2} + ((k + 2r)/2 + 1)\pi, 2\pi)$ $(k = 1, 2, \ldots; r = 0)$ for $t_0 = rT$.

For a pair of asymmetric period-4 motions with $\tau = 2.92$, initial conditions are $x_0 = 3.668104$, $y_0 = 1.737186$ (left); $x_0 = 2.577874$, $y_0 = -1.676010$ (right). $\tau/T \approx 0.8365$. Displacements for the two asymmetric period-4 motion are presented in Fig. 4(i) and (ii). The time for period-4 motion is doubled from the period-2 motion, and the period-4 motion almost repeats the period-1 motion responses four times. The two trajectories with time-delay for the two period-4 motions are shown in Fig. 4(iii) and (iv). The trajectories have eight small cycles on both sides plus four big circles to connect small cycles on the both sides. The harmonic amplitudes and phases for the period-4 motions are determined, in Fig. 4(v) and (vi), respectively. The harmonic magnitudes are $A_0 = a_0^{(4)R} = -a_0^{(4)L} \approx 0.2075$, $A_{1/4} \approx 0.0285$, $A_{1/2} \approx 0.0676$, $A_{3/4} \approx 1.3334\text{e-}3$, $A_1 \approx 2.8606$, $A_{5/4} \approx 1.8314\text{e-}3$, $A_{3/2} \approx 0.0121$, $A_{7/4} \approx 0.0113$, $A_2 \approx 0.2092$, $A_{9/4} \approx 0.0202$, $A_{5/2} \approx 0.0755$, $A_{11/4} \approx 5.0156\text{e-}3$, $A_3 \approx 0.4548$, $A_{13/4} \approx 4.0509\text{e-}3$, $A_{7/2} \approx 0.0220$, $A_{15/4} \approx 0.0125$, $A_4 \approx 0.2509$, $A_{17/4} \approx 0.0262$, $A_{9/2} \approx 0.1328$, $A_{19/4} \approx 0.0100$, $A_5 \approx 0.0720$, $A_{21/4} \approx 9.0113\text{e-}3$, $A_{11/2} \approx 0.0339$, $A_{23/4} \approx 0.0137$, $A_6 \approx 0.3141$, $A_{25/4} \approx 0.0378$, $A_{13/2} \approx 0.2079$, $A_{27/4} \approx 0.0120$, $A_7 \approx 0.5464$, $A_{29/4} \approx 0.0109$, $A_{15/2} \approx 0.0124$, $A_{31/4} \approx 0.0112$, $A_8 \approx 0.2633$, $A_{33/2} \approx 0.0263$, $A_{17/2} \approx 0.1267$, $A_{35/4} \approx 9.0150\text{e-}3$, $A_9 \approx 0.1907$, $A_{37/4} \approx 5.5408\text{e-}3$, $A_{19/2} \approx 8.0743\text{e-}3$, $A_{39/4} \approx 1.0027\text{e-}3$, $A_{10} \approx 0.0163$, $A_{41/4} \approx 1.3272\text{e-}3$, $A_{21/2} \approx 5.1506\text{e-}3$, $A_{43/4} \approx 7.1315\text{e-}4$, $A_{11} \approx 0.0170$, $A_{45/4} \approx 9.1745\text{e-}4$, $A_{23/2} \approx 8.4709\text{e-}3$, $A_{47/4} \approx 8.8042\text{e-}4$, $A_{12} \approx 5.9744\text{e-}3$, $A_{49/4} \approx 1.3697\text{e-}3$, $A_{25/2} \approx 3.5724\text{e-}3$,

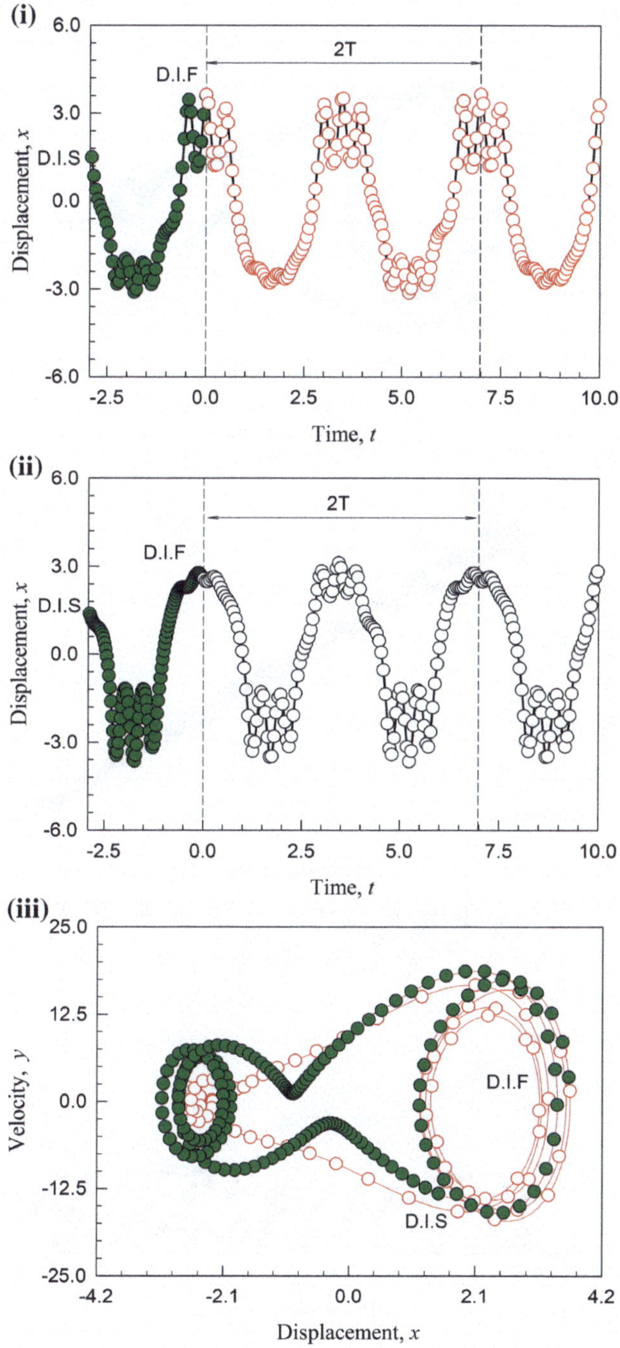

Fig. 3 Period-2 motion with large time-delay ($\tau = 2.9$): (i) displacement (left), (ii) displacement (right); (iii) trajectory (left), (iv) trajectory (right), (v) harmonic amplitudes, (vi) harmonic phases. (I.C.: $x_0 = 3.655843$, $y_0 = 1.633443$ (left); $x_0 = 2.571812$, $y_0 = -1.932056$ (right)) ($\alpha_1 = 10.0, \alpha_2 = 5.0, \beta = 10.0, \delta = 0.5, Q_0 = 200, \Omega = 1.8$)

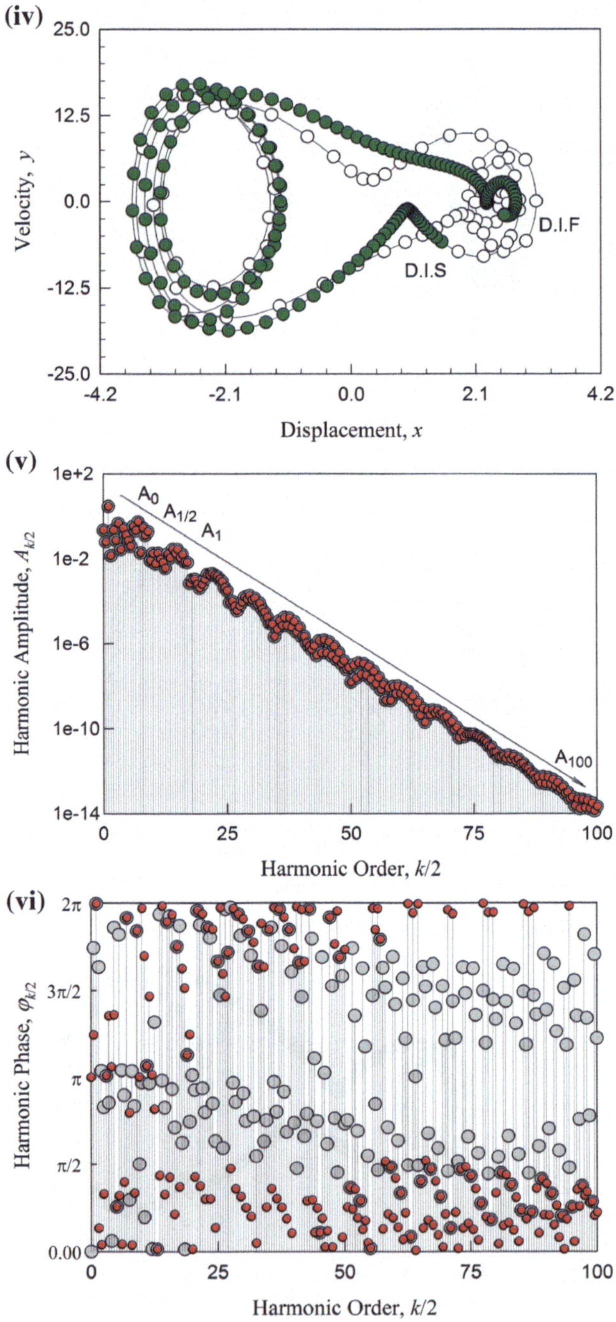

Fig. 3 (continued)

Fig. 4 Period-4 motion with large time-delay ($\tau = 2.92$): (i) displacement (left), (ii) displacement (right); (iii) trajectory (left), (iv) trajectory (right), (v) harmonic amplitudes, (vi) harmonic phases. (I.C.: $x_0 = 3.668104$, $y_0 = 1.737186$ (left); $x_0 = 2.577874$, $y_0 = -1.676010$ (right)) ($\alpha_1 = 10.0, \alpha_2 = 5.0, \beta = 10.0, \delta = 0.5, Q_0 = 200, \Omega = 1.8$)

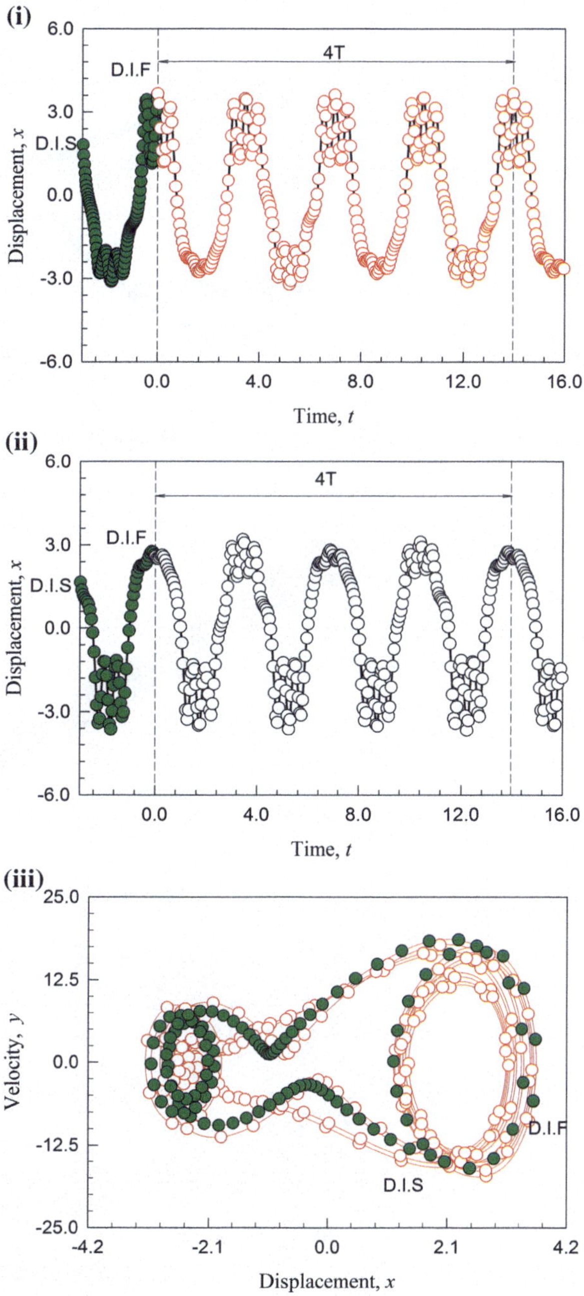

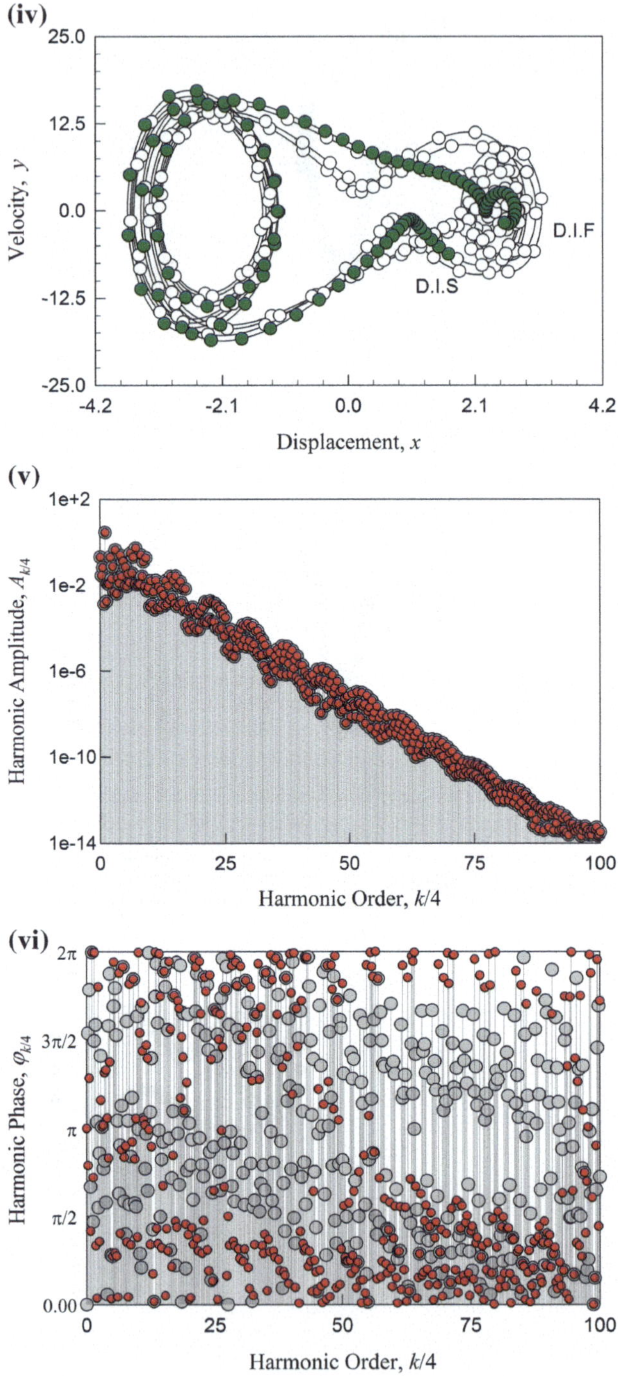

Fig. 4 (continued)

$A_{51/4} \approx 1.1010\text{e-}3$, $A_{13} \approx 0.0164$, $A_{53/4} \approx 1.3546\text{e-}3$, $A_{27/2} \approx 9.4996\text{e-}3$, $A_{55/4} \approx 2.1266\text{e-}3$, $A_{14} \approx 0.0266$, $A_{57/4} \approx 2.8443\text{e-}3$, $A_{29/2} \approx 0.0175$, $A_{59/4} \approx 1.4433\text{e-}3$, $A_{15} \approx 0.0271$, $A_{61/4} \approx 1.0085\text{e-}3$, $A_{31/2} \approx 9.8767\text{e-}3$, $A_{63/4} \approx 1.3299\text{e-}3$, $A_{16} \approx 0.0150$, $A_{65/4} \approx 1.6896\text{e-}3$, $A_{32/2} \approx 7.3809\text{e-}3$, $A_{67/4} \approx 7.2258\text{e-}4$, and $A_{17} \approx 6.8573\text{e-}3$. Other harmonic amplitudes lie in $A_{k/4} \in (10^{-15}, 10^{-3})$ $(k = 69, 70, \ldots, 400)$ and $A_{100} \approx 2.4579\text{e-}14$. With increasing harmonic orders, the harmonic amplitudes also decrease. Harmonic terms $A_{k/4}$ $(\mathrm{mod}(k, 4) \neq 0$ and $\mathrm{mod}(k, 2) \neq 0)$ for period-4 motion only is much smaller than $A_{k/4}$ $(\mathrm{mod}(k, 2) = 0$ or $\mathrm{mod}(k, 4) = 0)$. Thus, one can use 400 harmonic terms to approximate the two asymmetric period-4 motions. The relations of harmonic phases between the two asymmetric period-4 motions from 0 to 2π are with $\varphi^{L}_{k/4} = \mathrm{mod}(\varphi^{R}_{k/4} + ((k + 4r)/4 + 1)\pi, 2\pi)$ $(k = 1, 2, \ldots; r = 0)$ for $t_0 = rT$.

8 Concluding Remarks

The time-delay effects of periodic motions in a periodically forced, time-delayed, hardening Duffing oscillator were presented in this chapter. The bifurcation trees varying with time-delay for period-1 motions to chaos were presented, and the corresponding stability and bifurcation were also presented to show time-delay effects on periodic motions. From the analytical predictions of periodic motions, numerical results of period-1 to period-4 motions were illustrated to verify the analytical predictions. Period-1 motions of the time-delayed Duffing oscillator did not vary too much with $\tau/T \in [0, 1]$. With period-doubling bifurcation, period-2 and period-4 motions in the time-delayed Duffing oscillator were in the vicinity of the period-1 motions in $\tau/T \in [0, 1]$.

References

1. Hu, H.Y., Dowell, E.H., Virgin, L.N.: Resonance of harmonically forced Duffing oscillator with time-delay state feedback. Nonlinear Dyn. **15**(4), 311–327 (1998)
2. Hu, H.Y., Wang, Z.H.: Dynamics of Controlled Mechanical Systems with Delayed Feedback. Springer, Berlin (2002)
3. MacDonald, N.: Harmonic balance in delay-differential equations. J. Sound Vib. **186**(4), 649–656 (1995)
4. Leung, A.Y.T., Guo, Z.: Bifurcation of the periodic motions in nonlinear delayed oscillators. J. Vib. Control **20**, 501–517 (2014)
5. Liu, L., Kalmar-Nagy, T.: High-dimensional harmonic balance analysis for second-order delay-differential equations. J. Vib. Control **16**(7–8), 1189–1208 (2010)
6. Luo, A.C.J.: Analytical solutions of periodic motions in dynamical systems with/without time-delay. Int. J. Dyn. Control **1**, 330–359 (2013)
7. Luo, A.C.J., Jin, H.X.: Bifurcation trees of period-m motion to chaos in a time-delayed, quadratic nonlinear oscillator under a periodic excitation. Discontinuity Nonlinearity Complexity **3**, 87–107 (2014)

8. Luo, A.C.J., Jin, H.X.: Complex period-1 motions of a periodically forced Duffing oscillator with a time-delay feedback. Int. J. Dyn. Control **3**(4), 325–340 (2015)
9. Luo, A.C.J., Jin, H.X.: Period-*m* motions to chaos in a periodically forced Duffing oscillator with a time-delay feedback. Int. J. Bifurcat. Chaos **24**(10), article no: 1450126, 20 p (2014)
10. Luo, A.C.J.: Periodic flows in nonlinear dynamical systems based on discrete implicit maps. Int. J. Bifurcat. Chaos **25**(3), Article No: 1550044, 62 p (2015)
11. Luo, A.C.J., Guo, Yu.: A semi-analytical prediction of periodic motions in Duffing oscillator through mapping structures. Discontinuity Nonlinearity Complexity **4**(2), 121–150 (2015)
12. Luo, A.C.J., Xing, S.Y.: Symmetric and asymmetric period-1 motions in a periodically forced, time-delayed, hardening Duffing oscillator. Nonlinear Dyn. **85**, 1141–1186 (2016)
13. Luo, A.C.J., Xing, S.Y.: Multiple bifurcation trees of period-1 motions to chaos in a periodically forced, time-delayed, hardening Duffing oscillator. Nonlinear Dyn. **89**, 405–434 (2016)
14. Luo, A.C.J., Xing, S.Y.: Time-delay effects on periodic motions in a periodically forced, time-delayed, hardening Duffing oscillator. J. Vib. Test. Syst. Dyn. **1**(1), 73–91 (2017)

Nonchaos-Mediated Mixed-Mode Oscillations in a Prey-Predator Model with Predator Dormancy

Joana G. Freire, Marcia R. Gallas and Jason A.C. Gallas

1 Introduction

A well-known property of ecosystems is that the equilibrium of the classical prey-predator system is destabilized when the carrying capacity, i.e., the environment maximal load, of the preys increases [1]. Under this circumstance, a Hopf bifurcation occurs. Its effect is to destabilize the coexistence of prey and predator in favor of a stable periodic cycle. Then, as the carrying capacity further increases, the amplitude of the periodic orbits also increase, with the minimum value of the population density decreasing. This dynamical interplay acts so that population extinction due to stochastic environmental perturbations becomes more likely. This is known as the paradox of enrichment [2]. However, this does not need to be the only scenario. As discussed by Kuwamura et al. [3], the extinction of populations is not always observed in natural enriched environments. In other words, the destabilization effect induced by eutrophication, namely the enrichment of the environment with nutrients,

J.G. Freire · M.R. Gallas · J.A.C. Gallas (✉)
Instituto de Altos Estudos da Paraíba, Rua Silvino Lopes 419-2502,
João Pessoa, PB 58039-190, Brazil
e-mail: jason.gallas@gmail.com

J.G. Freire
e-mail: negracarvao@gmail.com

M.R. Gallas
e-mail: marciagallas@gmail.com

J.G. Freire · M.R. Gallas · J.A.C. Gallas
Departamento de Física, Universidade Federal da Paraíba,
João Pessoa 58051-970, Brazil

J.G. Freire
Faculdade de Ciências, Instituto Dom Luiz, Universidade de Lisboa,
1749-016 Lisbon, Portugal

M.R. Gallas · J.A.C. Gallas
Complexity Sciences Center, 9225 Collins Ave. #1208,
Surfside, FL 33154-3046, USA

M. Edelman et al. (eds.), *Chaotic, Fractional, and Complex Dynamics: New Insights and Perspectives*, Understanding Complex Systems,
https://doi.org/10.1007/978-3-319-68109-2_6

is somehow canceled in natural ecosystems. Many studies have attempted to clarify mechanisms for stabilizing the population dynamics in enriched environments. See, e.g., Refs. [3, 4] and several references therein.

This paper is motivated by a study of Kuwamura and Chiba [5] who proposed a mathematical framework to describe a stabilizing mechanism in enriched environment. The key idea of the approach is to consider the differentiation of the predator into active and dormant states. As indicated by a number of studies, dormancy of populations plays an important role in various biological studies [6–9]. According to Gyllström and Hansson [7], in comfortable environments zooplankton produces mainly subitaneous eggs. However, fertilized eggs (resting eggs, dormancy state) may be produced in order to escape periods of harsh environmental conditions. Experiments show [10] that the amplitude of prey-predator cycles of Daphnia and its algal prey in microcosms increases when a portion of ephippia-producing females is replaced by asexually-reproducing gravid females. This suggests that dormancy of predators may stabilize the population dynamics of Daphnia and its algal prey at high nutrient levels.

Kuwamura and Chiba [5] have shown that, under certain conditions, the dormancy of predators induces mixed-mode oscillations and chaos in the population dynamics of their prey-predator model. Using the theory of fast-slow systems, they argued mixed-mode oscillations and chaos to bifurcate from a coexisting equilibrium. These results are of interest because they may help to find experimental conditions under which one can demonstrate chaotic population dynamics in a simple phytoplankton-zooplankton (-resting eggs) community in a microcosm with a short duration.

The main finding reported here is the discovery of nonchaos-mediated cascades of mixed-mode oscillations observed in the prey-predator model with dormancy. Mixed-mode oscillations exist in two distinct flavors with easy to distinguish signatures [11]: Nonchaos-mediated cascades display *spike-adding sequences* while the more familiar chaos-mediated cascades involve *spikes-doubling sequences*. In contrast to the standard chaos-mediated cascades, the elusive nonchaos-mediated cascades were discovered quite recently and at present are known only for a handfull of cases, namely, for an enzyme reaction [11, 12], for familiar models of oscillators [13, 14], and for a centrifugal flywheel governor system [15]. Thus, the observation of nonchaos-mediated cascades in the rather different context of a biological model adds one more example to the list above.

The present prey-predator model shares a sigmoidal function and other similarities with situations known to lead to very rich dynamics [16, 17]. However, its equations of motion also contain contributions from different terms, which are of interest for our on-going quest of classifying complexity phases in continuous-time dynamical systems [12, 18], so that a closer investigation seems warranted. This fact is another motivation for the present investigation. Before starting, recall that there are no theoretical methods capable of locating stability phases for motions of arbitrary periodicity, going beyond time-honored fixed-point analysis. Therefore, all such studies are necessarily of a numerical nature.

2 Prey-Predator Model with Predator Dormancy

The model of dormancy effects in the prey-predator population dynamics studied here is defined by a set of three coupled ordinary differential equations [3, 5]:

$$\frac{dp}{dt} = r\left(1 - \frac{p}{k}\right) p - f(p)z, \tag{1}$$

$$\frac{dz}{dt} = k_1\mu(p)f(p)z + \alpha w - d_1 z, \tag{2}$$

$$\frac{dw}{dt} = \varepsilon\left[k_2\left(1 - \mu(p)\right) f(p)z - \alpha w - d_2 w\right]. \tag{3}$$

Here, p and z denote population densities of prey and predators, respectively, and w is the population density of predators with a dormancy state (resting eggs). Parameters r and k correspond to the intrinsic growth rate and the carrying capacity of prey, respectively. The function $f(p)$, represents a positive strictly monotone increasing bounded function, taken to be a Holling type II functional response, namely $f(p) = bp/(c + p)$, where b and c represent the maximum foraging rate and the half saturation constant, respectively. Here, ε is a small time-scale separation parameter used to control the speed in the system [5]. Parameters k_1 and k_2 denote the growth rates of predators in the active and dormant states. The function $\mu(p)$ is a switching function which controls the induction of dormancy. It is given the sigmoidal function [16]

$$\mu(p) = \frac{1}{2}\left[\tanh\left(\frac{p-\eta}{\sigma}\right) + 1\right], \tag{4}$$

where η and σ denote the switching level and the sharpness of the switching effect. This function implies that predators produce more resting eggs than subitaneous eggs when the prey density decreases below a certain level η. Parameters d_1 and d_2 denote the mortality rates of the active and dormant predators, respectively, while η is the hatching rate, i.e., resting eggs have a dormancy period with $1/\alpha$ on average. The model above is an extension of a prey-predator interaction-diffusion system based on the Bazykin model, known as the MacArthur-Rosenzweig model with intraspecific interaction (density-dependent inhibition) among predators, to which the effect of predator dormancy is incorporated. For details see Ref. [19].

Basically, Kuwamura and Chiba [5] considered two different situations of the model, $\varepsilon = 1$ and $\varepsilon = 0.2$, and studied how the bifurcation structure changes as a function k, the carrying capacity, and with variations of b and d_1. Table 1 collects the default parameter values used here, unless stated otherwise.

Stability diagrams were constructed by integrating numerically Eqs. (1)–(3) using a standard fourth-order Runge-Kutta with fixed time-step $h = 0.01$. Such integrations were performed horizontally, from left to right, starting from an arbitrarily chosen initial condition, $(p, z, w) = (0.6, 0.15, 1.5)$, and proceeding by 'following the attractor' [17], namely by using the values stored in the computer buffers as initial conditions when incrementing parameters infinitesimally. The first 0.6×10^6

Table 1 Parameters values used here, taken from Ref. [5], in the same units

$r = 0.5$	$\alpha = 0.02$	$c = 2$	$\sigma = 0.1$	$b = 2$
$\eta = 1.0$	$d_1 = 0.2$	$d_2 = 0.0001$	$k_1 = 0.6$	$k_2 = 0.12$

integration steps were disregarded as a transient time needed to come close to the attractor, with an additional of 12×10^6 steps used to compute the Lyapunov spectrum (not presented in this paper). To find the number of peaks per a period, subsequently to the computation of Lyapunov exponents, integrations were continued for 12×10^6 additional time-steps, recording up to 800 extrema (maxima and minima) of the three variables and, from the recorded extrema, determining whether or not pulses repeated.

3 Nonchaos-Mediated Cascades of Mixed-Mode Oscillations

Figure 1 shows typical stability diagrams computed for the prey-predator system with dormancy, Eqs. (1)–(3). Such diagrams, called *isospike diagrams*, use a palette of 17 colors, as indicated by the colorbars, to display the number of spikes per period of the stable oscillations. Patterns with more than 17 peaks are plotted by recycling the 17 basic colors modulo 17. Black represents "chaos" (i.e., lack of numerically detectable periodicity), and white marks non-zero amplitude fixed-points (non-oscillatory solutions). The three diagrams on the top row in Fig. 1 display from left to right the number of spikes as observed by following the temporal evolution of the three variables, p, z, w, respectively. The next row shows magnifications of the regions inside the yellow boxes seen on the top panels.

Figure 1 shows a number of interesting facts. The three panels show the precise location where the number of spikes change for every variable. Oscillations in p and z display a much larger variation of their number of spikes then oscillations of w which display one spike over extended intervals when b increases. The vertical white stripes seen on the left of the panels show that, independently of b, the maximum foraging rate, the fixed-point is not affected by the carrying capacity k. There is a dynamical threshold for the effects of k to start to be noticed in the system. Furthermore, the onset of chaotic oscillations occurs only for specific ranges of b and k. In particular, the onset occurs for considerably larger values of k when b decreases.

Figure 2 shows a companion of two cascades of mixed-mode oscillations found in Eqs. (1)–(3), recorded the parallel lines seen in Fig. 1d and defined by the equations

$$b = -0.483k + 20.483 \quad \text{and} \quad b = -0.483k + 26.035. \tag{5}$$

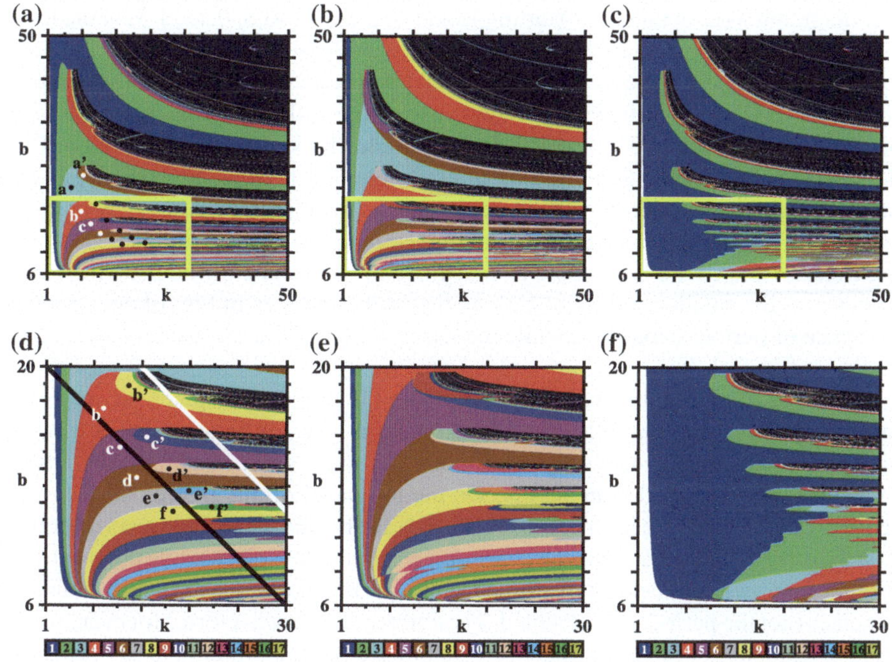

Fig. 1 Nonchaos-mediated cascades of mixed-mode oscillations, illustrated by sequences of domains not separated by chaos (shown in black), as a function of the maximum foraging rate b and the carrying capacity k. Panels (**a**)–(**c**) represent the number of spikes per period as measured for p, z, and w, respectively. The boxes in panels (**a**)–(**c**) are shown magnified in panels (**d**)–(**f**). The points marked in panels (**a**) and (**d**) are the same shown with more detail in Fig. 3, which is a magnification of panel (**d**). Bifurcations diagrams along the pair of lines in (**d**) are shown in Fig. 2. Here $\varepsilon = 0.2$ and other parameters as given in Table 1. Each individual panel displays the analysis of a mesh formed by 600×600 equally spaced parameter points

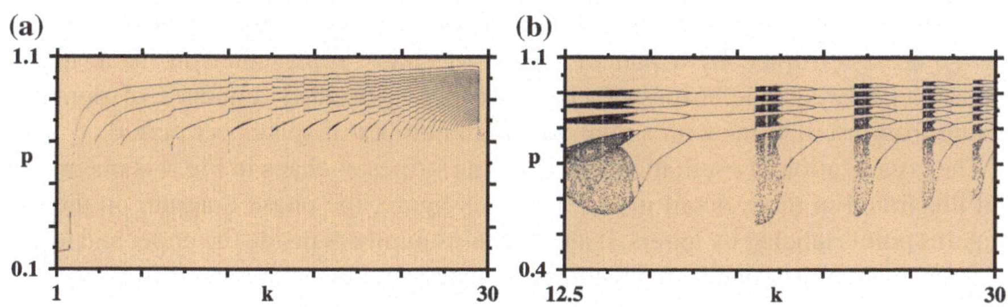

Fig. 2 Comparison of the rather distinct bifurcation diagrams underlying (**a**) nonchaos-mediated mixed-mode oscillations, and (**b**) the familiar chaos-mediated cascade. Panel (**a**) was recorded along the black line in Fig. 1d while (**b**) was recorded along the white line in the same figure. Nonchaos-mediated cascades display *spike-adding* sequences while chaos-mediated cascades show *spike-doubling*

The diagrams were obtained by starting from $(p, z, w) = (0.6, 0.15, 1.5)$ at the lowest value of k and proceeding by following the attractor until the highest k value. Figure 2a illustrates the new cascade of nonchaos-mediated mixed mode oscillations reported in this paper, while Fig. 2b shows an example of the chaos-mediated cascade found by Kuwamura and Chiba [5]. Both cascades look very different and reflect the distinct self-organization of oscillations seen in Fig. 1. Both cascades exist over relatively wide range of control parameters and, over limited intervals, may be even observed while varying just a single parameter, b, while keeping k fixed. Comparing Figs. 1d and 2b its is possible to recognize the existence of multistability: For instance, at the smallest values of k along the white line seen in Fig. 1d one sees the existence of period-3 solutions while the leftmost end of Fig. 2b displays chaos and no trace of period-3.

From Fig. 1d it is possible to recognize that the number of spikes increases by one as k grows along the black line. So, a natural question to ask is what exactly happens to the waveforms as they get more and more spikes continuously added to them. Figure 3 provides an answer. On the top panel of this figure one sees two sequences of points. The leftmost sequence, labeled by unprimed letters, corresponds to stability regions characterized by "primitive" number of spikes while points on the rightmost sequence, labeled with primed letters, lie inside domains where the number of spikes (not the period!) has doubled. The corresponding waveforms for every point in both sequences are shown under the stability diagram, while their coordinates, period, and number of spikes of their waveforms are collected in Table 2. In this Table, note that although the number of spikes doubles, the corresponding periods vary continuously, being not necessarily doubled. This lack of "period doubling" is a generic characteristic of continuous-time dynamical systems because the period varies *continuously* with parameters.

Figure 4 shows a remarkably complex "braided" self-organization of mixed-mode oscillations discovered for higher values of the carrying capacity k. In this region of the control parameter space one finds that the relatively regular sequences of nonchaos-mediated mixed-mode oscillations (seen in upper and lower portions of Fig. 4), are interrupted by a pair of stripes of chaos, represented in black in the diagrams. Between the stripes of chaos we find a braided sequence of domains arising from oscillations with a relatively high number of spikes per period.

The organization of oscillations between the stripes of chaos in Fig. 4 is magnified and illustrated in more detail in Fig. 5. In this figure, the phase diagram on the top contains points labeled by letters. It also contains numbers inside the upper and lower cascades of stability islands. Such numbers correspond to the number of spikes of the oscillations characteristic of each island. At about the center of the phase diagram there one sees four points labeled a, b, c, d, all of them lying inside the same oscillatory phase. As shown in the four panels in the lower part of Fig. 5, such points are all characterized by trains of periodic oscillations containing 38 spikes per period and, because 38 mod 17 = 4, are represented with the color corresponding to 4 in

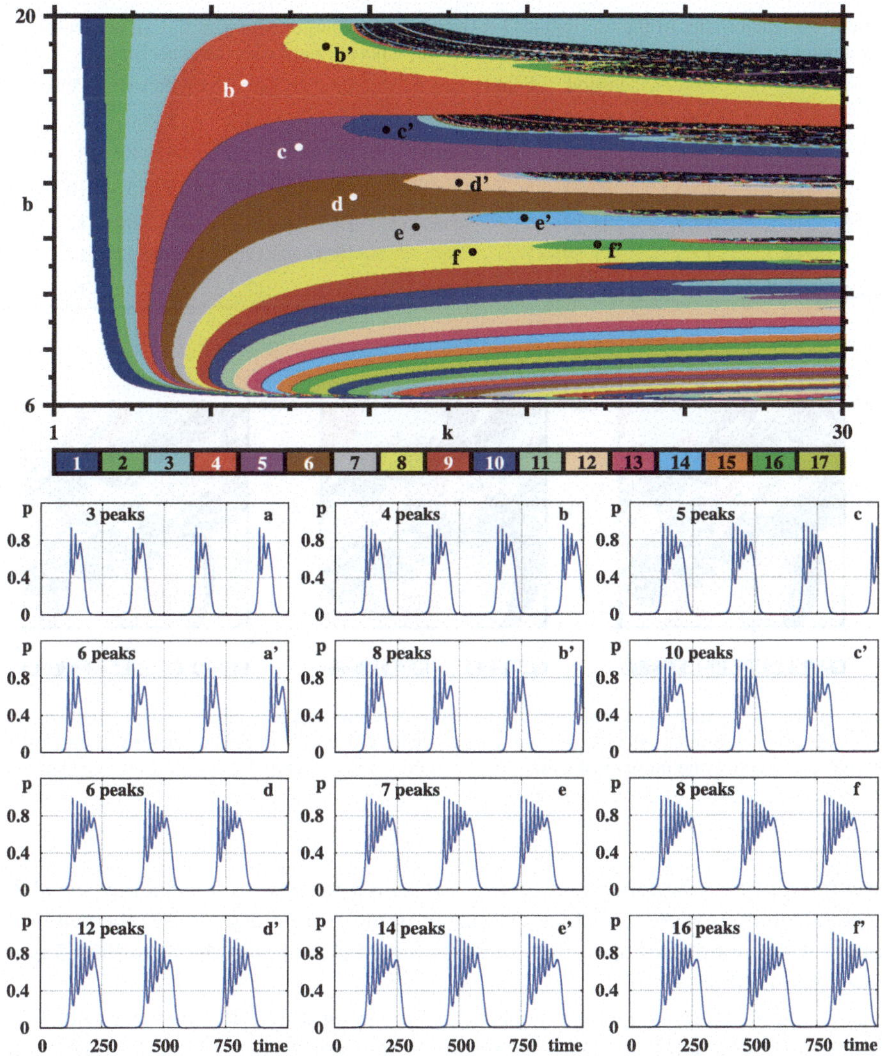

Fig. 3 Characteristic p waveforms along the nonchaos-mediated mixed-mode oscillation sequences indicated by the dots. Both sequences show cascades of spike additions and spike doublings. Here $\varepsilon = 0.2$. Evolutions start from the arbitrary initial condition $(p, z, w) = (0.6, 0.15, 1.5)$

Table 2 Characteristics of the oscillations for twelve time evolutions for $\varepsilon = 0.2$, shown in Fig. 3. Coordinates (k, b), oscillation period, and number of peaks per period of the p oscillations

	k	b	Period	Peaks		k	b	Period	Peaks
a	5.9	22.0	253.02	3	a'	8.4	24.3	555.07	6
b	8	17.6	264.54	4	b'	11	18.9	570.85	8
c	10	15.3	280.71	5	c'	13.2	15.9	592.3	10
d	12	13.5	294.14	6	d'	15.9	14.0	619.96	12
e	14.3	12.4	311.59	7	e'	18.3	12.72	649.71	14
f	16.4	11.5	327.54	8	f'	21	11.77	681.35	16

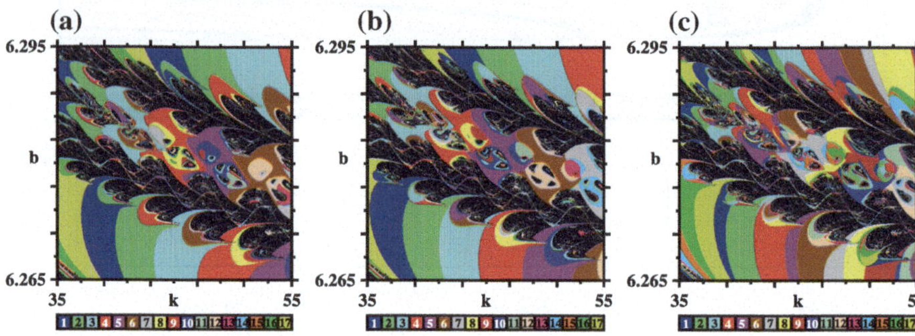

Fig. 4 Braided organization of mixed-mode oscillations, embedded between two stripes of chaos (in black), observed at higher values of the carrying capacity k. Panels (**a**)–(**c**) correspond to diagrams obtained by counting the number of spikes of p, z, and w, respectively. Details of these oscillations are given in Figs. 5 and 6. Here, $\varepsilon = 0.2$

the colorbar. From the four panels in Fig. 5 it is also possible to see that the period T of the oscillations increases clockwise, from point a to point d, inside the stability islands. The coordinates and characteristics for all points considered in Figs. 5 and 6 are given in Table 3.

Figures 5 and 6 show the time-evolution of the p variable. The other two variables show similar characteristics and, therefore, where not presented. The information recorded in Table 3 allow the time-evolution of the waveforms for all three variables to be easily recovered, if needed. An interesting point, however, is to clarify the nature of the reinjection loop responsible for starting every train of pulses shown in Figs. 5 and 6. As it is known, an important class of reinjection loops is associated with homoclinic bifurcations of a saddle-focus equilibrium state. In this case, the oscillatory part of the time-evolution corresponds to spiralling occurring essentially on a plane, with the reinjection happening perpendicular to it (see, e.g., Fig. 4 of Ref. [20]). Here, however, the spiralling occurs not on a plane but along a conical surface, as illustrated in Fig. 7. Furthermore, a close inspection of Fig. 7 reveals that the pair of spikes which appears between the pulse trains visible in Fig. 6 are responsible for a small loop that exists on the top of the cone in Fig. 7. We conjecture

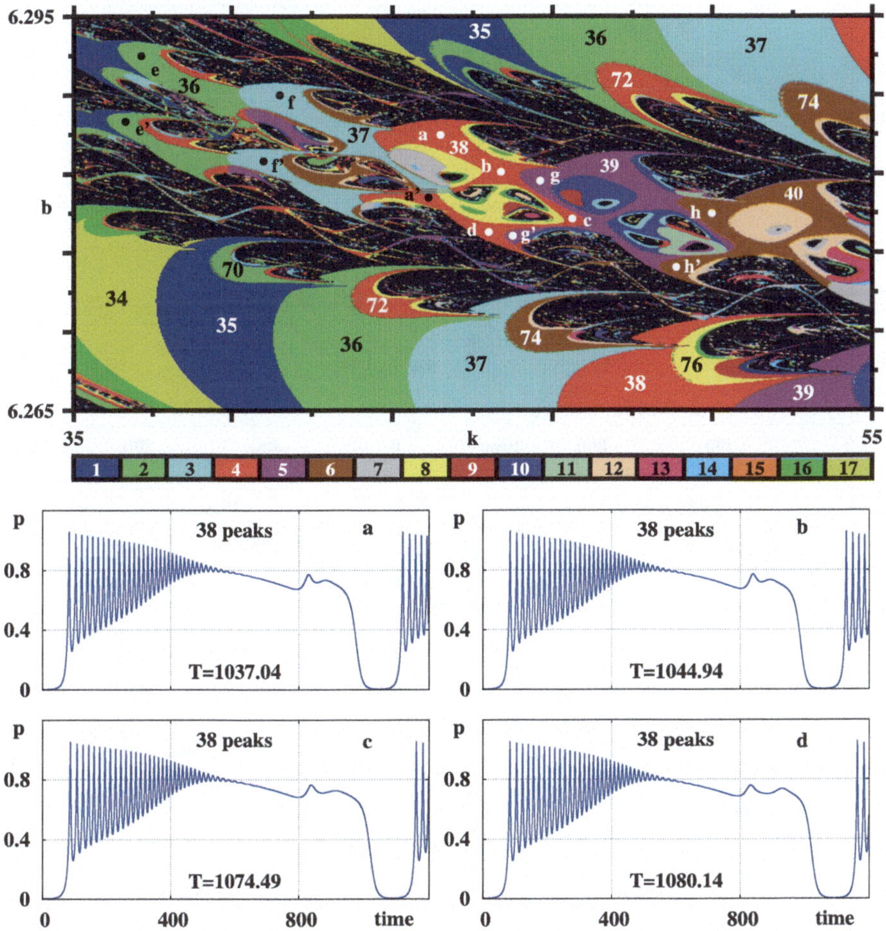

Fig. 5 Exceedingly complicated arrangement of periodic oscillations. Top panel: Magnification of Fig. 4a showing details of the braided organization found at higher values of the carrying capacity. Numbers indicate the number of spikes per period of the self-similar phases. The four bottom panels show how the waveform and period T of p pulses change when moving clockwise from points $a \to b$, $\to c$, $\to d$, defined in Table 3. Here, $\varepsilon = 0.2$. Evolutions start from $(p, z, w) = (0.6, 0.15, 1.5)$

that for other operation regimes of the model it should be possible to observe more complicated configurations in this region. We have not attempted to locate them since this requires investing considerable additional computer time.

So far, we discussed properties of nonchaos-mediated mixed-mode oscillations observed on the $k \times b$ control plane of the model. Is it possible to observe such oscillations in other control planes? Fig. 8 shows that it is not only possible to find them in other control planes but, in addition, that they exist over relatively wider parameter windows. In contrast to what was found in Fig. 1, note that the control-plane mosaic obtained by counting spikes of the w variable, Fig. 8c, is much different

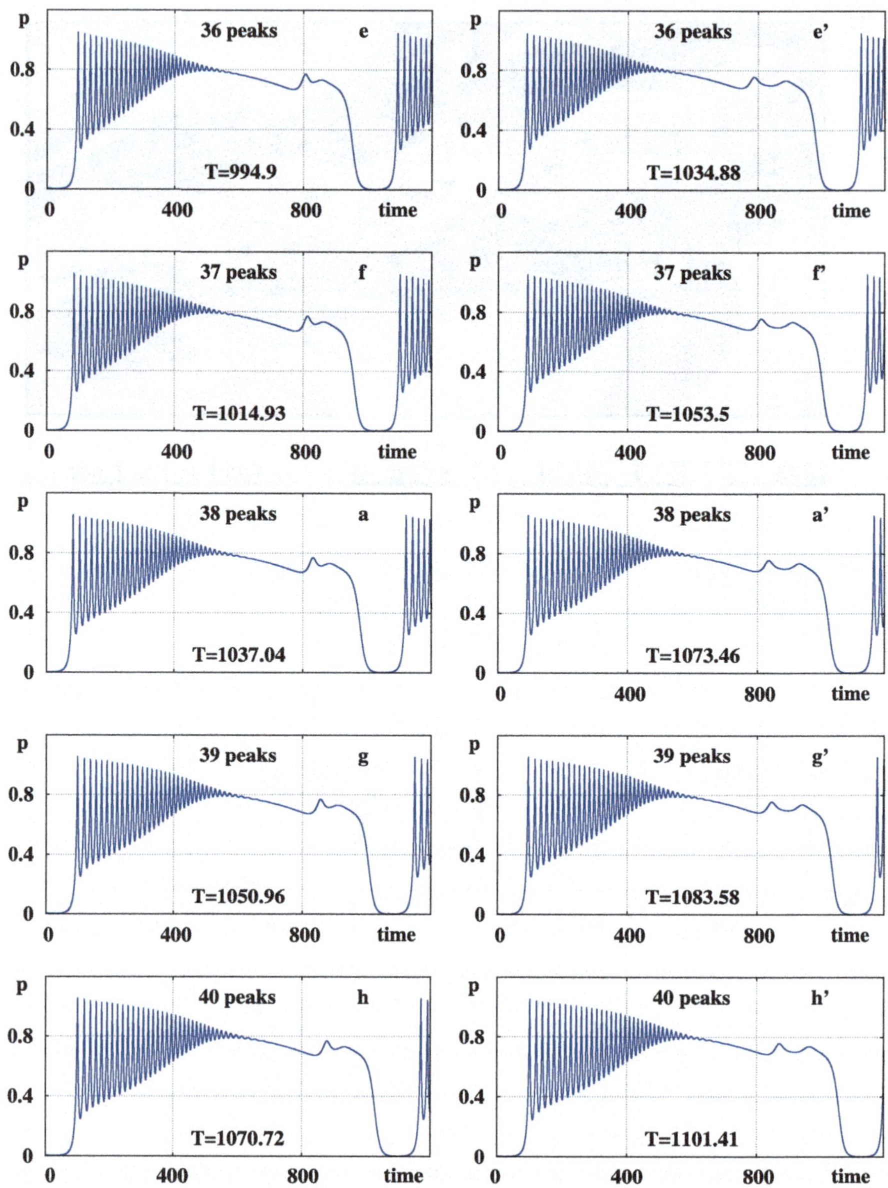

Fig. 6 Sequences of spike additions observed along the upper and lower parts of the braided structures in Fig. 5. Individual panels show the waveform of p for points labeled e, f, a, g, h and e', f', a', g', h' in Fig. 5. Note variations of the period T. All temporal evolutions start from the initial condition $(p, z, w) = (0.6, 0.15, 1.5)$

Table 3 Characteristics of the braided phases of periodic oscillations containing the points marked in Figs. 5 and 6. Coordinates (k, b), period T, and number of spikes per period of the p oscillations. Here, $\varepsilon = 0.2$

	k	b	Period	Peaks		k	b	Period	Peaks		k	b	Period	Peaks
a	44.2	6.286	1037.04	38						a’	43.9	6.2812	1073.46	38
b	45.7	6.2832	1044.94	38	e	36.7	6.292	994.9	36	e’	36.3	6.287	1034.88	36
c	47.5	6.2796	1074.49	38	f	40.2	6.289	1014.93	37	f’	39.8	6.284	1053.5	37
d	45.4	6.2786	1080.14	38	g	46.7	6.2825	1050.96	39	g’	46	6.2783	1083.58	39
					h	51	6.28	1070.72	40	h’	50.1	6.2759	1101.41	40

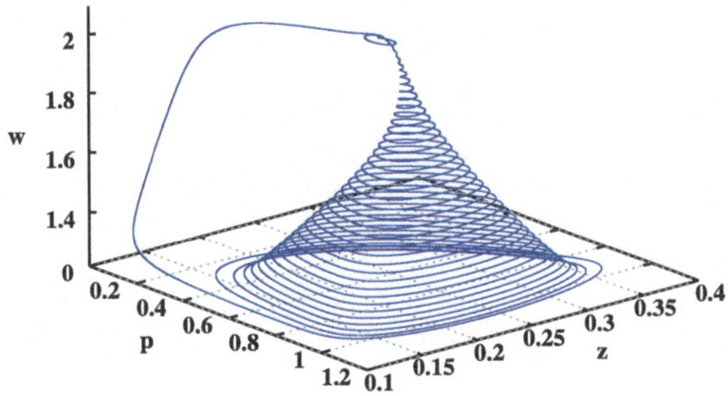

Fig. 7 Representation of the attractor corresponding to the point labeled a in Fig. 5, containing 38 spikes per period. The spiralling proceeds along a cone-like "witch hat" surface. All patterns in Figs. 5 and 6 produce similar *hats*

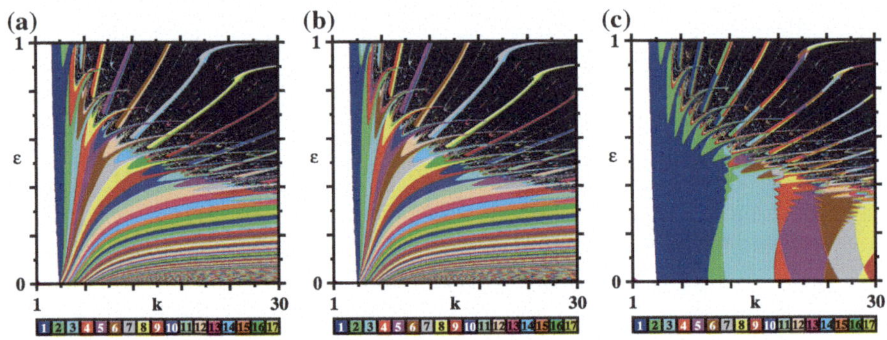

Fig. 8 Wide mosaics of nonchaos-mediated mixed-mode oscillations exist also o n the $k \times \varepsilon$ control plane. Panels (**a**)–(**c**) were obtained by counting spikes per period of p, z, w, respectively. Here, $b = 7$. Each panel displays the analysis of 600×600 parameter points

from the analogous mosaics obtained for variables p and z. We see no reason for nonchaos-mediated cascades of oscillations not to also exist for other combinations of parameters. However, a full exploration of all possible combinations is also a task demanding considerable additional computations.

4 Conclusions

This paper reported the discovery of abundant nonchaos-mediated sequences of oscillations, a novel and elusive type of mixed-mode oscillations, in an interesting prey-predator system including effects of predator dormancy, a strategy adopted in Nature to avoid extinction. As the carrying capacity increases, nonchaos-mediated sequences are found to appear well before the onset of chaos in the system, i.e., before the emergence of the more familiar chaos-mediated sequences. The observation of nonchaos-mediated sequences in the system is of interest from a dynamical point of view because such sequences have been reported only recently and, at present time, are known to exist only for a small number of systems. As seen in Fig. 8 when $\varepsilon \to 0$, nonchaos-mediated display intricate accumulation limits, which remain to be investigated, particularly to understand the interplay between the fast and slow time-scales governing the system. In addition, as seen in Figs. 4, 5, and 8, stability phases emerge in control parameter space self-organized regularly but arranged in exceedingly complicated ways which are best described by graphical means than by words. It would be nice if the intricate variations predicted here could be observed in real-life measurements.

Acknowledgements JGF was supported by a postdoctoral fellowship (SFRH/BPD/101760/2014), from the FCT, Portugal. JACG was supported by CNPq, Brazil. All phase diagrams were computed on the CESUP-UFRGS Supercomputing Center located in Porto Alegre, Brazil.

References

1. Rosenzweig, M.L., MacArthur, R.H.: Graphical representation and stability conditions of predator-prey interactions. Am. Nat. **47**, 209–223 (1963)
2. Rosenzweig, M.L.: Paradox of enrichment: destabilization of exploitation ecosystems in ecological time. Science **171**, 385–387 (1971)
3. Kuwamura, M., Nakazawa, T., Ogawa, T.: A minimum model of prey-predator system with dormancy of predators and the paradox of enrichment. J. Math. Biol. **58**, 459–479 (2009)
4. Jensen, C.X.J., Ginzburg, L.R.: Paradox or theoretical failures? The jury is still out. Ecol. Model. **188**, 314 (2005)
5. Kuwamura, M., Chiba, H.: Mixed-mode oscillations and chaos in a prey-pradator system with dormancy of predators. Chaos **19**, 043121 (2009)
6. Alekseev, V., Lampert, W.: Maternal control of resting-egg production in Dapnia. Nature **414**, 899901 (2001)
7. Gyllström, M., Hansson, K.-A.: Dormancy in freshwater zooplankton: induction, termination and the importance of benthic-pelagic coupling. Aquat. Sci. **66**, 274295 (2004)
8. Hairston Jr., N.G., Hansen, A.M., Schaffner, W.R.: The effect of diapause emergence on the seasonal dynamics of a zooplankton assemblage. Freshw. Biol. **45**, 133145 (2000)
9. Ricci, C.: Dormancy patterns in rorifers. Hydrobiologia **446**, 111 (2001)
10. McCauley, E., Nisbet, R.M., Murdoch, W.W., de Roos, A.M., Gurney, W.S.C.: Large-amplitude cycles of Daphnia and its algal prey in enriched environments. Nature **402**, 653656 (1999)
11. Hauser, M.J.B., Gallas, J.A.C.: Nonchaos-mediated mixed-mode oscillations in an enzyme reaction system. J. Phys. Chem. Lett. **5**, 4187–4193 (2014)
12. Gallas, M.R., Gallas, J.A.C.: Nested arithmetic progressions of oscillatory phases in Olsen's enzyme reaction model. Chaos **25**, 064603 (2015)
13. Freire, J.G., Gallas, M.R., Gallas, J.A.C.: Chaos-free oscillations. Europhys. Lett. **118**, 38003 (2017)
14. Freire, J.G., Gallas, M.R., Gallas, J.A.C.: Stability mosaics in a forced Brusselator: auto-organization of oscillations in control parameter space. Eur. Phys. J. Spec. Top. **226**, 1987–1995 (2017)
15. Rao, X., Chu, Y., Lu-Xu, Y.Chang, Zhang, J.: Fractal structures in centrifugal flywheel governor system. Commun. Nonlinear Sci. Numer. Simul. **50**, 330339 (2017)
16. Freire, J.G., Pöschel, T., Gallas, J.A.C.: Stern-Brocot trees in spiking and bursting of sigmoidal maps. Europhys. Lett. **100**, 48002 (2012)

17. Freire, J.G., Field, R.J., Gallas, J.A.C.: Relative abundance and structure of chaotic behavior: the nonpolynomial Belousov Zhabotinsky reaction kinetics. J. Chem. Phys. **131**, 044105 (2009)
18. Gallas, J.A.C.: Spiking systematics in some CO_2 laser models. Adv. Atom. Mol. Opt. Phys. **65**, 127–191 (2016)
19. Kuwamura, M.: Turing instabilities in prey-predator systems with dormancy of predators. J. Math. Biol. **71**, 125–149 (2015)
20. Vitolo, R., Glendinning, P., Gallas, J.A.C.: Global structure of periodicity hubs in Lyapunov phase diagrams of dissipative flows. Phys. Rev. E **84**, 016216 (2011)

Bifurcations and Stability Regions of Nonlinear Dynamical Systems

Luis F.C. Alberto, Fabiolo M. Amaral and Josaphat R.R. Gouveia Jr.

1 Stability Regions of Nonlinear Dynamical Systems

Attractors of nonlinear dynamical systems are rarely globally stable. Actually, there exists a subset of the state space, called *stability region*, composed of all the initial conditions that have trajectories approaching the attractor as time tends to infinity. Region of attraction, area of attraction and basin of attraction are other names commonly employed in the literature for stability region.

In this chapter, we will study stability regions of the following class of nonlinear dynamical systems:

$$\dot{x} = f(x), \tag{1}$$

where $f : \mathbb{R}^n \to \mathbb{R}^n$ is a C^1-function. We will assume that solutions of (1) are defined for all $t \in \mathbb{R}$ and the solution of (1) passing through x_o at time $t = 0$ is denoted by $\varphi(t, x_o)$.

Definition 1 (*Invarinat Set*) An invariant set γ is an attracting set of system (1) if there exists a neighborhood (open set) N of γ such that $\varphi(t, x) \to \gamma$ as $t \to \infty$ for all $x \in N$.

Definition 2 (*Stability Region*) The stability region $A(\gamma)$ of an attracting set γ of system (1) is the set:

$$A(\gamma) = \{x \in \mathbb{R}^n : \varphi(t, x) \to \gamma \text{ as } t \to \infty\}$$

L.F.C. Alberto (✉)
Sao Carlos Engineering School, University of Sao Paulo, Av. Trabalhador Sancarlense, 400, Sao Carlos, SP 13566-590, Brazil
e-mail: lfcalberto@usp.br

F.M. Amaral · J.R.R. Gouveia
Federal Institute of Bahia- College Eunápolis, Av. David Jonas Fadini, S/N, Rosa Neto, Eunápolis, BA 45823-431, Brazil
e-mail: fabiolo@ifba.edu.br

J.R.R. Gouveia
e-mail: josaphat@ifba.edu.br

M. Edelman et al. (eds.), *Chaotic, Fractional, and Complex Dynamics: New Insights and Perspectives*, Understanding Complex Systems,
https://doi.org/10.1007/978-3-319-68109-2_7

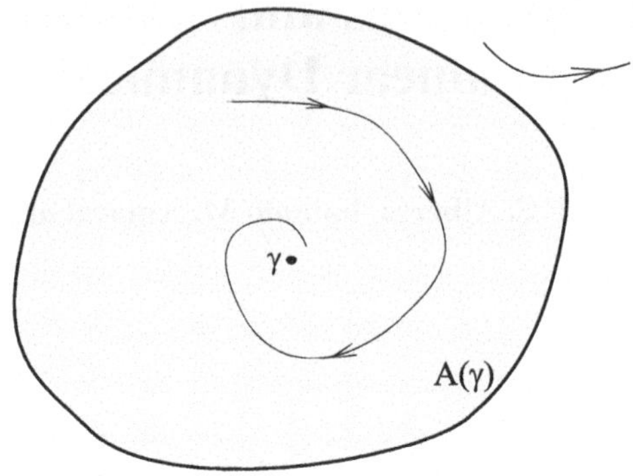

Fig. 1 Stability region of an attracting set γ

Asymptotically stable equilibrium points and asymptotically stable limit cycles are examples of attracting sets. Figure 1 illustrates the concept of stability region. The stability region $A(\gamma)$ is an open and invariant set. Its topological boundary will be called stability boundary and will be denoted $\partial A(\gamma)$. The stability boundary is a closed and invariant set [8].

Determining stability regions is relevant in many areas of sciences. For instance, stability regions plays an important role in the assessment of transient stability in electrical power systems [8] and in the problem of immunization in biological systems [13]. In the process of determining or estimating stability regions, the determination or estimation of the stability boundary is relevant. In the next subsections, the existing theory of characterization of stability regions and stability boundaries is reviewed. Invariant sets on the stability boundary play an important role in the theory of stability boundary characterization. In Sect. 1.1, the characterization of hyperbolic equilibrium points on the stability boundary is studied while Sect. 1.2 studies closed orbits on the stability boundary.

1.1 Hyperbolic Equilibrium Points on the Stability Boundary

In this section, a characterization of the stability boundary of a fairly large class of dynamical systems is developed. This class is composed of the dynamical systems that admit hyperbolic equilibrium points as the only type of critical element (minimal invariant set) on the stability boundary.

A key point to derive a characterization of the stability boundary is to understand the relationship between the critical elements and the stability region and its boundary. Theorem 1, proven in [6], establishes this relationship offering necessary and sufficient conditions for a hyperbolic equilibrium point lying on the stability boundary.

Theorem 1 (Equilibrium Points on the Stability Boundary) [6] *Let x^s be a hyperbolic asymptotically stable equilibrium point of (1) and $A(x^s)$ be its stability region. If assumptions:*

(A1) *All the equilibrium points on $\partial A(x^s)$ are hyperbolic,*
(A2) *The stable and unstable manifolds of equilibrium points on $\partial A(x^s)$ satisfy the transversality condition,*
(A3) *Every trajectory on $\partial A(x^s)$ approaches one of the equilibrium points as $t \to +\infty$,*

are satisfied and x^* ($x^* \neq x^s$) is a hyperbolic equilibrium point of (1). Then the following statements are equivalent:

(i) $x^* \in \partial A(x^s)$
(ii) $W^u(x^*) \cap A(x^s) \neq \emptyset$
(iii) $W^s(x^*) \subseteq \partial A(x^s)$.

Figure 2 illustrates the conclusions of Theorem 1. Next theorem, proven in [6], extends the characterization given in Theorem 1 by asserting the stability boundary $\partial A(x^s)$ is the union of the stable manifolds of the equilibrium points on $\partial A(x^s)$.

Theorem 2 (Stability Boundary Characterization) *Let x^s be a hyperbolic asymptotically stable equilibrium point of (1) and $A(x^s)$ be its stability region. If assumptions* (A1) *and* (A3) *are satisfied, then:*

$$\partial A(x^s) \subseteq \bigcup_i W^s(x^i),$$

where x^i, $i = 1, 2, \ldots$ are the equilibrium points on $\partial A(x^s)$. If, additionally, (A2) *is satisfied, then:*

$$\partial A(x^s) = \bigcup_i W^s(x^i).$$

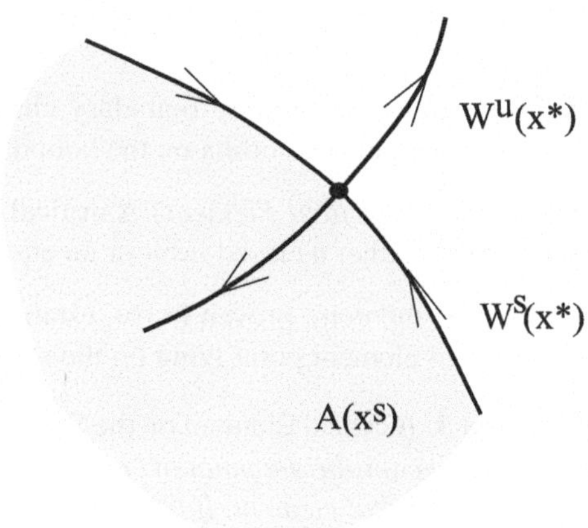

Fig. 2 The equilibrium x^* is on the stability boundary. Its stable manifold $W^s(x^*)$ lies on the stability boundary and the unstable manifold $W^u(x^*)$ has a non empty intersection with the stability region $A(x^s)$

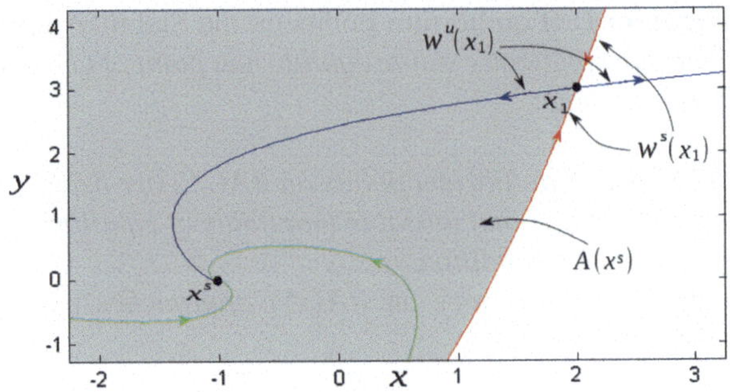

Fig. 3 Phase portrait of system (2). The stability boundary of $x^s = (-1, 0)$ is formed by the stable manifold $W^s(x_1)$ of the type-1 hyperbolic equilibrium point $x_1 = (2, 3)$

The following example illustrate the characterization of the stability boundary of Theorem 2.

Example Consider the following system of differential equations:

$$\begin{aligned} \dot{x} &= x^2 - y - 1 \\ \dot{y} &= x - y + 1 \end{aligned} \tag{2}$$

where $(x, y) \in \mathbb{R}^2$. System (2) possesses two equilibrium points, an asymptotically stable equilibrium point, $x^s = (-1, 0)$, and a type-1 hyperbolic equilibrium point $x_1 = (2, 3)$ on the stability boundary $\partial A(x^s)$. Since all equilibrium points are hyperbolic, then assumption (A1) is satisfied. In agreement with Theorem 1, the unstable manifold $W^u(x_1)$ of the hyperbolic equilibrium point $x_1 = (2, 3)$ intersects the stability region $A(x^s)$ and its stable manifold $W^s(x_1)$ is contained on the stability boundary $\partial A(x^s)$. The stability boundary $\partial A(x^s)$ is the stable manifold $W^s(x_1)$ of the type-1 hyperbolic equilibrium point $x_1 = (2, 3)$, in agreement with the results of Theorem 2. See Fig. 3.

1.2 Closed Orbits on the Stability Boundary

In this section, the stability boundary characterization of Sect. 1.1 is extended to accommodate periodic orbits on the boundary of stability regions.

Definition 3 (*Critical Element*) A critical element ϕ of the autonomous dynamic system (1) is either a closed orbit or an equilibrium point.

The next theorem, proven in [6], establishes necessary and sufficient conditions for a critical element point lying on the stability boundary.

Theorem 3 (Critical Element on the Stability Boundary) [6] *Let x^s be an asymptotically stable equilibrium point of (1) and $A(x^s)$ be its corresponding stability region. Let ϕ be a critical element. If assumptions:*

(B1) *All the critical elements of (1) on* $\partial A(x^s)$ *are hyperbolic,*
(B2) *The stable and unstable manifolds of critical elements of (1) on* $\partial A(x^s)$ *satisfy the transversality condition,*
(B3) *Trajectories on* $\partial A(x^s)$ *approach one of the critical elements of system (1) as* $t \to +\infty$,

are held, then the follwoing statements are equivalent:

(i) $\phi \subset \partial A(x^s)$
(ii) $W^u(\phi) \cap A(x^s) \neq \emptyset$
(iii) $W^s(\phi) \subset \partial A(x^s)$.

Theorem 3 is an extension of Theorem 1 and offers a local characterization of the boundary of the stability region in the neighborhood of critical elements. The following theorem, proven in [6], develops a global characterization of the stability boundary. Under assumptions (**B1**)–(**B3**), it asserts the stability boundary is the union of the stable manifolds of the hyperbolic critical elements on the boundary of the stability region.

Theorem 4 (Stability Boundary Characterization) [6] *Let* x^s *be an asymptotically stable equilibrium point of (1) and* $A(x^s)$ *its stability region. If assumptions* ***(B1)*** *and* ***(B3)*** *are held, then:*

$$\partial A(x^s) \subset \bigcup_i W^s(x_i) \bigcup_j W^s(\phi_j)$$

where x_i, $i = 1, 2, \ldots$ *are the equilibrium points and* ϕ_j, $j = 1, 2, \ldots$ *are the closed orbits in* $\partial A(x^s)$. *If, additionally, assumption* ***(B2)*** *is satisfied, then*

$$\partial A(x^s) = \bigcup_i W^s(x_i) \bigcup_j W^s(\phi_j).$$

1.3 Energy Functions and Stability Boundary Characterization

The characterizations of stability boundary given in Sects. 1.1 and 1.2 are given in terms of stable manifolds of critical sets. These manifolds are difficult to compute, specially in high dimensional systems. Despite that, level sets of energy functions provide concrete estimates of stability regions and stability boundaries, moreover, energy functions have important implications on the stability boundary characterization.

Consider the nonlinear dynamical system (1) and let $E := \{x \in \mathbb{R}^n : f(x) = 0\}$ be the set of all equilibrium points of (1). The following definition of energy function was firstly proposed in [7].

Definition 4 (*Energy Function*) A C^1-function $V : \mathbb{R}^n \to \mathbb{R}$ is an energy function of system (1) if the following conditions are satisfied:

(i) $\dot{V}(x) \leq 0$ for all $x \in \mathbb{R}^n$.
(ii) if $x_0 \notin E$, then the set $\{t \in \mathbb{R}_+ \ : \ \dot{V}(\varphi(t, x)) = 0\}$ has zero measure in $\mathbb{R}$.
(iii) if $V(\varphi(t, x_0))$ is bounded for $t \in \mathbb{R}_+$, then the trajectory $\varphi(t, x_0)$ is bounded for $t \in \mathbb{R}_+$.

The existence of an energy function guarantees that every bounded trajectory must approach an equilibrium point as $t \to +\infty$. As a consequence, complex behavior such as closed orbits and chaos cannot exist for systems that admit energy functions. Moreover, the existence of an energy function ensures that every trajectory on the stability boundary is bounded, although the stability boundary can be unbounded, and converges to an equilibrium point on the stability boundary as $t \to +\infty$ [6]. In other words, the existence of an energy function is a sufficient condition to guarantee assumption $(A3)$.

Next theorem, proven in [7], provides a complete characterization of the stability boundary for systems that admit energy functions.

Theorem 5 (Stability Boundary Characterization) [7] *Let x^s be a hyperbolic asymptotically stable equilibrium point of (1) and $A(x^s)$ be its stability region. If assumption $(A1)$ is satisfied and system (1) admits an energy function, then:*

$$\partial A(x^s) \subseteq \bigcup_i W^s(x^i)$$

where x^i, $i = 1, 2, \ldots$ are the hyperbolic equilibrium points on the stability boundary $\partial A(x^s)$.

2 Persistence of Stability Regions to Parameter Variation

Complete characterizations of stability regions and stability boundaries were proven in the literature [6, 8] and the main results of this theory were presented in Sect. 1. These characterizations are given in terms of the union of the stable manifolds of the critical elements on the stability boundary. However, systems are subjected to uncertainties and parameter changes and a natural question that pops up is how these characterizations are robust with respect to parameter variation. The answer to this question is crucial to ensure that estimates of the stability region obtained by means of these characterizations are robust to parameter changes.

In this chapter, we will study stability regions of the following class of nonlinear dynamical systems:

$$\dot{x} = f(x, \lambda) = f_\lambda(x), \tag{3}$$

where $f : \mathbb{R}^n \times \mathbb{R} \rightarrow \mathbb{R}^n$ is a $\mathscr{C}^1$-function and λ is a real parameter. The trajectory of system $\dot{x} = f_\lambda(x)$ passing through x_o at time $t = 0$ will be denoted $\varphi_\lambda(t, x_o)$.

Suppose $x^s_{\lambda_o}$ is a hyperbolic asymptotically stable equilibrium point of (3) for $\lambda = \lambda_o$ and let $A_{\lambda_o}(x^s_{\lambda_o})$ be its stability region. Hyperbolic equilibrium points persist to parameter changes. Consequently, it does make sense to study the stability region $A_\lambda(x^s_\lambda)$ of the perturbed equilibrium x^s_λ. Suppose assumptions (A1)–(A3) hold for $\lambda = \lambda_o$ and $x^i_{\lambda_o}$, $i = 1, \ldots, m$ are the unstable equilibrium points on the stability boundary. Then, according to Theorem 2, the stability boundary is given by:

$$\partial A_{\lambda_o}(x^s_{\lambda_o}) = \bigcup_i W^s_{\lambda_o}(x^i_{\lambda_o})$$

If assumptions (A1)–(A3) hold for all λ in a neighborhood of λ_0 and the number of equilibrium points on the boundary is finite, it can be proven, under reasonable conditions, that the stability region and the stability boundary do not suffer drastic changes. More precisely, if an unstable equilibrium point $x^u_{\lambda_o}$ belongs to the stability boundary $\partial A_{\lambda_o}(x^s_{\lambda_o})$ of the unperturbed system, then the perturbed unstable equilibrium point x^u_λ will persist on the stability boundary, i.e., $x^u_\lambda \in \partial A_\lambda(x^s_\lambda)$ for every λ sufficiently close to λ_o. Consequently,

$$\partial A_\lambda(x^s_\lambda) \subset \bigcup_i W^s_\lambda(x^i_\lambda),$$

indicating that the stability boundary does not suffer drastic changes for λ sufficient close to λ_o. However, with changes in the parameter λ, assumptions (A1) and (A2) may be violated. In these cases, drastic changes in the stability regions and stability boundaries may occur. In this chapter, we will study these changes when assumption (A1) is violated due to the appearance of two types of nonhyperbolic equilibrium points on the stability boundary: the saddle-node equilibrium point and the Hopf equilibrium point. We first develop characterizations of the stability boundary in the presence of these nonhyperbolic equilibrium points in Sect. 3 and then the stability region and stability boundary behavior due to changes in parameters in the neighborhood of saddle-node and Hopf bifurcations are studied in Sect. 4.

3 Non-hyperbolic Equilibrium Points on the Stability Boundary

Hyperbolicity of equilibrium points on the stability boundary is a fundamental property for the characterizations of the stability boundary developed in Sect. 1. Although the hyperbolicity of equilibrium points of a dynamical system is a generic property, i.e., it is satisfied for almost all dynamical systems, the violation of hyperbolicity condition of the equilibrium points on the stability boundary is very common when the system is subject to variation of parameters. In the analysis of voltage stability in electric power systems, for instance, the occurrence of saddle-node bifurcations on the stability boundary were reported, violating the hyperbolicity condition of the equilibrium points on the stability boundary [12].

In this section, we discuss the stability boundary characterization in the presence of two types of non-hyperbolic equilibrium points: the saddle-node and the Hopf equilibrium points. Exploring these characterizations, we also discuss on how to obtain estimates of the stability region for dynamical systems that admit energy functions.

3.1 Saddle-Node Equilibrium Points on the Stability Boundary

In Sect. 1.1, the properties of hyperbolic equilibriums on the stability boundary were studied. In this section, we develop necessary and sufficient conditions for a saddle-node, which is the simplest of the non-hyperbolic equilibrium points, belonging to the boundary of the stability region. In addition, we also develop a characterization of the stability boundary in the presence of saddle-node equilibrium points. The results developed in this section are a generalization of the ones proven in [3, 4]. They are also a generalization of the results presented in Sect. 1.1.

Definition 5 (*Saddle-Node Equilibrium Point*) [17] A non-hyperbolic equilibrium point $p \in \mathbb{R}^n$ of (1) is a *saddle-node equilibrium point* if the following conditions are satisfied:

(i) $D_x f(p)$ has a unique simple null eigenvalue and none of the other eigenvalues have real part equal to zero,
(ii) $w(D_x^2 f(p)(v, v)) \neq 0$,

with v as the right eigenvector and w as the left eigenvector associated with the null eigenvalue.

Saddle-node equilibrium points can be classified in types according to the number of eigenvalues of $D_x f(p)$ with positive real part.

Definition 6 (*Saddle-Node Equilibrium Type*) A saddle-node equilibrium point p of (1) is called a type-k saddle-node equilibrium point if $D_x f(p)$ has k eigenvalues with positive real part and $n - k - 1$ with negative real part.

If p is a saddle-node equilibrium point of (1), then there exist invariant local manifolds $W^s_{loc}(p)$, $W^{cs}_{loc}(p)$, $W^c_{loc}(p)$, $W^u_{loc}(p)$ and $W^{cu}_{loc}(p)$ of class C^r, tangent to the eigenspaces E^s, $E^c \oplus E^s$, E^c, E^u and $E^c \oplus E^u$ at p, respectively [14, 18]. These manifolds are respectively called stable, stable center, center, unstable and unstable center manifolds. The stable and unstable manifolds are unique, but the stable center, center and unstable center manifolds may not be. Dynamic properties of these manifolds can be found in [17, 18].

3.1.1 Stability Boundary Characterization

In the presence of non-hyperbolic equilibrium points on the stability boundary, assumption $(A1)$ is violated and the stability boundary characterization given in Theorem 2 is not valid. In this section, a characterization of the stability boundary in the presence of hyperbolic and saddle-node equilibrium points is developed. This characterization is developed in two steps. In the first step, a local characterization of equilibrium points on the stability boundary is developed and then a global characterization of the stability boundary in terms of manifolds of equilibrium points is developed.

Let x^s be an asymptotically stable equilibrium point of (1) and let $A(x^s)$ be its stability region. Consider the following assumptions:
(**A1**$'$) All the equilibrium points on $\partial A(x^s)$ are hyperbolic or saddle-node equilibrium points.
(**A2**$'$) The following transversality conditions are satisfied:

(i) The stable and unstable manifolds of equilibrium points on $\partial A(x^s)$ satisfy the transversality condition.
(ii) The unstable manifolds of equilibrium points and the stable component of the stable center manifolds of the type-k saddle-node equilibrium points, with $1 \leq k \leq n-2$, on $\partial A(x^s)$ satisfy the transversality condition.
(iii) The unstable manifolds of equilibrium points and the stable component of the center manifolds of the type-$(n-1)$ saddle-node equilibrium points on $\partial A(x^s)$ satisfy the transversality condition.
(iv) The stable manifolds of equilibrium points and the unstable component of the center manifolds of the type-0 saddle-node equilibrium points on $\partial A(x^s)$ satisfy the transversality condition.
(v) The stable component of the stable center manifolds of the type-k saddle-node equilibrium points, with $1 \leq k \leq n-2$, and the unstable component of the center manifolds of the type-0 saddle-node equilibrium points on $\partial A(x^s)$ satisfy the transversality condition.
(vi) The stable component of the center manifolds of the type-$(n-1)$ saddle-node equilibrium points and the unstable component of the center manifolds of the type-0 saddle-node equilibrium points on $\partial A(x^s)$ satisfy the transversality condition.

Assumptions $(A1')$ and $(A2')$ are generic properties of dynamical systems [16]. Under assumptions $(A1')$, $(A2') - (iv)$, (v), (vi) and $(A3)$, next theorem, proven in [1], offers necessary and sufficient conditions to guarantee that a type-0 saddle-node equilibrium point lies on the stability boundary of a nonlinear autonomous dynamical system.

Theorem 6 (Type-0 Saddle-Node Equilibrium Points on the Stability Boundary) [1] *Let x^s be an asymptotically stable equilibrium point of (1) and let $A(x^s)$ be its stability region. Suppose that assumptions $(A1^{'})$, $(A2^{'})$-(iv), (v), (vi) and $(A3)$ are satisfied. If p is a type-0 saddle-node equilibrium point, then the following statements are equivalent:*

(a) $p \in \partial A(x^s)$
(b) $W^{c^+}(p) \cap A(x^s) \neq \emptyset$
(c) $W^s(p) \subseteq \partial A(x^s)$

Theorem 6 offers necessary and sufficient conditions to guarantee that a type-0 saddle-node equilibrium point p belongs to the stability boundary. More precisely, it shows that the non empty intersection of the unstable component $W^{c^+}(p)$ of the center manifold with the stability region guarantees that the saddle-node equilibrium point p lies on the stability boundary.

In this sense, we observe that $W^{c^+}(p)$ plays in Theorem 6 the same role of $W^u(p)$ in Theorem 1. Consequently, one can check if an equilibrium point p lies on the stability boundary by checking if the unstable manifold intersects the stability region in the case of a hyperbolic equilibrium point and if the unstable component of the center manifold intersects the stability region in the case of a type-0 saddle-node equilibrium point. Figure 4 illustrates the results of Theorem 6.

Next theorem, proven in [1], offers necessary and sufficient conditions to guarantee that a hyperbolic or a type-r saddle-node equilibrium point, with $r \geq 1$, lies on the stability boundary of a nonlinear autonomous dynamical system.

Theorem 7 (Hyperbolic and Type-r Saddle-Node Equilibrium Points, with $r \geq 1$ on the Stability Boundary) [1] *Let x^s be an asymptotically stable equilibrium point of (1) and let $A(x^s)$ be its stability region. Suppose that assumptions $(A1^{'})$, $(A2^{'})$ and $(A3)$ are satisfied. Then:*

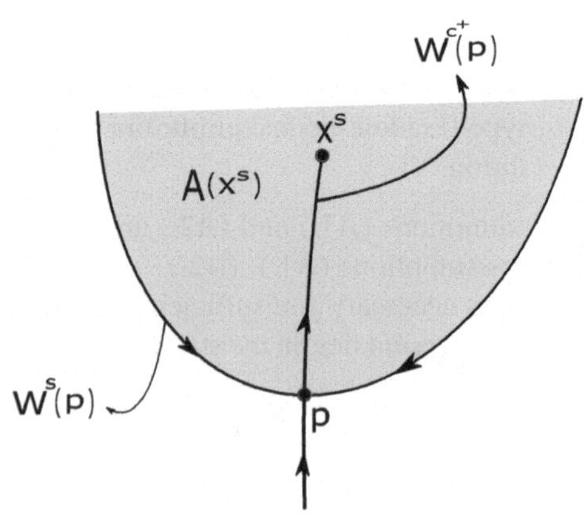

Fig. 4 The type-0 saddle-node equilibrium point p is on the stability boundary. Its stable manifold $W^s(p)$ lies on the stability boundary $\partial A(x^s)$ and the unstable component of the center manifold $W^{c^+}(p)$ has a non empty intersection with the stability region $A(x^s)$

(i) If p^* is a hyperbolic equilibrium point or a type-r saddle-node equilibrium point, with $r \geq 1$, of (1) and $(W^u(p^*)\text{-}\{p^*\}) \cap \overline{A(x^s)} \neq \emptyset$, then the following statements are equivalent:

(a) $p^* \in \partial A(x^s)$

(b) $W^u(p^*) \cap A(x^s) \neq \emptyset$

(c) $\begin{cases} W^s(p^*) \subseteq \partial A(x^s) & \text{if } p^* \text{ is a hyperbolic equilibrium point} \\ W^{cs^-}(p^*) \subseteq \partial A(x^s) & \text{if } p^* \text{ is a type-}r \text{ saddle-node equilibrium point, } r \leq n-2 \\ W^{c^-}(p^*) \subseteq \partial A(x^s) & \text{if } p^* \text{ is a type-}(n\text{-}1) \text{ saddle-node equilibrium point.} \end{cases}$

(ii) If p is a type-r saddle-node equilibrium point, with $r \geq 1$, of (1) and $(W^u(p) - \{p\}) \cap \overline{A(x^s)} = \emptyset$ then the following statements are equivalent:

(a) $p \in \partial A(x^s)$

(b) $W^{c^+}(p) \cap A(x^s) \neq \emptyset$ *for some unstable component* $W^{c^+}(p)$ *of the center manifold.*

(c) $W^s(p) \subseteq \partial A(x^s)$.

Admitting the existence of non hyperbolic saddle-node equilibrium points on the stability boundary, generalizing the transversality condition and exploring assumption $(A3)$, Theorem 7 extends the results of Theorem 1. Observe that the same equivalences proven in Theorem 1 are still valid for hyperbolic equilibrium points even in the presence of saddle-node equilibrium points on the stability boundary.

Theorem 7 offers necessary and sufficient conditions for a type-r saddle-node equilibrium point, with $r \geq 1$, lying on the stability boundary $\partial A(x^s)$. For saddle-node equilibrium points, two different situations can occur. Therefore, two cases are separately treated in Theorem 7, the case (i), in which $(W^u(p) - \{p\}) \cap \overline{A(x^s)} \neq \emptyset$, and the case (ii), in which $(W^u(p) - \{p\}) \cap \overline{A(x^s)} = \emptyset$.

Next theorem, proven in [1], combines the results of Theorems 6 and 7 to offer a complete characterization of the stability boundary of a nonlinear autonomous dynamical system in the presence of saddle-node equilibrium points on the stability boundary $\partial A(x^s)$.

Theorem 8 (Stability Boundary Characterization) [1] *Let x^s be an asymptotically stable equilibrium point of (1) and $A(x^s)$ be its stability region. Suppose that assumptions $(A1')$, $(A2')$ and $(A3)$ are satisfied. Then:*

$$\partial A(x^s) = \bigcup_i W^s(x_i) \bigcup_j W^s(p_j) \bigcup_l W^{cs^-}(z_l) \bigcup_t W^s(z_t) \bigcup_m W^{c^-}(q_m)$$

where x_i are the hyperbolic equilibrium points on $\partial A(x^s)$, p_j the type-0 saddle-node equilibrium points on $\partial A(x^s)$, z_l the type-k saddle-node equilibrium points on $\partial A(x^s)$, with $1 \leq k \leq n-2$, and $(W^u(z_l) - \{z_l\}) \cap \overline{A(x^s)} \neq \emptyset$, z_t the type-d saddle-node equilibrium points on $\partial A(x^s)$, with $d \geq 1$ and $(W^u(z_t) - \{z_t\}) \cap \overline{A(x^s)} = \emptyset$ and q_m the type-$(n-1)$ saddle-node equilibrium points on $\partial A(x^s)$, with $(W^u(q_m) - \{q_m\}) \cap \overline{A(x^s)} \neq \emptyset$, $i, j, l, t, m = 1, 2, \ldots$.

Figure 5 which was presented in [1], shows an example of a dynamical system in $\mathbb{R}^3$, where a type-1 saddle-node equilibrium point p lies on the stability boundary $\partial A(x^s)$ of an asymptotically stable equilibrium point x^s, but $(W^u(p) - \{p\}) \cap \overline{A(x^s)} = \emptyset$. The stability boundary $\partial A(x^s)$ is formed, according to Theorem 8, as the union of the stable manifold $W^s(p)$ and the stable manifolds $W^s(x_1)$, $W^s(x_2)$ of the unstable hyperbolic equilibrium points x_1 and x_2 that belong to the stability boundary $\partial A(x^s)$.

Figure 6 which was presented in [1], shows an example of a dynamic system in $\mathbb{R}^3$, where $(W^u(p) - \{p\}) \cap \overline{A(x^s)} \neq \emptyset$. The stability boundary $\partial A(x^s)$ is formed, according to Theorem 8, as the union of the stable component $W^{cs^-}(p)$ of the stable center manifold and the stable manifold $W^s(x_1)$ of the unstable hyperbolic equilibrium point x_1 that belongs to the stability boundary $\partial A(x^s)$.

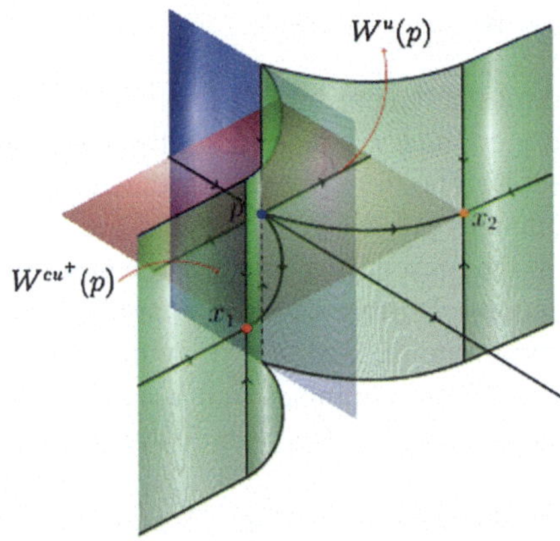

Fig. 5 Example of a dynamical system on $\mathbb{R}^3$ where the unstable component of the unstable center manifold $W^{cu^+}(p)$ of the type-1 saddle-node equilibrium point p lying on the stability boundary $\partial A(x^s)$ intersects the closure of the stability region $\overline{A(x^s)}$ and $(W^u(p) - \{p\}) \cap \overline{A(x^s)} = \emptyset$. Reprinted from [1]

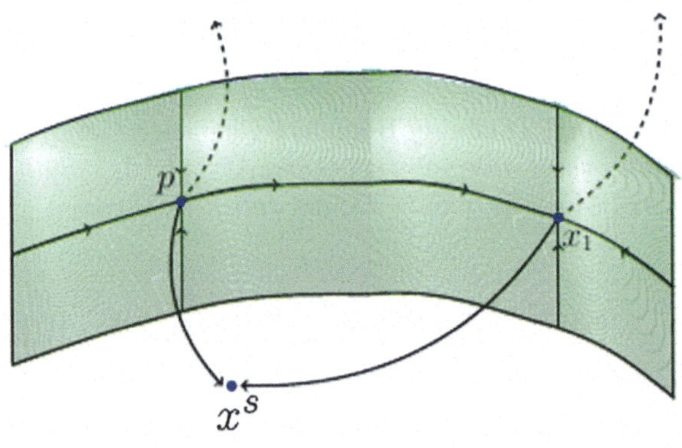

Fig. 6 Example of a dynamical system on $\mathbb{R}^3$ where the unstable component of the unstable center manifold $W^{cu^+}(p)$ of the type-1 saddle-node equilibrium point p lying on the stability boundary $\partial A(x^s)$ intersects the closure of the stability region $\overline{A(x^s)}$ and $(W^u(p) - \{p\}) \cap \overline{A(x^s)} \neq \emptyset$. Reprinted from [1]

3.1.2 Stability Region Estimation

In this section, we derive a scheme to obtain optimal estimates of stability regions via level sets of a given energy function even in the presence of saddle-node equilibrium points on the stability boundary.

The first theorem of this section was proven in [7]. It guarantees that every local minimum of the energy function on the stability boundary is attained at an equilibrium point.

Theorem 9 (Energy Functions and Equilibrium Points I) [7] *Let x_s be an asymptotically stable equilibrium point of the nonlinear dynamical system (1) and $A(x_s)$ be its stability region. If system (1) admits an energy function, then, the point on the stability boundary $\partial A(x_s)$ at which the energy function attains the minimum value must be an equilibrium point.*

The point of minimum energy on the stability boundary may not be unique. However, since the property that all equilibrium points of system (1) have distinct energy function values is generic, we can affirm that the point of minimum energy on the stability boundary is generically unique. In other words, the uniqueness of the point with minimum energy is almost always guaranteed.

Theorem 10, proven in [5], gives a characterization for the equilibrium points at which the global minimum of energy is attained over the stability boundary.

Theorem 10 (Energy Functions and Equilibrium Points II) [5] *Let x_s be an asymptotically stable equilibrium point of the nonlinear dynamical system (1) and $A(x_s)$ be its stability region. Suppose that system (1) admits an energy function. If x^* is the equilibrium point with the minimum value of the energy function over the stability boundary $\partial A(x_s)$, then*

(i) if x^* is a hyperbolic equilibrium point, then x^* is of the type-one;
(ii) if x^* is a saddle-node equilibrium point, then x^* is of the type-zero.

Theorem 11, proven in [5], gives a dynamical characterization of this equilibrium in terms of its invariant manifolds. Note that this theorem holds without the transversality condition.

Theorem 11 (Dynamical Characterization) [5] *Let x_s be an asymptotically stable equilibrium point of the nonlinear dynamical system (1) and $A(x_s)$ be its stability region. Suppose that system (1) admits an energy function. If x^* is the equilibrium point with the minimum value of the energy function over the stability boundary $\partial A(x_s)$, then*

(i) if x^* is hyperbolic, then $W^u(x^*) \cap A(x_s) \neq \emptyset$;
(ii) if x^* is a type-zero saddle-node equilibrium point, then $W^{c^+}(x^*) \cap A(x^s) \neq \emptyset$

Next theorem provides a scheme to obtain the best estimate of the stability region, via level sets of a particular given energy function, even in the presence of a saddle-node equilibrium point on the stability boundary. Its proof can be found in [5].

Theorem 12 (Stability Region Estimation) [5] *Let $A(x^s)$ be the stability region of the asymptotically stable equilibrium point x^s of (1). Suppose also that system (1) admits an energy function V. If $L = \min_{x \in E \cap \partial A(x^s)} V(x)$, then:*

(i) the connected component $D(L)$ of the level set $\{x \in \mathbb{R}^n : V(x) < L\}$ containing the equilibrium x^s is inside the stability region $A(x^s)$.
(ii) the connected component $D(B)$ of the level set $\{x \in \mathbb{R}^n : V(x) < B\}$ containing the equilibrium x^s has a nonempty intersection with the complement of the stability region $A^c(x^s)$ for any number $B > L$.

Theorem 12 ensures that, calculating all type-one hyperbolic and type-zero saddle-node equilibrium points on the stability boundary, we can obtain the best estimate of the stability region, in the form of a level set of the energy function V, by picking the level set with a level value that equals the value of the energy of the equilibrium point on the stability boundary which has the lowest value of energy. The choice L is optimal in the sense that any level set with an energy level greater than L is not contained in $A(x^s)$.

Theorem 12 generalizes the results in [7] by allowing the existence of saddle-node equilibrium points on the stability boundary. It suggests the following conceptual algorithm, which is also a generalization of the one proposed in [7], to obtain the optimal estimate of the stability region in the form of level sets of a given energy function V:

Conceptual Algorithm for Stability Region Estimation

Step 1 Compute all the equilibrium points on $\partial A(x^s)$.
Step 2 Identify the equilibrium point $x^{\min}$ that possesses the lowest energy among them. Let $L = V(x^{\min})$.
Step 3 The connected component $D(L)$ of the level set $\{x \in \mathbb{R}^n : V(x) < L\}$ containing x^s is the largest estimate of the stability region $A(x^s)$ in the form of a level set of V.

Example This example, proposed in [5], illustrates the application of the conceptual algorithm for stability region estimation. Consider the nonlinear autonomous dynamical system

$$\begin{aligned} \dot{x} &= -y \\ \dot{y} &= -x^4 + x^2 - y \end{aligned} \tag{4}$$

where $(x, y) \in \mathbb{R}^2$. Function $V(x, y, \lambda) = -\frac{x^5}{5} + \frac{x^3}{3} + \frac{y^2}{2}$ is an energy function for system (4). System (4) possesses three equilibrium points; they are $x^s = (-1, 0)$, a

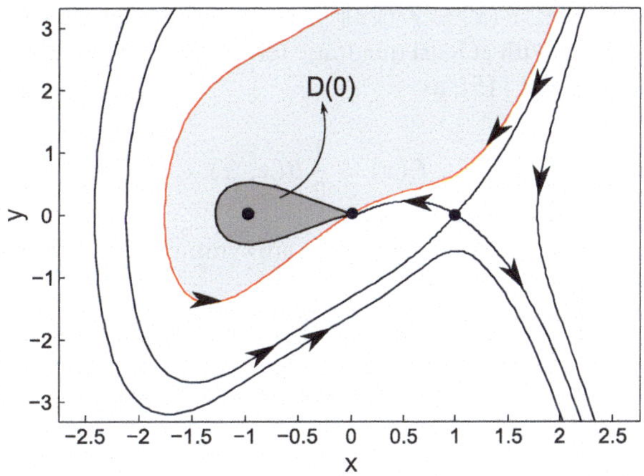

Fig. 7 The phase portrait of system (4). The type-zero saddle-node equilibrium point (0, 0) is on the stability boundary $\partial A(-1, 0)$. The region in light gray is the stability region of $(-1, 0)$ and the dark gray region is the largest estimate obtained via level set of the energy function $V(x, y, \lambda) = -\frac{x^5}{5} + \frac{x^3}{3} + \frac{y^2}{2}$. Reprinted from [5]

hyperbolic asymptotically stable equilibrium point, $p = (0, 0)$, a type-zero saddle-node equilibrium point and $x^* = (1, 0)$, a type-one hyperbolic equilibrium point. The stability boundary $\partial A(-1, 0)$ is depicted in Fig. 7. It is formed of the stable manifold of the type-zero saddle-node equilibrium point (0, 0). The minimum of the energy function V on the stability boundary $\partial A(-1, 0)$ is attained at the type-zero saddle-node equilibrium point p, the unique equilibrium on $\partial A(-1, 0)$. The energy function value at p is $L = V(0, 0) = 0$. The connected component $D(0)$ of the level set $\{x \in \mathbb{R}^2 : V(x, y) < 0\}$ containing the asymptotically stable equilibrium point $x^s = (-1, 0)$ is completely contained in $A(-1, 0)$, see Fig. 7, and it is the largest estimate that can be obtained in the form of a level set of V.

3.2 Hopf Equilibrium Points on the Stability Boundary

In this section, we study the properties of another type of non-hyperbolic equilibrium point on the stability boundary, the so called Hopf equilibrium point. In particular, we develop necessary and sufficient conditions for a Hopf equilibrium point belonging to the stability boundary. Moreover, we develop a complete characterization of the stability boundary in the presence of Hopf equilibrium points. The results of this section generalize the characterization of stability boundary given in Sect. 1 and are a compilation of the results presented in [9–11].

Consider the nonlinear dynamical system (1). We can always perform a change of coordinates in system (1), shifting the equilibrium point to origin. Thus, without loosing generality, system (1) can be rewritten as

$$\dot{x} = Ax + F(x), x \in \mathbb{R}^n, \tag{5}$$

where F, $F(x) = \mathcal{O}(\|x\|^2)$, is a smooth function that has Taylor expansion in x starting with at least quadratic terms. We can also write function $F(x)$, following the notation of [15], as

$$F(x) = \frac{1}{2}B(x,\ x) + \frac{1}{6}C(x,\ x,\ x) + \mathcal{O}(\|x\|^4)$$

where $B(x,\ y)$ and $C(x,\ y,\ z)$ are symmetric multilinear vector functions of $x,\ y,\ z \in \mathbb{R}^n$ such that

$$B_i(x,\ y) = \sum_{j,\,k=1}^{n} \left.\frac{\partial^2 F_i(\xi)}{\partial \xi_j \partial \xi_k}\right|_{\xi=0} x_j y_k,\ i = 1,\ \ldots,\ n$$

and

$$C_i(x,\ y,\ z) = \sum_{j,\,k,\,l=1}^{n} \left.\frac{\partial^3 F_i(\xi)}{\partial \xi_j \partial \xi_k \partial \xi_l}\right|_{\xi=0} x_j y_k z_l,\ i = 1,\ \ldots,\ n.$$

Definition 7 (*Hopf Equilibrium Point*) A non-hyperbolic equilibrium point $p \in \mathbb{R}^n$ of (1) is called a Hopf equilibrium point if the following conditions are satisfied:

(i) $A = D_x f(p)$ has a simple pair of purely imaginary eigenvalues, $\pm i\omega$, and no other eigenvalue with null real part;

(ii) $l_1 \neq 0$ where l_1 is the first Lyapunov coefficient, which can be computed by the formula:

$$l_1 = \frac{1}{2\omega}\Re\left[\langle u,\ C(v,\ v,\ \overline{v})\rangle - 2\langle u,\ B(v,\ A^{-1}B(v,\ \overline{v}))\rangle + \langle u,\ B(\overline{v},\ (2i\omega I - A)^{-1}B(v,\ v))\rangle\right]$$

where v is the complex eigenvector associated with the imaginary eigenvalue $i\omega$, u is the complex adjoint eigenvector of the transposed matrix A associated with its eigenvalue $-i\omega$ and satisfying the normalization condition:

$$\langle u,\ v\rangle = 1,$$

where $\langle x,\ y\rangle = \sum_{i=1}^{n} \overline{x}_i y_i$ represents the inner product in $\mathbb{C}^n$, and $\Re$ is the real part of a complex number.

Hopf equilibrium points can be classified according to the sign of the first Lyapunov coefficient.

Definition 8 (*Supercritical and Subcritical Hopf Equilibrium Point*) A Hopf equilibrium point $p \in \mathbb{R}^n$ of (1) is called a supercritical Hopf equilibrium point if the first Lyapunov coefficient $l_1 < 0$. A Hopf equilibrium point $p \in \mathbb{R}^n$ of (1) is called a subcritical Hopf equilibrium point if the first Lyapunov coefficient $l_1 > 0$.

Lyapunov coefficients are related to the asymptotic behavior of the system on the central manifold. Supercritical Hopf equilibrium points attract orbits on the central manifold while subcritical Hopf equilibrium points repel them. Furthermore, Hopf equilibrium points can be also classified in types according to the number of eigenvalues of $D_x f(p)$ with positive real part.

Definition 9 (*Type-k Hopf Equilibrium Point*) A Hopf equilibrium point p of (1) is called a type-k Hopf equilibrium point if $D_x f(p)$ has k ($k \leq n-2$) eigenvalues with positive real part and $n-k-2$ with negative real part.

Figure 8 illustrates the invariant manifolds for a type-0 supercritical Hopf equilibrium point in $\mathbb{R}^3$ and Fig. 9 illustrates these invariant manifolds for a type-1 supercritical Hopf equilibrium point in $\mathbb{R}^3$.

Figure 10 illustrates the invariant manifolds for a type-1 subcritical Hopf equilibrium point in $\mathbb{R}^3$ and Fig. 11 illustrates these invariant manifolds for a type-0 subcritical Hopf equilibrium point in $\mathbb{R}^3$.

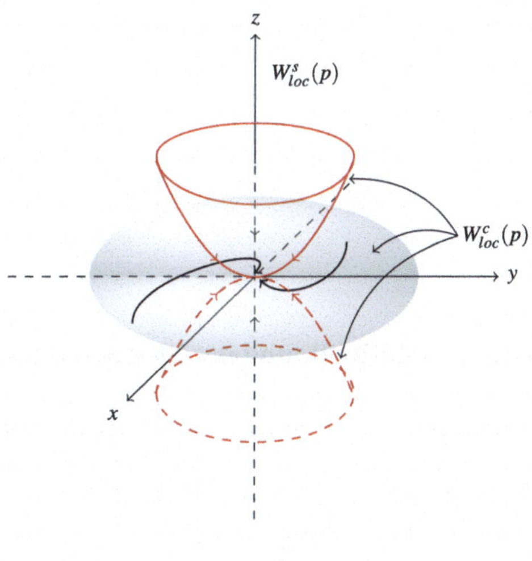

Fig. 8 Manifolds $W^c_{loc}(p)$ and $W^s_{loc}(p)$ for a type-0 supercritical Hopf equilibrium point p of system (1) in $\mathbb{R}^3$. $W^c_{loc}(p)$ is not unique. Three choices of $W^c_{loc}(p)$ are displayed in this figure. Reprinted from [9]

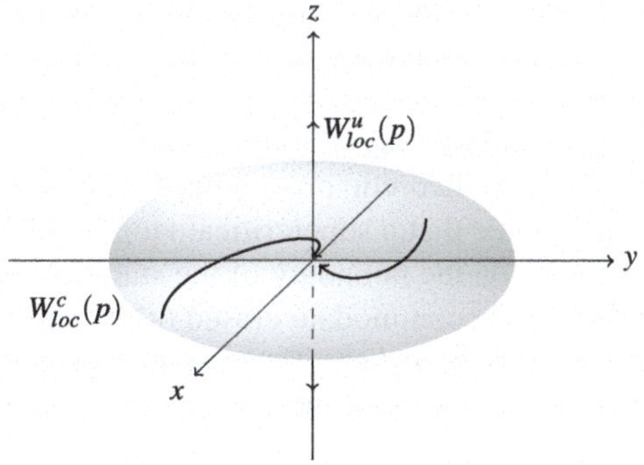

Fig. 9 Manifolds $W^c_{loc}(p)$ and $W^u_{loc}(p)$ for a type-1 supercritical Hopf equilibrium point p of system (1) in $\mathbb{R}^3$. In this case, $W^c_{loc}(p)$ is unique. Reprinted from [9]

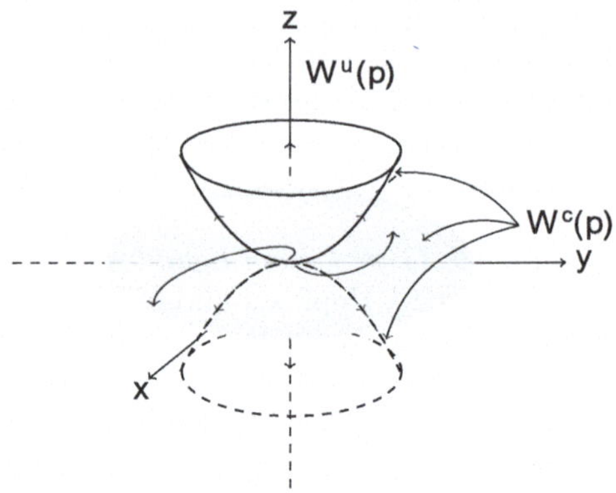

Fig. 10 Manifolds $W^c(p)$ and $W^s(p)$ for a type-1 subcritical Hopf equilibrium point p of system (1) in $\mathbb{R}^3$. $W^c(p)$ is not unique. Three choices of $W^c(p)$ are displayed in this figure. Reprinted from [10]

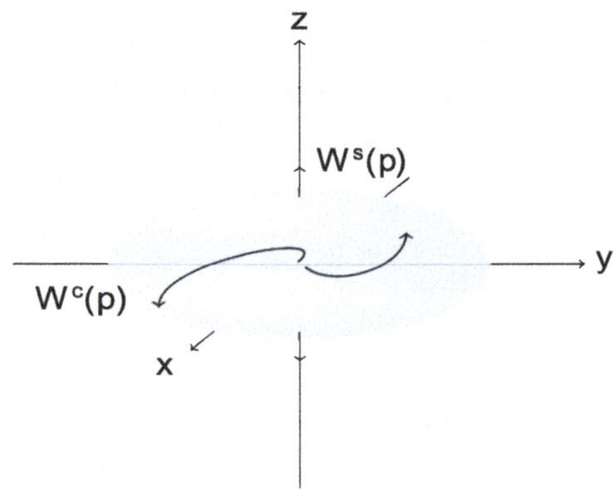

Fig. 11 Manifolds $W^c(p)$ and $W^u(p)$ for a type-0 subcritical Hopf equilibrium point p of system (1) in $\mathbb{R}^3$. In this case, $W^c(p)$ is unique. Reprinted from [10]

3.2.1 Stability Boundary Characterization

In this section, a complete characterization of the stability boundary in the presence of a Hopf equilibrium point on the stability boundary is developed. This characterization is developed in two steps. First we study a local characterization of the stability boundary by studying and characterizing the equilibrium points and closed orbits that belong to the stability boundary, then a global characterization is developed.

Initially, a characterization of the stability boundary considering supercritical Hopf equilibrium points is developed and then the result is extended to consider subcritical Hopf equilibrium points.

The next theorem, proven in [9, 11], provides necessary and sufficient conditions to guarantee that a supercritical Hopf equilibrium point or a hyperbolic critical element lies on the boundary of the stability region. It extends the results of Theorems 1 and 3 to accommodate closed orbits and supercritical Hopf equilibrium points on the stability boundary. These conditions are expressed in terms of the properties of their stable, unstable and center-stable manifolds.

Theorem 13 (Critical Elements on the Stability Boundary) [9, 11] *Let $A(x^s)$ be the stability region of an asymptotically stable equilibrium point x^s of (1). Let p be a type-k supercritical Hopf equilibrium point, with $1 \leq k \leq n-2$, and let ϕ be a type-k' hyperbolic critical element, with $k' \leq n$, of (1). If the assumptions:*

(B1') *All the critical elements on $\partial A(x^s)$ are hyperbolic critical elements or supercritical Hopf equilibrium points;*
(B2') *The stable, center-stable and/or center and unstable manifolds of the critical elements on $\partial A(x^s)$ satisfy the transversality condition;*
(B3) *Trajectories on $\partial A(x^s)$ approach one of the critical elements as $t \to \infty$ are held, then:*

(i) If ϕ is a type-k' critical element, with $1 \leq k' \leq n$, then

$$\phi \subset \partial A(x^s) \Longleftrightarrow W^u(\phi) \cap A(x^s) \neq \emptyset$$
$$\phi \subset \partial A(x^s) \Longleftrightarrow W^s(\phi) \subset \partial A(x^s)$$

(ii) If p is a type-k supercritical Hopf equilibrium point, with $1 \leq k \leq n-3$, then

$$p \in \partial A(x^s) \Longleftrightarrow W^u(p) \cap A(x^s) \neq \emptyset$$
$$p \in \partial A(x^s) \Longleftrightarrow W^{cs}(p) \subset \partial A(x^s)$$

(iii) If p is a type-$(n-2)$ supercritical Hopf equilibrium point, then

$$p \in \partial A(x^s) \Longleftrightarrow W^u(p) \cap A(x^s) \neq \emptyset$$
$$p \in \partial A(x^s) \Longleftrightarrow W^c(p) \subset \partial A(x^s)$$

The next theorem offers a complete characterization of the stability boundary when a supercritical Hopf equilibrium points lies on $\partial A(x^s)$. It is a generalization of Theorems 2 and 4 that allows the existence of closed orbits and supercritical Hopf equilibrium points on the stability boundary.

Theorem 14 (Characterization of the Stability Boundary for Critical Elements) [9, 11] *Let x^s be an asymptotically stable equilibrium point of (1) and let $A(x^s)$ be its stability region. If assumptions **(B1')** and **(B3)** are held, then:*

$$\partial A(x^s) \subset \bigcup_i W^s(\phi_i) \bigcup_j W^{cs}(p_j) \bigcup_l W^c(q_l)$$

*where ϕ_i are the hyperbolic critical elements, p_j the type-k supercritical Hopf equilibrium points, with $1 \leq k \leq n-3$, and q_l the type-$(n-2)$ supercritical Hopf equilibrium points on $\partial A(x^s)$, $i, j, l = 1, 2, \ldots$. If assumption **(B2')** is additionally satisfied, then*

$$\partial A(x^s) = \bigcup_i W^s(\phi_i) \bigcup_j W^{cs}(p_j) \bigcup_l W^c(q_l).$$

Theorem 14 states that the stability boundary is composed of the stable manifolds of the critical elements on the stability boundary union with the center-stable manifold of the supercritical Hopf equilibrium points on the stability boundary.

Now, we will construct the same characterization of the stability boundary when the non-hyperbolic equilibrium point on the stability boundary is a subcritical Hopf equilibrium point.

The next theorem provide necessary and sufficient conditions to guarantee that a subcritical Hopf equilibrium point or a hyperbolic critical element lies on the boundary of the stability region. These conditions are expressed in terms of the properties of its stable, center-unstable and center manifolds.

Theorem 15 (Critical Elements on the Stability Boundary) [11] *Let $A(x^s)$ be the stability region of an asymptotically stable equilibrium point x^s of (1). Let p be a type-k subcritical Hopf equilibrium point, with $1 \leq k \leq n-3$, and let ϕ be a type-k' hyperbolic critical element, with $k' \leq n$, of (1). If the assumptions:*

- ***(B1")*** *All the critical elements on $\partial A(x^s)$ are hyperbolics critical elements or subcritical Hopf equilibrium points;*
- ***(B2")*** *The stable and unstable, center-unstable and/or center manifolds of the critical elements on $\partial A(x^s)$ satisfy the transversality condition;*
- ***(B3)*** *Trajectories on $\partial A(x^s)$ approach one of the critical elements as $t \to \infty$ are held, then:*

(i) *If ϕ is a type-k' critical element, with $1 \leq k' \leq n$, then*
$\phi \in \partial A(x^s) \Longleftrightarrow W^u(\phi) \cap A(x^s) \neq \emptyset$
$\phi \in \partial A(x^s) \Longleftrightarrow W^s(\phi) \subset \partial A(x^s)$

(ii) *If p is a type-0 subcritical Hopf equilibrium point, then*
$p \in \partial A(x^s) \Longleftrightarrow W^c(p) \cap A(x^s) \neq \emptyset$
$p \in \partial A(x^s) \Longleftrightarrow W^s(p) \subset \partial A(x^s)$

(iii) *If p is a type-k subcritical Hopf equilibrium point, with $1 \leq k \leq n-3$, then*
$p \in \partial A(x^s) \Longleftrightarrow W^{cu}(p) \cap A(x^s) \neq \emptyset$
$p \in \partial A(x^s) \Longleftrightarrow W^s(p) \subset \partial A(x^s)$

(iv) *If p is a type-$(n-2)$ subcritical Hopf equilibrium point, then*
$p \in \partial A(x^s) \Longleftrightarrow W^{cu}(p) \cap A(x^s) \neq \emptyset$

The next theorem provides a complete characterization of the boundary of the stability region when there are subcritical Hopf equilibrium points in $\partial A(x^s)$. It is a generalization of Theorems 2 and 4 that allows the presence of closed orbits and subcritical Hopf equilibrium points on the stability boundary.

Theorem 16 (Characterization of the Stability Boundary for Critical Elements) *Let x^s be an asymptotically stable equilibrium point of (1) and let $A(x^s)$ be its stability region. If assumptions **(B1")** and **(B3)** are held, then:*

$$\partial A(x^s) \subset \bigcup_i W^s(\phi_i) \bigcup_j W^{cs}(p_j)$$

*where ϕ_i are the hyperbolic critical elements and p_j the subcritical Hopf equilibrium points, with $1 \le k \le n-2$, on $\partial A(x^s)$, $i = 1, 2, \ldots$. If, additionally, assumption (**B2'**) is satisfied, then*

$$\partial A(x^s) = \bigcup_i W^s(\phi_i) \bigcup_j W^{cs}(p_j).$$

Next example illustrates the results and characterizations developed in this section.

Example Consider the autonomous nonlinear dynamical system proposed in [9]:

$$\begin{cases} \dot{x} = -xz^2 - y - x(x^2+y^2); \\ \dot{y} = -yz^2 + x - y(x^2+y^2); \\ \dot{z} = -z(z-3)(8-z); \end{cases} \tag{6}$$

where $(x, y, z) \in \mathbb{R}^3$.

System (6) has three equilibrium points, they are an asymptotically stable equilibrium point $x^s = (0, 0, 3)$, a type-1 hyperbolic equilibrium point $x = (0, 0, 8)$ and a type-1 supercritical Hopf equilibrium point $p = (0, 0, 0)$. Consequently, assumptions (**B1**$'$) and (**B2**$'$) are satisfied.

Function $V(x, y, z) = \frac{x^2}{2} + \frac{y^2}{2} - \frac{z^4}{4} + \frac{11}{3}z^3 - -12z^2$ is an energy function for system (6). For instance, $\dot{V} = -(x^2+y^2)z^2 - (x^2+y^2)^2 - z^2(z-3)^2(8-z)^2 \le 0$ and assumption (**E1**) holds. The derivative of V equals zero only at equilibrium points, consequently assumption (**E2**) holds. And finally, if a solution $\varphi(t, x_0)$ is unbounded for $t \ge 0$, then $V(\varphi(t, x_0))$ is also unbounded for $t \ge 0$. As a consequence, assumption (**E3**) is satisfied and V is an energy function for system (6).

The existence of an energy function implies that assumption (**B3**) is held. Consequently the assumptions of Theorems 13 and 14 are satisfied and the complete characterization of stability boundary developed in Theorem 14 also holds. The unstable manifold of the type-1 supercritical Hopf equilibrium point $p = (0, 0, 0)$ intersects the stability region of $x^s = (0, 0, 3)$, consequently, according to Theorem 13, p lies on the stability boundary $\partial A(x^s)$ and the center manifold is contained in the boundary of the stability region of $x^s = (0, 0, 3)$, see Fig. 12. The unstable manifold of the type-1 hyperbolic equilibrium point $x = (0, 0, 8)$ also intersects the stability region of $x^s = (0, 0, 3)$ and therefore x lies on the stability boundary of x^s and the stable manifold is contained in the boundary of the stability region of $x^s = (0, 0, 3)$, according to the Theorem 13, see Fig. 12.

Figure 12 illustrates the boundary of the stability region of the asymptotically stable equilibrium point $x^s = (0, 0, 3)$. The boundary is formed, according to Theorem 14, of the union of the stable manifold of the type-1 hyperbolic equilibrium point $x = (0, 0, 8)$, the highest shaded surface passing by x at Fig. 12, with the center manifold of the type-1 supercritical Hopf equilibrium point $p = (0, 0, 0)$, the lowest shaded surface passing by p at Fig. 12.

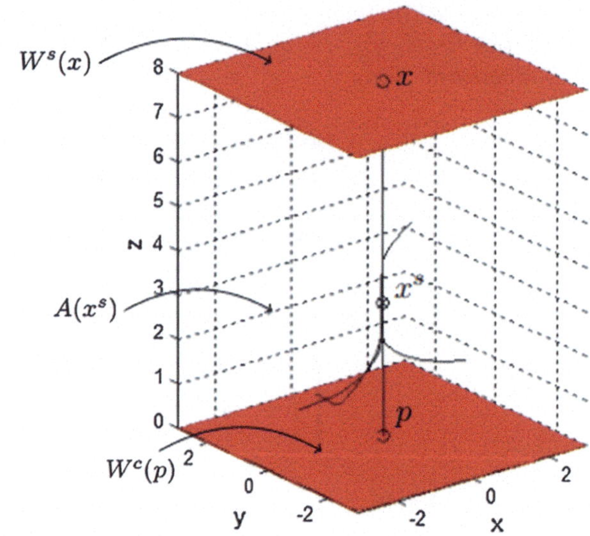

Fig. 12 The stability boundary of the asymptotically stable equilibrium point $x^s = (0, 0, 3)$ of system (6) is composed of two surfaces, the stable manifold of the type-1 hyperbolic equilibrium point $x = (0, 0, 8)$ and the center manifold of the type-1 supercritical Hopf equilibrium point $p = (0, 0, 8)$. Reprinted from [9]

4 Stability Region Bifurcations

The characterization of stability boundaries derived in Sect. 1 were developed under assumptions $(A1) - (A3)$. Under parameter variation, bifurcations may occur on the stability boundary and assumptions $(A1)$ or $(A2)$ may be violated at bifurcation points. Studying the characterization of the stability boundary at these bifurcation points is of fundamental importance to understanding how the stability region behaves under parameter variation. In this section, we study bifurcations of the stability boundary that are induced by local bifurcations of critical elements on the stability boundary and, in particular, by local bifurcation of equilibrium points. It will be shown that drastic changes in the size of the stability region might occur.

4.1 Sadde-Node Bifurcation

Consider the nonlinear dynamical system (3) and let $f : \mathbb{R}^n \times \mathbb{R} \to \mathbb{R}^n$ be a vector field of class C^r, with $r \geq 2$.

Definition 10 (*Saddle-Node Bifurcation Point*) The point $(p_{\lambda_0}, \lambda_0) \in \mathbb{R}^n \times \mathbb{R}$ is called a saddle-node bifurcation point of system (3) if $p_{\lambda_0} \in \mathbb{R}^n$ is a non-hyperbolic equilibrium point of (3) for the fixed parameter $\lambda = \lambda_0$ and the following conditions are satisfied:

(SN1) $D_x f_{\lambda_0}(p_{\lambda_0})$ has a unique simple eigenvalue equal to 0 with v as an eigenvector to the right and w to the left.

(SN2) $w(D_x^2 f_{\lambda_0}(p_{\lambda_0})(v, v)) \neq 0$.
(SN3) $w((\partial f_\lambda/\partial\lambda)(p_{\lambda_0}, \lambda_0)) \neq 0$.

In other words, $(p_{\lambda_0}, \lambda_0) \in \mathbb{R}^n \times \mathbb{R}$ is a saddle-node bifurcation point of system (3) if $p_{\lambda_0} \in \mathbb{R}^n$ is a saddle-node equilibrium point of (3) for a fixed parameter $\lambda = \lambda_0$ and the transversality condition $(SN2)$ is satisfied. A saddle-node bifurcation point $(p_{\lambda_0}, \lambda_0)$ will be of type k if the non-hyperbolic equilibrium point p_{λ_0} is a type-k saddle-node equilibrium point. The parameter λ_0 will also be called a type-k saddle-node bifurcation value.

Next theorem, proven in [17], studies the dynamical behavior of system (3) in the neighborhood of a saddle-node bifurcation point.

Theorem 17 (Saddle-Node Bifurcation) [17] *Let $(p_{\lambda_0}, \lambda_0)$ be a saddle-node bifurcation point of (3). Then there exist a neighborhood N of p_{λ_0} and $\delta > 0$ such that, depending on the signs of the expressions in (SN2) and (SN3), there is no equilibrium point on N when $\lambda \in (\lambda_0 - \delta, \lambda_0)[\lambda \in (\lambda_0, \lambda_0 + \delta)]$ and two equilibrium points p_λ^k and p_λ^{k+1} in N for each $\lambda \in (\lambda_0, \lambda_0 + \delta)[\lambda \in (\lambda_0 - \delta, \lambda_0)]$. The two equilibrium points on N are hyperbolic, more specifically p_λ^k is of type-k and p_λ^{k+1} is of type-$k+1$, $k \in \mathbb{N}$. Moreover, the stable manifold of the type-k equilibrium point and the unstable manifold of the type-$k+1$ equilibrium point intersect along an one-dimensional manifold.*

4.2 Saddle-Node Bifurcation on the Stability Boundary

In this section, we develop results that describe the behavior of the stability region and stability boundary in the neighborhood of a saddle-node bifurcation value. These results generalize the results of [4], which explore the behavior of the stability region and stability boundary in the neighborhood of only a type-zero saddle-node bifurcation value.

Next theorem, proven in [2], describes the local behavior of the stability boundary in the neighborhood of a type-k saddle-node equilibrium point.

Theorem 18 (Stability Boundary Behavior Near a Saddle-Node) [2] *Let p_{λ_0} be a type-k saddle-node equilibrium point lying on the stability boundary $\partial A_{\lambda_0}(x_{\lambda_0}^s)$ of the hyperbolic asymptotically stable equilibrium point $x_{\lambda_0}^s$ of (3) for $\lambda = \lambda_0$. If assumptions $(A1) - (A3)$ are satisfied in an open interval containing λ_0, except at the type-k saddle-node bifurcation value, with $k \geq 0$, where assumptions $(A1')$, $(A2')$ and $(A3)$ are satisfied, and the number of equilibrium points on $\partial A_{\lambda_0}(x_{\lambda_0}^s)$ is finite, then:*

(i) If $(p_{\lambda_0}, \lambda_0)$ is a type-zero saddle-node bifurcation point, with p_{λ_0} lying on the stability boundary $\partial A_{\lambda_0}(x_{\lambda_0}^s)$, then there is $\beta > 0$ such that, for all $\lambda \in (\lambda_0 - \beta, \lambda_0)$, we have that

$$p_{\lambda^0} \notin \partial A_\lambda(x_\lambda^s) \;\; and \;\; p_{\lambda^1} \in \partial A_\lambda(x_\lambda^s)$$

where p_{λ^0} and p_{λ^1} are the hyperbolic equilibrium points originated from the type-zero saddle-node bifurcation.

(ii) *If $(x_{\lambda_0}, \lambda_0)$ is a type-r saddle-node bifurcation point, with $r \geq 1$, with x_{λ_0} lying on the stability boundary $\partial A_{\lambda_0}(x^s_{\lambda_0})$, then there is $\beta > 0$ such that, for all $\lambda \in (\lambda_0 - \beta, \lambda_0)$, we have that*

$$y_{\lambda^r} \in \partial A_\lambda(x^s_\lambda) \ \text{ and } \ y_{\lambda^{r+1}} \in \partial A_\lambda(x^s_\lambda)$$

where p_{λ^r} and $p_{\lambda^{r+1}}$ are the unstable hyperbolic equilibrium points originated from the type-r saddle-node bifurcation, with $r \geq 1$.

Theorem 18 shows that, in the occurrence of a type-r saddle-node bifurcation, with $r \geq 1$, on the stability boundary, necessarily the two hyperbolic equilibrium points that coalesce and disappear at the bifurcation saddle-node belong to the stability boundary. Otherwise, the generic assumption of transversality would be violated.

The following corollary offers a complete characterization of the stability boundary in the neighborhood of a type-k saddle-node bifurcation value, with $k \geq 0$.

Corollary 1 (Characterization of the Stability Boundary in the Neighborhood of a Type-k Saddle-Node Bifurcation Value, with $k \geq 0$) *Let p_{λ_0} be a type-k saddle-node equilibrium point lying on the stability boundary $\partial A_{\lambda_0}(x^s_{\lambda_0})$ of the hyperbolic asymptotically stable equilibrium point $x^s_{\lambda_0}$ of (3) for $\lambda = \lambda_0$. If assumptions $(A1) - (A3)$ are satisfied in an open interval containing λ_0, except at the type-k saddle-node bifrucation value, with $k \geq 0$, where assumptions $(A1')$, $(A2')$ and $(A3)$ are satisfied, and the number of equilibrium points on $\partial A_{\lambda_0}(p_{\lambda_0})$ is finite, then:*
(i) For $\lambda = \lambda_0$ we have that

$$\partial A_{\lambda_0}(x^s_{\lambda_0}) = \bigcup_i W^s_{\lambda_0}(w^i_{\lambda_0}) \bigcup_j W^s_{\lambda_0}(p^j_{\lambda_0}) \bigcup_l W^{cs^-}_{\lambda_0}(z^l_{\lambda_0}) \bigcup_m W^{c^-}_{\lambda_0}(q^m_{\lambda_0})$$

where $w^i_{\lambda_0}$ are the hyperbolic equilibrium points in $\partial A_{\lambda_0}(x^s_{\lambda_0})$, $p^j_{\lambda_0}$ are the type-zero saddle-node equilibrium points, $z^l_{\lambda_0}$ the type-k saddle-node equilibrium points, with $1 \leq k \leq n-2$ and $q^m_{\lambda_0}$ the type-$(n-1)$ saddle-node equilibrium points in $\partial A_{\lambda_0}(x^s_{\lambda_0})$, $i, j, l, m = 1, 2, \ldots$.
(ii) There is $\varepsilon > 0$ such that, for all $\lambda \in (\lambda_0 - \varepsilon, \lambda_0)$,

$$\partial A_\lambda(x^s_\lambda) = \bigcup_i W^s_\lambda(w^i_\lambda) \bigcup_j W^s_\lambda(y^j_{\lambda^k}) \bigcup_j W^s_\lambda(y^j_{\lambda^{k+1}})$$

where w^i_λ are the perturbed hyperbolic equilibrium points in $\partial A_\lambda(x^s_\lambda)$, $y^j_{\lambda^k}$ and $y^j_{\lambda^{k+1}}$ are the unstable hyperbolic equilibrium points originated from the type-k saddle-node bifurcation, with $k \geq 0$, that also belong to $\partial A_\lambda(x^s_\lambda)$, $i, j, = 1, 2, \ldots$.
(iii) There is $\varepsilon > 0$ such that, for all $\lambda \in (\lambda_0, \lambda_0 + \varepsilon)$,

$$\partial A_\lambda(x_\lambda^s) = \bigcup_i W_\lambda^s(w_\lambda^i)$$

where w_λ^i are the perturbed hyperbolic equilibrium points in $\partial A_\lambda(x_\lambda^s)$, $i = 1, 2, \ldots$.

Example Consider the following system of differential equations from [2]:

$$\begin{aligned} \dot{x} &= 1 - \lambda sen(x) - 2sen(x-y) \\ \dot{y} &= 1 - 3sen(y) - 2sen(y-x) \\ \dot{z} &= \qquad\qquad -z \end{aligned} \tag{7}$$

where $(x; y; z) \in \mathbb{R}^3$ and $\lambda \in \mathbb{R}$.

System (7) possesses for $\lambda_0 = 2, 84$, a hyperbolic asymptotically stable equilibrium point $x_{\lambda_0}^s = (0, 35; 0, 34; 0)$ and a type-1 saddle-node equilibrium point $x_{\lambda_0} = (1, 42; 3, 39; 0)$. The type-1 saddle-node equilibrium point belongs to the stability boundary $\partial_{\lambda_0}(0, 35; 0, 34; 0)$. For $\lambda = 2, 87$, system (7) possesses a hyperbolic asymptotically stable equilibrium point $x_\lambda^s = (0, 33; 32; 0)$, a type-1 hyperbolic equilibrium point $p_{\lambda^1} = (1, 14; 3, 34; 0)$ and a type-2 hyperbolic equilibrium point $p_{\lambda^2} = (1, 48; 3, 43; 0)$. The equilibrium points p_{λ^1} and p_{λ^2} are originated from the type-1 saddle-node equilibrium point in a type-1 saddle-node bifurcation. Moreover, $p_{\lambda^1} \in \partial A_\lambda(0, 33; 32; 0)$ and $p_{\lambda^2} \in \partial A_\lambda(0, 33; 32; 0)$, according to Theorem 18, see Fig. 13.

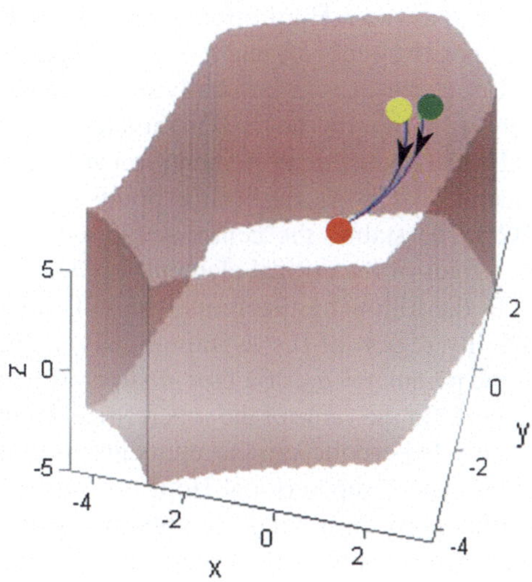

Fig. 13 The surface in this figure is the stability boundary of the stability region of the asymptotically stable equilibrium point $x_\lambda^s = (0, 33; 32; 0)$ of system (7) for $\lambda = 2, 87$. The unstable equilibrium points $p_{\lambda^1} = (1, 14; 3, 34; 0)$ and $p_{\lambda^2} = (1, 48; 3, 43; 0)$, originated from the type-1 saddle-node bifurcation, belong to the stability boundary $\partial_\lambda(0, 33; 32; 0)$. Reprinted from [2]

4.3 Hopf Bifurcation on the Stability Boundary

In this section, a characterization of the stability boundary in a small neighborhood of the parameter μ_0 of a Hopf bifurcation of type-k, with $k \geq 1$, is developed. We begin the section establishing some concepts of the Hopf bifurcation theory.

Consider the autonomous dynamic system dependent on a parameter

$$\dot{x} = f(x, \mu), \; x \in \mathbb{R}^n, \; \mu \in \mathbb{R} \tag{8}$$

where $f : \mathbb{R}^n \times \mathbb{R} \to \mathbb{R}^n$ is a $\mathscr{C}^1$-vector field .

Definition 11 (*Hopf Bifurcation Point*) A non-hyperbolic equilibrium point $x_{\mu_0} \in \mathbb{R}^n$ of (8), for a fixed parameter $\mu = \mu_0$, is called a Hopf equilibrium point and (x_{μ_0}, μ_0) a Hopf bifurcation point if the following conditions are satisfied:

(i) $D_x f(x_{\mu_0})$ has a simple pair of purely imaginary eigenvalues, $\pm i\omega$, and no other eigenvalue with null real part; and
(ii) $l_1 \neq 0$, where l_1 is the first coefficient of Lyapunov, see [9].

Hopf bifurcation points can also be classified according to the sign of the first Lyapunov coefficient. A Hopf bifurcation point (x_{μ_0}, μ_0) of (8) is called a supercritical Hopf bifurcation point if the first Lyapunov coefficient $l_1 < 0$ and is called a subcritical Hopf bifurcation point if the first Lyapunov coefficient $l_1 > 0$.

Hopf bifurcation points can also be classified in types according to the number of eigenvalues of $D_x f(x_{\mu_0})$ with positive real part. The Hopf bifurcation point x_{μ_0} of (8) is called a type-k Hopf bifurcation point if $D_x f(x_{\mu_0})$ has k ($k \leq n-2$) eigenvalues with positive real part and $n-k-2$ with negative real part.

Let $x^s_{\mu_0}$ be an asymptotically stable equilibrium point of (8) and let $A_{\mu_0}(x^s_{\mu_0})$ be its stability region for the fixed parameter $\mu = \mu_0$. Consider the following assumptions:
(B1') All the critical elements on $\partial A_{\mu_0}(x^s_{\mu_0})$ are hyperbolic critical elements or supercritical Hopf equilibrium points.
(B2') The stable, the center-stable and/or the center manifolds and the unstable manifolds of the critical elements on $\partial A_{\mu_0}(x^s_{\mu_0})$ satisfy the transversality condition.

In the following theorems, we will explore the behavior of the boundary of the stability region of the asymptotically stable equilibrium in a small neighborhood of the parameter μ_0 of a type-k supercritical Hopf bifurcation, with $k \geq 1$. We will assume, for the value of the supercritical Hopf bifurcation parameter μ_0, the existence of only hyperbolic critical elements of system (8) at $\mu = \mu_0$, with the exception of the type-k supercritical Hopf nonhyperbolic equilibrium point, with $k \geq 1$, x_{μ_0}. Furthermore, in a small neighborhood of the parameter μ_0, we will assume the existence of only critical elements that are the perturbed critical elements of the original system (8) at $\mu = \mu_0$. Initially, we will establish the behavior of the boundary of the stability region in the neighborhood of a type-k supercritical Hopf equilibrium point with $k \geq 1$ and, then we will present a global characterization of the boundary in that neighborhood.

Theorem 19 (Boundary of the Stability Region in the Neighborhood of a Type-k Supercritical Hopf Bifurcation Point with $k \geq 1$) [11] *Let (μ_0, x_{μ_0}) be a type-k supercritical Hopf bifurcation point, with $k \geq 1$, of (8) for $\mu = \mu_0$. Suppose that the type-k supercritical Hopf bifurcation point x_{μ_0} belongs to the boundary of the stability region $\partial A_{\mu_0}(x^s_{\mu_0})$ of an asymptotically stable hyperbolic equilibrium point $x^s_{\mu_0}$ of (8) for $\mu = \mu_0$. Admit that assumptions **(B1)**, **(B2)** and **(B3)** are satisfied for all μ belonging to an open interval I containing $\mu = \mu_0$, except in μ_0 where assumptions **(B1')** and **(B2')** are satisfied. Furthermore, assume that x_{μ_0} is the only nonhyperbolic equilibrium point in $\mu = \mu_0$. Suppose also that for all $\mu \in I$, all the critical elements of the perturbed system $\dot{x} = f(x, \mu)$ are perturbed critical elements originated from the system $\dot{x} = f(x, \mu_0)$. Then there is a neighborhood U of x_{μ_0} and $\varepsilon_1 \geq \varepsilon > 0$ such that:*

(i) There is a hyperbolic equilibrium point x^H_μ of type-k, with $1 \leq k \leq n-2$, in U for all $\mu \in (\mu_0 - \varepsilon_1, \mu_0)$ and there are a hyperbolic closed orbit Ω^H_μ of type-k, with $1 \leq k \leq n-2$, and a hyperbolic equilibrium point x^H_μ of type-$k+2$, with $1 \leq k \leq n-2$, in U for all $\mu \in (\mu_0, \mu_0 + \varepsilon_1)$.
(ii) For $\mu \in (\mu_0, \mu_0 + \varepsilon)$ we have that $\Omega^H_\mu \in \partial A_\mu(x^s_\mu)$ and $x^H_\mu \in \partial A_\mu(x^s_\mu)$.
(iii) For $\mu \in (\mu_0 - \varepsilon, \mu_0)$ we have that $x^H_\mu \in \partial A_\mu(x^s_\mu)$.

Theorem 19 ensures, for $\mu \in (\mu_0 - \varepsilon, \mu_0)$, the hyperbolic equilibrium point x^H_μ of type-k, with $1 \leq k \leq n-2$, in the neighborhood U, belongs to the stability boundary of x^s_μ. At $\mu = \mu_0$, the equilibrium point loses hyperbolicity, leading to the emergence of a type-k supercritical Hopf equilibrium point, with $k \geq 1$. The supercritical Hopf equilibrium point is on the stability boundary of x^s_μ. For values of $\mu > \mu_0$, the hyperbolic equilibrium point x^H_μ of type-$k+2$, with $1 \leq k \leq n-2$ in U loses stability and a hyperbolic closed orbit Ω^H_μ of type-$(k+1)$ arises, with $1 \leq k \leq n-2$, on the stability boundary of x^s_μ. Theorem 19 states that both the stability region as the stability boundary undergo changes when the parameter changes in the interval $(\mu_0 - \varepsilon, \mu_0 + \varepsilon)$. The next result establishes the characterization of the boundary of the stability region in a small neighborhood of the type-k supercritical Hopf bifurcation parameter value, with $k \geq 1$.

Theorem 20 (Characterization of the Stability Boundary in the Neighborhood of a Type-k Supercritical Hopf Equilibrium Point with $k \geq 1$) [11] *Let (μ_0, x_{μ_0}) be a type-k supercritical Hopf bifurcation point, with $k \geq 1$, of (8) for $\mu = \mu_0$. Suppose that the type-k supercritical Hopf bifurcation point x_{μ_0} belongs to the boundary of the stability region $\partial A_{\mu_0}(x^s_{\mu_0})$ of an asymptotically stable hyperbolic equilibrium point $x^s_{\mu_0}$ of (8) for $\mu = \mu_0$. Admit that assumptions **(B1)**, **(B2)** and **(B3)** are satisfied for all μ in an open interval I containing $\mu = \mu_0$, except at μ_0, where assumptions **(B1')** and **(B2')** are satisfied. Furthermore, assume that x_{μ_0} is the only nonhyperbolic equilibrium point in $\mu = \mu_0$. Suppose also that for all $\mu \in I$, all the critical elements of the perturbed system $\dot{x} = f(x, \mu)$ are perturbed critical elements originated from the system $\dot{x} = f(x, \mu_0)$. If $r^i_{\mu_0}$ are the critical elements in $\partial A_{\mu_0}(x^s_{\mu_0})$, $i = 1, \ldots, k$, then:*

(i) *For* $\mu = \mu_0$ *we have* $\partial A_{\mu_0}(x^s_{\mu_0}) = \bigcup_i W^s_{\mu_0}(r^i_{\mu_0}) \bigcup W^c_{\mu_0}(x_{\mu_0})$.

(ii) *There is* $\varepsilon > 0$ *such that, for all* $\mu \in (\mu_0 - \varepsilon,\ \mu_0)$, $\partial A_\mu(x^s_\mu) = \bigcup_i W^s_\mu(r^i_\mu) \bigcup W^s_\mu(x^H_\mu)$ *where* r^i_μ, $i = 1, 2, \ldots, k$ *are the perturbed hyperbolic critical elements in* $\partial A_\mu(x^s_\mu)$ *and* x^H_μ *is the type-k hyperbolic equilibrium point, with* $1 \le k \le n - 2$, *originated from the type-k supercritical Hopf bifurcation,* $k \ge 1$.

(iii) *There is* $\varepsilon > 0$ *such that, for all* $\mu \in (\mu_0,\ \mu_0 + \varepsilon)$, $\partial A_\mu(x^s_\mu) = \bigcup_i W^s_\mu(r^i_\mu) \bigcup W^s_\mu(x^H_\mu) \bigcup W^s_\mu(\Omega^H_\mu)$ *where* r^i_μ, $i = 1, 2, \ldots, k$ *are the perturbed hyperbolic critical elements in* $\partial A_\mu(x^s_\mu)$ *and* x^H_μ *and* Ω^H_μ *are the type-*$(k + 2)$ *hyperbolic equilibrium point, with* $1 \le k \le n - 2$, *and the type-k periodic orbit, with* $1 \le k \le n - 2$, *respectively, originated from the type-k supercritical Hopf bifurcation,* $k \ge 1$.

In the next two theorems, we will present the behavior of the stability boundary of an asymptotically stable equilibrium point in a small neighborhood of the parameter μ_0 of a type-k subcritical Hopf bifurcation, with $k \ge 1$. We will assume, for the value of the subcritical Hopf bifurcation parameter μ_0, the existence of only hyperbolic critical elements of the system (8), with the exception of the type-k subcritical Hopf non-hyperbolic equilibrium point, with $k \ge 1, x_{\mu_0}$. Furthermore, in a small neighborhood of the parameter μ_0, we will assume the existence of only critical elements that are the disturbed critical elements of the original system (8) in $\mu = \mu_0$. Proceeding in the same way we did in the occurrence of a supercritical Hopf bifurcation, we will establish the behavior of the boundary of the stability region in the neighborhood of a type-k subcritical Hopf equilibrium point with $k \ge 1$ and, then we will present a global characterization of the boundary in that neighborhood.

Theorem 21 (Stability Boundary in the Neighborhood of a Type-k Subcritical Hopf Bifurcation Point with $k \ge 1$) [11] *Let* (μ_0, x_{μ_0}) *be a type-k subcritical Hopf bifurcation point, with* $k \ge 1$, *of (8) for* $\mu = \mu_0$. *Suppose that the type-k subcritical Hopf bifurcation point* x_{μ_0} *belongs to the boundary of the stability region* $\partial A_{\mu_0}(x^s_{\mu_0})$ *of an asymptotically stable hyperbolic equilibrium point* $x^s_{\mu_0}$ *of (8) for* $\mu = \mu_0$. *Admit that assumptions (**B1**), (**B2**) and (**B3**) are satisfied for all* μ *belonging to an open interval I containing* $\mu = \mu_0$, *except in* μ_0 *where assumptions (**B1"**) and (**B2"**) are satisfied. Furthermore, assume that* x_{μ_0} *is the only nonhyperbolic equilibrium point in* $\mu = \mu_0$. *Suppose also that for all* $\mu \in I$, *all the critical elements of the perturbed system* $\dot{x} = f(x,\ \mu)$ *are perturbed critical elements originated from the system* $\dot{x} = f(x,\ \mu_0)$. *Then there is a neighborhood of* x_{μ_0} *and* $\varepsilon_1 \ge \varepsilon > 0$ *such that:*

(i) *There is a hyperbolic closed orbit* Ω^H_μ *of type-*$(k + 1)$, *with* $1 \le k \le n - 2$, *and a hyperbolic equilibrium point* x^H_μ *of type-k, with* $1 \le k \le n - 2$, *in U for all* $\mu \in (\mu_0 - \varepsilon_1,\ \mu_0)$ *and a hyperbolic equilibrium point* x^H_μ *of type-*$(k + 2)$, *with* $1 \le k \le n - 2$, *in U for all* $\mu \in (\mu_0,\ \mu_0 + \varepsilon_1)$.

(ii) *For* $\mu \in (\mu_0 - \varepsilon,\ \mu_0)$ *we have that*

$$\Omega^H_\mu \in \partial A_\mu(x^s_\mu) \quad \text{and} \quad x^H_\mu \in \partial A_\mu(x^s_\mu).$$

(iii) For $\mu \in (\mu_0,\ \mu_0+\varepsilon)$ we have that

$$x_\mu^H \in \partial A_\mu(x_\mu^s).$$

Theorem 21 ensures, for $\mu \in (\mu_0-\varepsilon,\ \mu_0)$, the hyperbolic periodic orbit Ω_μ^H of type-$(k+1)$, with $1 \leq k \leq n-2$, and the hyperbolic equilibrium point x_μ^H of type-k, with $1 \leq k \leq n-2$, in the neighborhood U, belongs to the stability boundary of x_μ^s. As the parameter μ grows, the amplitude of the closed orbit decreases and approaches the type-k hyperbolic equilibrium point, with $1 \leq k \leq n-2$ in U. At $\mu = \mu_0$, the periodic orbit coalesces with the hyperbolic equilibrium point in U, resulting in the emergence of a type-k subcritical Hopf equilibrium point, with $k \geq 1$. The subcritical Hopf equilibrium point is on the stability boundary of x_μ^s. For values of $\mu > \mu_0$, we have a hyperbolic equilibrium point x_μ^H in U, which belongs to the stability boundary. Theorem 21 states that both the stability region as the stability boundary undergo changes when the parameter changes in the interval $(\mu_0-\varepsilon,\ \mu_0+\varepsilon)$.

The next result establishes the characterization of the stability boundary in a small neighborhood of the parameter value of a type-k subcritical Hopf bifurcation, with $k \geq 1$.

Theorem 22 (Characterization of the Stability Boundary in the Neighborhood of a Type-k Subcritical Hopf Equilibrium Point with $k \geq 1$) [11] *Let (μ_0, x_{μ_0}) be a type-k subcritical Hopf bifurcation point, with $k \geq 1$, of (8) for $\mu = \mu_0$. Suppose that the type-k subcritical Hopf bifurcation point x_{μ_0} belongs to the stability boundary $\partial A_{\mu_0}(x_{\mu_0}^s)$ of an asymptotically stable hyperbolic equilibrium point $x_{\mu_0}^s$ of (8) for $\mu = \mu_0$. Admit that assumptions **(B1)**, **(B2)** and **(B3)** are satisfied for all μ belonging to an open interval I containing $\mu = \mu_0$, except in μ_0 where assumptions **(B1")** and **(B2")** are satisfied. Furthermore, assume that x_{μ_0} is the only nonhyperbolic equilibrium point in $\mu = \mu_0$. Suppose also that for all $\mu \in I$, all the critical elements of the perturbed system $\dot{x} = f(x,\ \mu)$ are perturbed critical elements originated from the system $\dot{x} = f(x,\ \mu_0)$. If $r_{\mu_0}^i$ are the critical elements in $\partial A_{\mu_0}(x_{\mu_0}^s)$, $i = 1,\ \ldots,\ k$, then:*

(i) For $\mu = \mu_0$ we have

$$\partial A_{\mu_0}(x_{\mu_0}^s) = \bigcup_i W_{\mu_0}^s(r_{\mu_0}^i) \bigcup W_{\mu_0}^s(x_{\mu_0})$$

(ii) There is $\varepsilon > 0$ such that, for all $\mu \in (\mu_0-\varepsilon,\ \mu_0)$,

$$\partial A_\mu(x_\mu^s) = \bigcup_i W_\mu^s(r_\mu^i) \bigcup W_\mu^s(x_\mu^H) \bigcup W_\mu^s(\Omega_\mu^H)$$

where r_μ^i, $i = 1,\ 2,\ \ldots,\ k$ are the perturbed hyperbolic critical elements in $\partial A_\mu(x_\mu^s)$ and x_μ^H and Ω_μ^H are the type-k hyperbolic equilibrium point, with $1 \leq k \leq n-2$, and a type-$(k+1)$ periodic orbit, with $1 \leq k \leq n-2$, respec-

tively, originated from the type-k subcritical Hopf bifurcation, with $k \geq 1$.

(*iii*) *There is* $\varepsilon > 0$ *such that, for all* $\mu \in (\mu_0,\ \mu_0 + \varepsilon)$,

$$\partial A_\mu(x^s_\mu) = \bigcup_i W^s_\mu(r^i_\mu) \bigcup W^s_\mu(x^H_\mu)$$

where r^i_μ, $i = 1,\ 2,\ \ldots,\ k$ *are the perturbed hyperbolic critical elements in* $\partial A_\mu(x^s_\mu)$ *and* x^H_μ *is the type-*$(k+2)$ *hyperbolic equilibrium point, with* $1 \leq k \leq n-2$, *originated from the type-k subcritical Hopf bifurcation,* $k \geq 1$.

Example Consider the nonlinear dynamical system from [9]:

$$\begin{cases} \dot{x} = (-z + \mu)x - y - x(x^2 + y^2); \\ \dot{y} = (-z + \mu)y + x - y(x^2 + y^2); \\ \dot{z} = -0.1(z + 0.5(x^2 + y^2))(z - 3)(8 - z); \end{cases} \tag{9}$$

where $(x,\ y,\ z) \in \mathbb{R}^3$ and $\mu \in \mathbb{R}$. For $\mu_0 = 0$, system (9) has three equilibrium points, they are: a type-1 hyperbolic equilibrium point, $x_1 = (0,\ 0,\ 8)$, a type-1 supercritical Hopf equilibrium point, $x^H_{\mu_0} = (0,\ 0,\ 0)$, and an asymptotically stable equilibrium point, $x^s_\mu = (0,\ 0,\ 3)$. The boundary of the stability region of $x^s_{\mu_0} = (0,\ 0,\ 3)$ is formed by the union of the stable manifold of the type-1 hyperbolic equilibrium point $x_1 = (0,\ 0,\ 8)$ with the center manifold of the type-1 supercritical

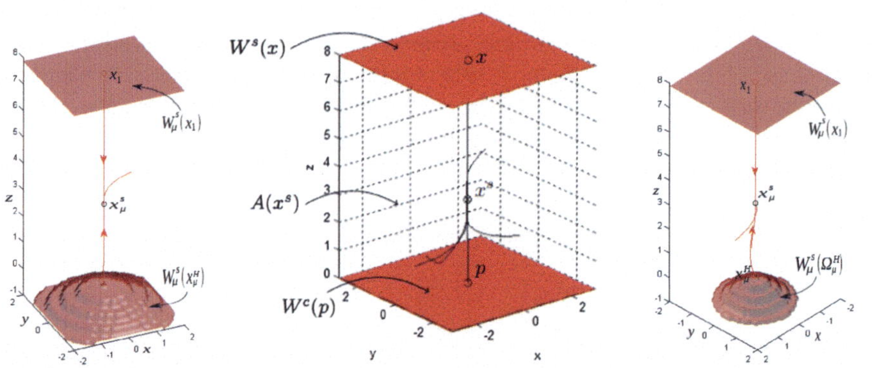

(a) Phase picture of the system for $\mu = -0.5$.

(b) Phase picture of the system for $\mu_0 = 0$.

(c) Phase picture of the system for $\mu = 0.5$.

Fig. 14 **a** The boundary of the stability region of $x^s_\mu = (0,\ 0,\ 3)$ is formed by the union of the stable manifold $W^s_\mu(x_1)$ with the stable manifold $W^s_\mu(x^H_\mu)$. **b** The boundary of the stability region of $x^s_{\mu_0} = (0,\ 0,\ 3)$ is formed by the union of the stable manifold $W^s_{\mu_0}(x_1)$ with the center manifold $W^c_{\mu_0}(x_{\mu_0})$. **c** The boundary of the stability region of $x^s_\mu = (0,\ 0,\ 3)$ is formed by the union of the stable manifold $W^s_\mu(x_1)$ with the stable manifold $W^s_\mu(\phi^H_\mu)$. Reprinted from [9]

Hopf equilibrium point, $x^H_{\mu_0} = (0, 0, 0)$, see Fig. 14b. For $\mu = -0.5$, the system has three equilibrium points, they are: two type-1 hyperbolic equilibrium points, $x^H_\mu = (0, 0, 0)$ and $x_1 = (0, 0, 8)$, and an asymptotically stable equilibrium point, $x^s_\mu = (0, 0, 3)$. The equilibrium point x^H_μ is originated from the type-1 supercritical Hopf equilibrium point in a type-1 supercritical Hopf bifurcation. The hyperbolic equilibrium points $x^H_\mu = (0, 0, 0)$ and $x_1 = (0, 0, 8)$ belong to the boundary of the stability region $\partial A_\mu(x^s_\mu)$, according to Theorem 19, see Fig. 14a. For $\mu = 0.5$, the system has four critical elements, they are: a type-3 hyperbolic equilibrium point, $x^H_\mu = (0, 0, 0)$, a type-1 hyperbolic equilibrium point, $x_1 = (0, 0, 8)$, an asymptotically stable equilibrium poin, $x^s_\mu = (0, 0, 3)$, and a type-1 hyperbolic periodic orbit ϕ^H_μ. The critical elements x^H_μ and ϕ^H_μ were originated from the type-zero supercritical Hopf equilibrium point in a type-zero supercritical Hopf bifurcation. The hyperbolic equilibrium points x_1 and ϕ^H_μ belong to the boundary of the stability region $\partial A_\mu(x^s_\mu)$, according to Theorem 19, see Fig. 14c.

5 Concluding Remarks

In this chapter, the body of the existing theory regarding the study of changes in the stability region due to parameter variation has been presented. These changes might be very complex and we have studied in this chapter only the ones triggered by two types of local bifurcation on the stability boundary: the saddle-node bifurcation and the Hopf bifurcation. It has been shown that these bifurcations may induce drastic changes in the "size" of the stability region, impacting on the stability of practical systems. There are many open issues to investigate to understand how stability region of general nonlinear systems behave as a consequence of parameter variation. Other types of local bifurcations on the stability boundary and global bifurcations are examples of potential themes for future research.

References

1. Amaral, F.M., Alberto, L.F.C., Gouveia, J.R.R.: Stability boundary and saddle-node equilibrium points on the stability boundary of nonlinear autonomous dynamical systems. Dyn. Syst. **01**, 1–23 (2017)
2. Amaral, F.M., Alberto, L.F.C.: Bifurcações Sela-Nó da Região de Estabilidade de Sistemas Dinâmicos Autônomos Não Lineares. Trends Appl. Comput. Math. **1**, 71–80 (2016)
3. Amaral, F.M., Alberto, L.F.C.: Stability boundary characterization of nonlinear autonomous dynamical systems in the presence of saddle-node equilibrium points. Trends Appl. Comput. Math. **2**, 143–154 (2012)
4. Amaral, F.M., Alberto, L.F.C.: Stability region bifurcations of nonlinear autonomous dynamical systems: type-zero saddle-node bifurcations. Int. J. Robust Nonlinear Control **21**, 591–612 (2011)

5. Amaral, F.M., Alberto, L.F.C.: Type-zero saddle-node bifurcations and stability region estimation of nonlinear autonomous dynamical systems. Int. J. Bifurcat. Chaos Appl. Sci. Eng. **22**(1), 1250020–1 (2012)
6. Chiang, H.D., Hirsch, M.W., Wu, Felix, F.: Stability region of nonlinear autonomous dynamical systems. IEEE Trans. Autom. Control **33**, 16–27 (1988)
7. Chiang, H.-D., Wu, F., Varaiya, P.: Foundations of direct methods for power system transient stability analysis. Circ. Syst. **34**, 160–173 (1987)
8. Chiang, H.-D: Alberto, L.F.C.: Stability Regions of Nonlinear Dynamical Systems, Theory, Estimation and Applications. Cambridge University Press (2015)
9. Gouveia, J.R.R., Amaral, F.M., Alberto, L.F.C.: Stability boundary characterization of nonlinear autonomous dynamical systems in the presence of a supercritical Hopf equilibrium point. Int. J. Bifurcat. Chaos **23**(12), 1350196 (2013)
10. Gouveia, J.R.R., Amaral, F.M., Alberto, L.F.C.: Subcritical hopf equilibrium points in the boundary of the stability region. TEMA São Carlos **17**(2), 211–224 (2016)
11. Gouveia, R.R.: Stability region bifurcations induced by local bifurcations of the type Hopf. Thesis - EESC, University of Sao Paulo (2015)
12. Guedes, R.B, Alberto, L.F.C., Bretas, N.G.: Power system low-voltage solutions using an auxiliary gradient system for voltage collapse purposes. Power Syst. **20**, 1528–1537 (2005)
13. Gunther, N., Hoffmann, G.W.: Qualitative dynamics of a network model of regulation of the immune system: a rationale for the IgM to IgG switch. J. Theor. Biol. **94**(4), 815–855 (1982)
14. Hirsch, M.W., Pugh, C.C., Shub, M.: Invariant manifolds. Bull. Am. Math. Soc. **76**(9), 1015–101 (1970)
15. Kuznetsov, Y.A.: Elements of Applied Bifurcation Theory. Springer, New York (1995)
16. Palis, J.: On morse-smale dynamical systems. Topology **8**, 385–405
17. Sotomayor, J.: Generic bifurcations of dynamical systems. Dyn. Syst. 549 (1973)
18. Wiggins, S.: Introduction to Applied Nonlinear Dynamical Systems and Chaos. Springer, New York (1989)

Universality in Systems with Power-Law Memory and Fractional Dynamics

Mark Edelman

1 Introduction

Sir Robert M. May's paper "Simple mathematical models with very complicated dynamics" published 41 years ago in "Nature" [35] is one of the most cited papers—3057 citations are registered by the Web of Science at the moment I am writing this sentence. In this review the author, using the logistic map as an example, described the universal behavior typical for all nonlinear systems: transition to chaos through the period-doubling cascade of bifurcations. The main applications considered by the author are the biological (even the variable used in the text was treated as "the population"), economic, and social sciences. The major steps in the development of the notion of universality in non-linear dynamics are gathered in the reprinted selection of papers compiled by Predrag Cvitanovic [10], and applications of universality encompass all areas of science.

The logistic map is a very simple discrete non-linear model of dynamical evolution. More realistic models of biological, economic, and social systems are more complicated. One of the features, not reflected in this equation but which is present in all the abovementioned systems, is memory. Evolution of any social or biological system depends not only on the current state of a system but on the whole history of its development. In majority of cases this memory obeys the power law. There

M. Edelman (✉)
Department of Physics, Stern College at Yeshiva University,
245 Lexington Ave, New York, NY 10016, USA
e-mail: edelman@cims.nyu.edu

M. Edelman
Courant Institute of Mathematical Sciences, New York University 251 Mercer St,
New York, NY 10012, USA

M. Edelman
Department of Mathematics, BCC, CUNY 2155 University Avenue,
Bronx, NY 10453, USA

M. Edelman et al. (eds.), *Chaotic, Fractional, and Complex Dynamics: New Insights and Perspectives*, Understanding Complex Systems,
https://doi.org/10.1007/978-3-319-68109-2_8

are many reviews on power-law distributions and memory in various social systems (see, e.g., [33]). In papers [21, 29, 52, 53, 56, 62] the power-law adaptation has been used to describe the dynamics of biological systems. The impotence and origin of the memory in biological systems can be related to the presence of memory at the level of individual cells: it has been shown recently that processing of external stimuli by individual neurons can be described by fractional differentiation [31, 32, 43]. The orders of fractional derivatives α derived for different types of neurons fall within the interval [0, 1], which implies power-law memory $\sim t^{\beta}$ with power $\beta = \alpha - 1$, $\beta \in [-1, 0]$. For neocortical pyramidal neurons the order of the fractional derivative is quite small: $\alpha \approx 0.15$. At the level of a human individual as a whole the power law appears in the study of human memory: forgetting—the accuracy on memory tasks decays as a power law with $\beta \in [-1, 0]$ [25, 45, 54–56]; learning—the reduction in reaction times that comes with practice is a power function of the number of training trials [2]. Power-law memory appears in the study of the human organ tissues due to their viscoelastic properties (see, e.g., references in [18]). This leads to their description by fractional differential equations with time fractional derivatives which implies the power-law memory. In most of the biological systems with the power-law behavior the power β is between -1 and 1 ($0 < \alpha < 2$).

It is much easier to investigate general properties of discrete systems with power-law memory than properties of integro-differential equations with power-law kernel. In Sect. 2 we review different ways to introduce or derive maps with power-law memory and their relation to fractional differential/difference equations. Periodic sinks and their stability (the stability of fixed points and asymptotic period two ($T = 2$) sinks) in fractional systems are discussed in Sect. 3. In Sect. 4 we consider various forms (non-linearity parameter, two-dimensional, and memory parameter) of bifurcation diagrams and transition to chaos in discrete fractional systems. In the conclusion we discuss perspectives and application of the research on the universality in fractional systems.

2 Maps with Power-Law Memory and Fractional Maps

In this section we consider various ways to introduce maps with power-law memory and fractional maps following [11–20, 48–51, 57, 58].

In the following we will use two definitions of fractional derivatives. They are based on the fractional integral introduced by Liouville, which is a generalization of the Cauchy formula for the n-fold integral

$$ {}_aI_t^p x(t) = \frac{1}{\Gamma(p)} \int_a^t \frac{x(\tau)d\tau}{(t-\tau)^{1-p}}, \tag{1}$$

where p is a real number, $\Gamma()$ is the gamma function and we will assume $a = 0$.

The first one is the left-sided Riemann-Liouville fractional derivative ${}_0D_t^{\alpha}x(t)$ defined for $t > 0$ [28, 42, 46] as

$$ {}_0D_t^\alpha x(t) = D_t^n \, {}_0I_t^{n-\alpha} x(t) = \frac{1}{\Gamma(n-\alpha)} \frac{d^n}{dt^n} \int_0^t \frac{x(\tau)d\tau}{(t-\tau)^{\alpha-n+1}}, \quad (2) $$

where $n-1 \leq \alpha < n, n \in \mathbb{Z}$, $D_t^n = d^n/dt^n$.

The second one is the left-sided Caputo derivative, in which the order of integration and differentiation in Eq. (2) is switched [28]

$$ {}_0^C D_t^\alpha x(t) = {}_0 I_t^{n-\alpha} \, D_t^n x(t) = \frac{1}{\Gamma(n-\alpha)} \int_0^t \frac{D_\tau^n x(\tau)d\tau}{(t-\tau)^{\alpha-n+1}} \quad (n-1 < \alpha \leq n). \quad (3) $$

2.1 Direct Introduction of Maps with Power-Law Memory

The direct way to introduce maps with power-law memory is to define them as convolutions according to the formula (see [18, 47])

$$ x_n = \sum_{k=0}^{n-1} (n-k)^{\alpha-1} G_K(x_k, h), \quad (4) $$

where K is a parameter and h is a constant time step between time instants t_n and t_{n+1}. For a physical interpretation of this formula we consider a system which state is defined by the variable x and evolution by the function $G_K(x)$. The value of the state variable at the time t_n is a weighted total of the functions $G_K(x_k)$ from the values of this variable at past time instants t_k, $0 < k < n$, $t_k = kh$. The weights are the times between time instants t_n and t_k to the fractional power $\alpha - 1$.

The more general form of this map considered in [18, 47] (see, e.g., Eq. (73) from [18]) is

$$ x_n = \sum_{k=1}^{\lceil\alpha\rceil-1} \frac{c_k}{\Gamma(\alpha-k+1)} (nh)^{\alpha-k} + \sum_{k=0}^{n-1} (n-k)^{\alpha-1} G_K(x_k, h), \quad (5) $$

where $\alpha \in \mathbb{R}$. If we assume

$$ G_K(x, h) = \frac{1}{\Gamma(\alpha)} h^\alpha G_K(x), \quad (6) $$

where $G_K(x)$ is continuous, and

$$ x = x(t), \quad x_k = x(t_k), \quad t_k = a + kh, \quad nh = t - a \quad (7) $$

for $0 \leq k \leq n$, then this equation can be written as

$$x_n = \sum_{k=1}^{\lceil\alpha\rceil-1} \frac{c_k}{\Gamma(\alpha-k+1)}(nh)^{\alpha-k} + \frac{h^\alpha}{\Gamma(\alpha)}\sum_{k=0}^{n-1}(n-k)^{\alpha-1}G_K(x_k). \tag{8}$$

Eq. (8) in the limit $h \to 0+$ will yield the Volterra integral equation of the second kind

$$x(t) = \sum_{k=1}^{\lceil\alpha\rceil-1} \frac{c_k}{\Gamma(\alpha-k+1)}(t-a)^{\alpha-k} + \frac{1}{\Gamma(\alpha)}\int_a^t \frac{G_K(\tau, x(\tau))d\tau}{(t-\tau)^{1-\alpha}}, \quad (t > a). \tag{9}$$

This equation is equivalent to the fractional differential equation with the Riemann-Liouville or Grünvald-Letnikov fractional derivative [18, 26, 27]

$${}_a^{RL/GL}D_t^\alpha x(t) = G_k(t, x(t)), \;\; 0 < \alpha \tag{10}$$

with the initial conditions

$$({}_a^{RL/GL}D_t^{\alpha-k}x)(a+) = c_k, \quad k = 1, 2, ..., \lceil\alpha\rceil. \tag{11}$$

For $\alpha \notin \mathbb{N}$ we assume $c_{\lceil\alpha\rceil} = 0$, which corresponds to a finite value of $x(a)$.

2.2 Universal Map with Power-Law Memory From fractional Differential Equations of Systems with Periodic Delta-Function Kicks

The universal map and its particular form, the standard map, play an important role in the study of regular dynamical systems. Their fractional generalizations can be obtained in a way similar to the way in which the regular universal map is derived from the differential equation of a periodically (with the period h) kicked system in regular dynamics (see, e.g., [60]). The two-dimensional fractional universal map obtained from the differential equation of the order $1 < \alpha \le 2$ was introduced in [51], extended to any real $\alpha > 1$ in [48–50], and then to any $\alpha \ge 0$ in [12, 13, 15].

To derive the equations of the fractional universal map, which we'll call the universal α-family of maps (α-FM) for $\alpha \ge 0$, we start with the differential equation

$$\frac{d^\alpha x}{dt^\alpha} + G_K(x(t-\Delta h)) \sum_{k=-\infty}^{\infty} \delta\Big(\frac{t}{h} - (k+\varepsilon)\Big) = 0, \tag{12}$$

where $\varepsilon > \Delta > 0$, $\alpha \in \mathbb{R}$, $\alpha > 0$, and consider it as $\varepsilon \to 0$. The initial conditions should correspond to the type of fractional derivative used in Eq. (12). The case $\alpha = 2$, $\Delta = 0$, and $G_K(x) = KG(x)$ corresponds to the equation whose integration yields the regular universal map.

Integration of Eq. (12) with the Riemann-Liouville fractional derivative ${}_0D^{\alpha}_t x(t)$ and the initial conditions

$$({}_0D^{\alpha-k}_t x)(0+) = c_k, \tag{13}$$

where $k = 1, ..., N$ and $N = \lceil \alpha \rceil$, yields the Riemann-Liouville universal α-FM

$$x_{n+1} = \sum_{k=1}^{N} \frac{c_k}{\Gamma(\alpha - k + 1)} h^{\alpha-k}(n+1)^{\alpha-k} - \frac{h^{\alpha}}{\Gamma(\alpha)} \sum_{k=0}^{n} G_K(x_k)(n-k+1)^{\alpha-1}. \tag{14}$$

As in the Sect. 2.1, for $\alpha \notin \mathbb{N}$ boundedness of $x(t)$ at $t = 0$ requires $c_N = 0$ and $x(0) = 0$ (see [28, 42, 46]). Obtained in Sect. 2.1 Eq. (8) is identical to Eq. (14).

Integration of Eq. (12) with the Caputo fractional derivative ${}^C_0D^{\alpha}_t x(t)$ and the initial conditions $(D^k_t x)(0+) = b_k$, $k = 0, ..., N-1$, yields the Caputo universal α-FM

$$x_{n+1} = \sum_{k=0}^{N-1} \frac{b_k}{k!} h^k (n+1)^k - \frac{h^{\alpha}}{\Gamma(\alpha)} \sum_{k=0}^{n} G_K(x_k)(n-k+1)^{\alpha-1}. \tag{15}$$

Later in this paper we'll refer to the maps Eqs. (8) and (14), the RL universal α-FM, as the Riemann-Liouville universal map with power-law memory or the Riemann-Liouville universal fractional map; we'll call the Caputo universal α-FM, Eq. (15), the Caputo universal map with power-law memory or the Caputo universal fractional map.

In the case of integer α the universal map converges to

$$x_n = 0 \ \text{ for } \ \alpha = 0 \ \text{ and } \ x_{n+1} = x_n - hG_K(x_n) \ \text{ for } \ \alpha = 1, \tag{16}$$

and for $\alpha = N = 2$ with $p_{n+1} = (x_{n+1} - x_n)/h$

$$\begin{cases} p_{n+1} = p_n - hG_K(x_n), & n \geq 0, \\ x_{n+1} = x_n + hp_{n+1}, & n \geq 0. \end{cases} \tag{17}$$

N-dimensional, with $N \geq 2$, universal maps are investigated in [13], where it is shown that they are volume preserving.

2.3 *Universal Fractional Difference Map*

The fractional sum ($\alpha > 0$)/difference ($\alpha < 0$) operator introduced by Miller and Ross in [36]

$${}_a\Delta^{-\alpha}_t f(t) = \frac{1}{\Gamma(\alpha)} \sum_{s=a}^{t-\alpha} (t-s-1)^{(\alpha-1)} f(s) \tag{18}$$

can be considered as a fractional generalization of the n-fold summation formula [17, 24]

$$
{}_a\Delta_t^{-n} f(t) = \frac{1}{(n-1)!}\sum_{s=a}^{t-n}(t-s-1)^{(n-1)}f(s) = \sum_{s^0=a}^{t-n}\sum_{s^1=a}^{s^0}\dots\sum_{s^{n-1}=a}^{s^{n-2}} f(s^{n-1}), \quad (19)
$$

where $n \in \mathbb{N}$ and s^i, $i = 0, 1, ...n-1$, are the summation variables. In Eq. (18) f is defined on $\mathbb{N}_a$ and ${}_a\Delta_t^{-\alpha}$ on $\mathbb{N}_{a+\alpha}$, where $\mathbb{N}_t = \{t, t+1, t+2, \dots\}$. The falling factorial $t^{(\alpha)}$ is defined as

$$
t^{(\alpha)} = \frac{\Gamma(t+1)}{\Gamma(t+1-\alpha)}, \quad t \neq -1, -2, -3, ... \quad (20)
$$

and is asymptotically a power function:

$$
\lim_{t\to\infty}\frac{\Gamma(t+1)}{\Gamma(t+1-\alpha)t^{\alpha}} = 1, \quad \alpha \in \mathbb{R}. \quad (21)
$$

For $\alpha > 0$ and $m-1 < \alpha < m$ the fractional (left) Riemann-Liouville difference operator is defined (see [4, 5]) as

$$
{}_a\Delta_t^{\alpha} x(t) = \Delta_a^m\, \Delta_t^{-(m-\alpha)} x(t) = \frac{1}{\Gamma(m-\alpha)}\Delta^m \sum_{s=a}^{t-(m-\alpha)} (t-s-1)^{(m-\alpha-1)} x(s) \quad (22)
$$

and the fractional (left) Caputo-like difference operator (see [3]) as

$$
{}_a^C\Delta_t^{\alpha} x(t) = {}_a\Delta_t^{-(m-\alpha)} \Delta^m x(t) = \frac{1}{\Gamma(m-\alpha)}\sum_{s=a}^{t-(m-\alpha)} (t-s-1)^{(m-\alpha-1)} \Delta^m x(s), \quad (23)
$$

where Δ^m is the m-th power of the forward difference operator defined as $\Delta x(t) = x(t+1) - x(t)$. Due to the fact that ${}_a\Delta_t^{\lambda}$ in the limit $\lambda \to 0$ approaches the identity operator (see [17, 36]), the definition Eq. (23) can be extended to all real $\alpha \geq 0$ with ${}_a^C\Delta_t^m x(t) = \Delta^m x(t)$ for $m \in \mathbb{N}_0$.

Fractional h-difference operators, which are generalizations of the fractional difference operators, were introduced and investigated in [7, 8, 22, 37–40]. The h-sum operator is defined as

$$
({}_a\Delta_h^{-\alpha} f)(t) = \frac{h}{\Gamma(\alpha)}\sum_{s=\frac{a}{h}}^{\frac{t}{h}-\alpha}(t-(s+1)h)_h^{(\alpha-1)} f(sh), \quad (24)
$$

where $\alpha \geq 0$, $({}_a\Delta_h^0 f)(t) = f(t)$, f is defined on $(h\mathbb{N})_a$, and ${}_a\Delta_h^{-\alpha}$ on $(h\mathbb{N})_{a+\alpha h}$. $(h\mathbb{N})_t = \{t, t+h, t+2h, \dots\}$. The h-factorial $t_h^{(\alpha)}$ is defined as

$$t_h^{(\alpha)} = h^\alpha \frac{\Gamma(\frac{t}{h}+1)}{\Gamma(\frac{t}{h}+1-\alpha)} = h^\alpha \Big(\frac{t}{h}\Big)^{(\alpha)}, \quad \frac{t}{h} \neq -1, -2, -3, ... \tag{25}$$

With $m = \lceil \alpha \rceil$ the Riemann-Liouville (left) h-difference is defined as

$$\begin{aligned}({}_a\Delta_h^\alpha x)(t) &= (\Delta_h^m ({}_a\Delta_h^{-(m-\alpha)} x))(t) \\ &= \frac{h}{\Gamma(m-\alpha)} \Delta_h^m \sum_{s=\frac{a}{h}}^{\frac{t}{h}-(m-\alpha)} (t-(s+1)h)_h^{(m-\alpha-1)} x(sh)\end{aligned} \tag{26}$$

and the Caputo (left) h-difference is defined as

$$\begin{aligned}({}_a\Delta_{h,*}^\alpha x)(t) &= ({}_a\Delta_h^{-(m-\alpha)} (\Delta_h^m x))(t) \\ &= \frac{h}{\Gamma(m-\alpha)} \sum_{s=\frac{a}{h}}^{\frac{t}{h}-(m-\alpha)} (t-(s+1)h)_h^{(m-\alpha-1)} (\Delta_h^m x)(sh),\end{aligned} \tag{27}$$

where $(\Delta_h^m x))(t)$ is the mth power of the forward h-difference operator

$$(\Delta_h x)(t) = \frac{x(t+h)-x(t)}{h}. \tag{28}$$

As it has been noted in [7, 22, 23], due to the convergence of solutions of fractional Riemann-Liouville h-difference equations when $h \to 0$ to solutions of the corresponding differential equations, they can be used to solve fractional Riemann-Liouville differential equations numerically. A proof of the convergence (as $h \to 0$) of fractional Caputo h-difference operators to the corresponding fractional Caputo differential operators for $0 < \alpha \leq 1$ can be found in [38] (Proposition 17).

In what follows, we will consider fractional Caputo difference maps—the only fractional difference maps which behavior has been investigated. The following theorem [9, 16, 17, 58] is essential to derive the universal fractional difference map.

Theorem 1 *For $\alpha \in \mathbb{R}$, $\alpha \geq 0$ the Caputo-like difference equation*

$${}_0^C\Delta_t^\alpha x(t) = -G_K(x(t+\alpha-1)), \tag{29}$$

where $t \in \mathbb{N}_m$, with the initial conditions

$$\Delta^k x(0) = c_k, \quad k = 0, 1, \ldots, m-1, \quad m = \lceil \alpha \rceil \tag{30}$$

is equivalent to the map with falling factorial-law memory

$$x_{n+1} = \sum_{k=0}^{m-1} \frac{\Delta^k x(0)}{k!}(n+1)^{(k)}$$
$$-\frac{1}{\Gamma(\alpha)} \sum_{s=0}^{n+1-m} (n-s-m+\alpha)^{(\alpha-1)} G_K(x_{s+m-1}), \tag{31}$$

where $x_k = x(k)$, which is called the fractional difference Caputo universal α-family of maps.

To consider h-differences, we will extend this theorem using the property (see [22])

$$({}_0\Delta^{\alpha}_{h,*}x)(t) = h^{-\alpha}{}^C_0\Delta^{\alpha}_t \bar{x}\left(\frac{t}{h}\right), \tag{32}$$

where x is defined on $(h\mathbb{N})_a$, ${}_a\Delta^{\alpha}_{h,*}$ on $(h\mathbb{N})_{a+\alpha h}$, and $\bar{x}(s) = x(sh)$. It is easy to show that the following theorem is a generalization of Theorem 1.

Theorem 2 *For $\alpha \in \mathbb{R}$, $\alpha \geq 0$ the Caputo-like h-difference equation*

$$({}_0\Delta^{\alpha}_{h,*}x)(t) = -G_K(x(t+(\alpha-1)h)), \tag{33}$$

where $t \in (h\mathbb{N})_m$, with the initial conditions

$$({}_0\Delta^k_h x)(0) = c_k, \quad k = 0, 1, \ldots, m-1, \quad m = \lceil \alpha \rceil \tag{34}$$

is equivalent to the map with h-factorial-law memory

$$x_{n+1} = \sum_{k=0}^{m-1} \frac{c_k}{k!}((n+1)h)^{(k)}_h$$
$$-\frac{h^{\alpha}}{\Gamma(\alpha)} \sum_{s=0}^{n+1-m} (n-s-m+\alpha)^{(\alpha-1)} G_K(x_{s+m-1}), \tag{35}$$

where $x_k = x(kh)$, which is called the h-difference Caputo universal α-family of maps.

In the case of integer α the fractional difference universal map converges to

$$x_{n+1} = -G_K(x_n) \text{ for } \alpha = 0, \text{ to } x_{n+1} = x_n - hG_K(x_n) \text{ for } \alpha = 1, \tag{36}$$

and for $\alpha = N = 2$, with $p_{n+1} = (x_{n+1} - x_n)/h$, to

$$\begin{cases} p_{n+1} = p_n - hG_K(x_n), & n \geq 1, \quad p_1 = p_0, \\ x_{n+1} = x_n + hp_{n+1}, & n \geq 0. \end{cases} \tag{37}$$

N-dimensional, with $N \geq 2$, difference universal maps are investigated in [16]. They are volume preserving (as well as the N-dimensional universal maps of Sect. 2.2).

All the above considered universal maps in the case $\alpha = 2$ yield the standard map if $G_K(x) = K \sin(x)$ (harmonic nonlinearity) and we'll call them the standard α-families of maps. When $G_K(x) = x - Kx(1-x)$ (quadratic nonlinearity) in the one-dimensional case all maps yield the regular logistic map and we'll call them the logistic α-families of maps.

3 Periodic Sinks and Their Stability

As in regular dynamics, the notion of universality and transition to chaos in fractional dynamics is related to the dependence of the phase space structure of fixed and periodic points (sinks) on systems' parameters. Presence of power-law memory leads to some new features that appear in fractional dynamics.

- In addition to the dependence on nonlinearity parameters, the phase space structure of fractional systems depends on a memory (an order of a fractional derivative) parameter.
- Periodic points in fractional dynamics exist in the asymptotic sense. As it has been shown in [61], effects of memory on the phase space structure of fractional systems of the order $\alpha \in (1,2)$ are similar to the effects of dissipation. But in fractional systems periodic sinks have their basins of attraction to which they themselves may not belong [11, 19, 20]. In the latter case a trajectory that starts from a sink jumps out of the sink and may end up in a different sink.
- Evolution of systems with memory, in general, follows cascade of bifurcations type trajectories (CBTT). Two examples of CBTT are presented in Fig. 1. As time (number of iterations n) increases, the trajectory bifurcates and may end as a periodic sink (Fig. 1a) or as a chaotic trajectory (Fig. 1b).
- Not only the time of convergence of trajectories to the periodic sinks but also the way in which convergence occurs depends on the initial conditions. As $n \to \infty$, all trajectories in Fig. 2 converge to the same period two ($T = 2$) sink (as in Fig. 2c), but for small values of initial conditions x_0 all trajectories first converge to a $T = 1$ trajectory which then bifurcates and turns into the $T = 2$ sink converging to its limiting value. As x_0 increases, the bifurcation point n_{bif} gradually evolves from the right to the left (Fig. 2a). Ignoring this feature may result (as, e.g., in [57, 58]) in very messy bifurcation diagrams.

3.1 $T = 2$ Sinks and Stability of Fixed Points

Stability of fixed points in fractional dynamical systems was investigated in multiple publications, see, e.g., articles [1, 30, 34], Chap. 4 in book [41], and review [44]; for stability in discrete fractional systems see, e.g., [6, 19, 59]. There are various ways to define stability and various methods and criteria to analyze it. In this paper we

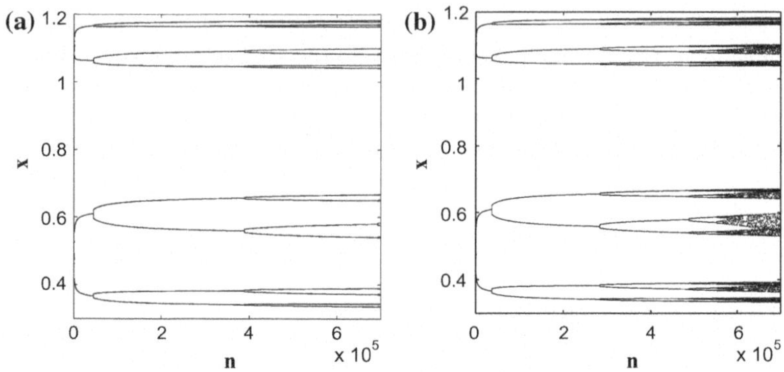

Fig. 1 Two examples of cascade of bifurcations type trajectories in the Caputo logistic α-family of maps (Eq. (15) with $T = 1$ and $G_K(x) = x - Kx(1-x)$) with $\alpha = 0.1$ and $x_0 = 0.001$: **a** for the nonlinearity parameter $K = 22.39$ the last bifurcation from period $T = 16$ to period $T = 32$ occurs after approximately 7×10^5 iterations; **b** when $K = 22.416$ the trajectory becomes chaotic after approximately 6×10^5 iterations

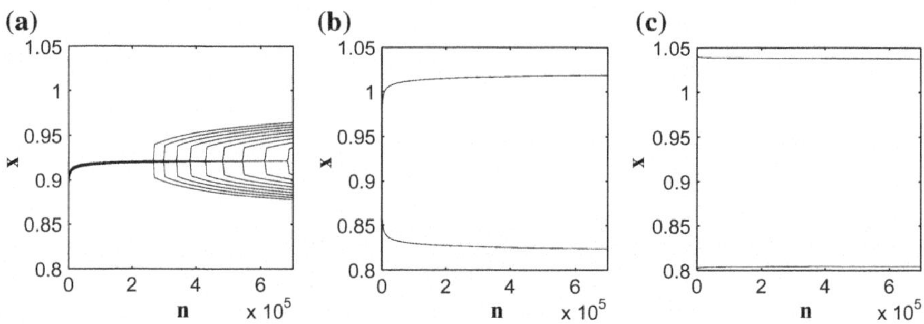

Fig. 2 Asymptotically period two trajectories for the Caputo logistic α-family of maps with $\alpha = 0.1$ and $K = 15.5$: **a** nine trajectories with initial conditions x_0 from 0.29 (the rightmost bifurcation) to 0.37 (the leftmost bifurcation) with the step 0.01; **b** $x_0 = 0.74$; **c** $x_0 = 0.94$

consider an asymptotic stability of periodic points. A periodic point is asymptotically stable if there exists an open set such that all trajectories with initial conditions from this set converge to this point as $t \to \infty$. It is known from the study of the ordinary nonlinear dynamical systems that as a nonliniearity parameter increases the system bifurcates. This means that at the point (value of the parameter) of birth of the $T = 2^{n+1}$ sink, the $T = 2^n$ sink becomes unstable. In this section we will investigate the $T = 2$ sinks of discrete fractional systems and apply our results to analyze stability of the systems' fixed points. As all published results on the existence and stability of the $T = 2$ point were obtained for $h = 1$, in this section we assume $h = 1$.

All published results on the asymptotic stability of the stable fixed point and $T = 2$ sink were obtained for the fractional and fractional difference standard and logistic α-families of maps.

3.1.1 Fractional Standard Map ($1 < \alpha < 2$)

First results on the first bifurcation and stability of the fixed point in discrete fractional systems were obtained in [11, 19, 20] for the Riemann-Liouville standard α-family of maps ($G_K(x) = K\sin(x)$) for $1 < \alpha < 2$. In this case the map Eq. (14) can be written as a two-dimensional map considered on a cylinder

$$p_{n+1} = p_n - K \sin x_n, \tag{38}$$

$$x_{n+1} = \frac{1}{\Gamma(\alpha)} \sum_{i=0}^{n} p_{i+1} V_\alpha^1(n-i+1), \quad (\mathrm{mod}\ 2\pi), \tag{39}$$

where

$$V_\alpha^k(m) = m^{\alpha-k} - (m-1)^{\alpha-k} \tag{40}$$

and the momentum $p(t)$ is defined as

$$p(t) = {}_0D_t^{\alpha-1}x(t). \tag{41}$$

The Caputo standard α-family of maps from Eq. (15) can be considered on a torus and written as

$$p_{n+1} = p_n - \frac{K}{\Gamma(\alpha-1)} \left[\sum_{i=0}^{n-1} V_\alpha^2(n-i+1)\sin x_i + \sin x_n\right], \ (\mathrm{mod}\ 2\pi), \tag{42}$$

$$x_{n+1} = x_n + p_0 - \frac{K}{\Gamma(\alpha)} \sum_{i=0}^{n} V_\alpha^1(n-i+1)\sin x_i, \ (\mathrm{mod}\ 2\pi). \tag{43}$$

Both maps have the fixed point in the origin (0, 0). Numerical simulations show that both maps also have two $T = 2$ sinks: the antisymmetric sink, with

$$p_{n+1} = -p_n, \ \ x_{n+1} = -x_n, \tag{44}$$

and the π-shift sink, with

$$p_{n+1} = -p_n, \ \ x_{n+1} = x_n - \pi. \tag{45}$$

For the Riemann-Liouville family of maps there are two types of convergence of the trajectories to the fixed point and the $T = 2$ sinks: fast (from the basins of attraction) with

$$\delta x_n \sim n^{-1-\alpha}, \ \ \delta p_n \sim n^{-\alpha} \tag{46}$$

and slow with

$$\delta x_n \sim n^{-\alpha}, \quad \delta p_n \sim n^{1-\alpha}. \tag{47}$$

For the Caputo family of maps

$$\delta x_n \sim n^{1-\alpha}, \quad \delta p_n \sim n^{1-\alpha}. \tag{48}$$

The antisymmetric $T = 2$ sink (x_l, p_l) and $(-x_l, -p_l)$, Eq. (44), can be found considering the $n \to \infty$ limit in Eqs. (38) and (39):

$$p_l = \frac{K}{2} \sin(x_l), \tag{49}$$

$$x_l = \frac{K}{2\Gamma(\alpha)} V_{\alpha l} \sin(x_l), \tag{50}$$

where

$$V_{\alpha l} = \sum_{k=1}^{\infty} (-1)^{k+1} V_{\alpha}^{1}(k). \tag{51}$$

A high accuracy algorithm for calculating the slow converging series in Eq. (51) can be found in the Appendix section of [13]. Equation (50) has a solution and the $T = 2$ sink exists when

$$K \geq K_{s1}(\alpha) = \frac{2\Gamma(\alpha)}{V_{\alpha l}}. \tag{52}$$

The opposite condition, as found in [19], is the condition of the stability of the (0, 0) fixed point. The same condition can be shown for the Caputo standard α-family of maps. It is used to plot the part $1 < \alpha \leq 2$ of the bottom thin line in Fig. 5a, which is a two-dimensional $(\alpha - K)$ bifurcation diagram. The fixed point (0, 0) is stable below this line.

π-shift $T = 2$ sink

$$p_n = (-1)^n p_l, \quad x_n = x_l - \frac{\pi}{2}[1 - (-1)^n] \tag{53}$$

can be found plugging the asymptotic expression for x_n from Eq. (53) and

$$p_n = (-1)^n p_l + A n^{1-\alpha} \tag{54}$$

into Eqs. (38) and (39) and considering $n \to \infty$ limit. This gives

$$p_l = K/2 \sin(x_l), \quad A = \frac{2x_l - \pi}{2\Gamma(2-\alpha)}, \tag{55}$$

$$\sin(x_l) = \frac{\pi\Gamma(\alpha)}{KV_{\alpha l}}, \tag{56}$$

which has solutions for

$$K > K_{s2} = \frac{\pi\Gamma(\alpha)}{V_{\alpha l}} = \frac{\pi}{2}K_{s1}. \tag{57}$$

$K_{s2}(\alpha)$ is used to plot the part $1 < \alpha \leq 2$ of the middle thin line in Fig. 5a.

3.1.2 Fractional Logistic Map ($1 < \alpha \leq 2$)

A fractional generalization of the logistic map became possible after a small time delay was introduced into the differential equation describing a periodically kicked system Eq. (12) (see [13]). The logistic Riemann-Liouville α-family of maps ($G_K(x) = x - Kx(1-x)$) can be written as

$$p_{n+1} = p_n + Kx_n(1-x_n) - x_n, \tag{58}$$

$$x_{n+1} = \frac{1}{\Gamma(\alpha)}\sum_{i=0}^{n} p_{i+1}V_\alpha^1(n-i+1). \tag{59}$$

Numerical simulations show that for $0 < K \leq 1$ all converging trajectories converge to the fixed point $(0, 0)$ as $x \sim n^{-\alpha-1}$, $p \sim n^{-\alpha}$. For $1 < K < K_{l1}$ the only stable periodic sink is the fixed point $((K-1)/K, 0)$. The rate of convergence to this fixed point is $\delta x \sim n^{-\alpha}$, $p \sim n^{-\alpha+1}$. At $K = K_{l1}$ the fixed point becomes unstable and the stable antisymmetric in p period two sink appears. From the results of numerical simulations [13], the asymptotic behavior of converging to the $T = 2$ sink trajectories follows the power law

$$p_n = p_l(-1)^n + \frac{A}{n^{\alpha-1}}. \tag{60}$$

Substituting this expression for p_n into Eq. (59) and considering even values of n, we obtain

$$\begin{aligned} x_{lo} &= \lim_{n\to\infty} x_{2n+1} = \frac{p_l}{\Gamma(\alpha)}\lim_{n\to\infty}\sum_{k=1}^{2n+1}(-1)^k V_\alpha^1(k) \\ &+ \frac{A}{\Gamma(\alpha)}\lim_{n\to\infty}\sum_{k=1}^{2n-1}\frac{\alpha-1}{k^{\alpha-1}(2n-k)^{2-\alpha}} = -\frac{p_l}{\Gamma(\alpha)}V_{\alpha l} \\ &+ \frac{(\alpha-1)A}{\Gamma(\alpha)}\int_0^1\frac{x^{1-\alpha}dx}{(1-x)^{2-\alpha}} = -\frac{p_l}{\Gamma(\alpha)}V_{\alpha l} + A\Gamma(2-\alpha). \end{aligned} \tag{61}$$

Here we took into account that the sum on the second line of the last equation is the Riemann sum for the integral on the third line, which is equal to the Beta-function $B(2-\alpha,\alpha-1)=\Gamma(2-\alpha)\Gamma(\alpha-1)$. Similarly,

$$x_{le} = \lim_{n\to\infty} x_{2n} = \frac{p_l}{\Gamma(\alpha)}V_{\alpha l} + A\Gamma(2-\alpha). \tag{62}$$

Then, in the limit $n\to\infty$, Eq. (58) gives

$$-2p_l = Kx_{le}(1-x_{le}) - x_{le}, \tag{63}$$

$$2p_l = Kx_{lo}(1-x_{lo}) - x_{lo}. \tag{64}$$

Two fixed points, $x_{lo}=x_{le}=p_l=A=0$ and $x_{lo}=x_{le}=x_l=(K-1)/K$, $p_l=0$, $A=x_l/\Gamma(2-\alpha)$, are the two expected solutions of the system of four equations Eqs. (61)–(64). For the two remaining solutions with $x_{lo}\neq x_{le}$

$$A = \frac{K-1+\frac{2\Gamma(\alpha)}{V_{\alpha l}}}{2K\Gamma(2-\alpha)} \tag{65}$$

and the quadratic equation defining x_{le} and x_{lo} can be written as

$$x^2_{lo,le} - \Big(\frac{2\Gamma(\alpha)}{V_{\alpha l}K} + \frac{K-1}{K}\Big)x_{lo,le} + \frac{2\Gamma^2(\alpha)}{(V_{\alpha l}K)^2} + \frac{(K-1)\Gamma(\alpha)}{V_{\alpha l}K^2} = 0. \tag{66}$$

The solutions of this equation

$$x_{lo,le} = \frac{K_{s1}+K-1\pm\sqrt{(K-1)^2-K^2_{s1}}}{2K} \tag{67}$$

are defined when

$$K \ge 1 + \frac{2\Gamma(\alpha)}{V_{\alpha l}} = 1+K_{s1} \ \text{ or } \ K \le 1 - \frac{2\Gamma(\alpha)}{V_{\alpha l}} = 1-K_{s1}. \tag{68}$$

From $V_{\alpha l}<1$ and $\Gamma(\alpha)>0.885$ for $\alpha>0$ follows that $2\Gamma(\alpha)/V_{\alpha l}>1$ and, considering only $K>0$, we may ignore the second of the inequalities in Eq. (68). We may also note that the fixed point $x=(K-1)/K$ is stable when

$$1 \le K < K_{l1} = 1 + \frac{2\Gamma(\alpha)}{V_{\alpha l}} = 1+K_{s1}. \tag{69}$$

$K_{l1}(\alpha)$ is used to plot the part $1<\alpha\le 2$ of the bottom thin line in Fig. 5b.

3.1.3 Fractional and Fractional Difference Standard α-families of Maps for $0 < \alpha < 1$

In this and the next sections we will follow the results obtained in [16]. For $0 < \alpha < 1$ fractional and fractional difference maps, Eq. (15) and Eq. (35), can be written (with $h = 1$ and $G_K(x) = K\sin(x)$) in the universal form

$$x_n = x_0 - \frac{K}{\Gamma(\alpha)} \sum_{k=0}^{n-1} W_\alpha(n-k)\sin(x_k), \tag{70}$$

where $W_\alpha(s) = s^{\alpha-1}$ for the fractional map and $W_\alpha(s) = \Gamma(s+\alpha-1)/\Gamma(s)$ for the fractional difference map.

For $\alpha = 0$ the Caputo fractional standard map is identically zero and fractional difference is the sine map

$$x_{n+1} = -K\sin(x_n), \quad (\text{mod } 2\pi). \tag{71}$$

For $\alpha = 1$ both maps converge to the circle map with zero driven phase

$$x_{n+1} = x_n + K\sin(x_n), \quad (\text{mod } 2\pi). \tag{72}$$

In the sine map and in the circle map with zero driven phase, when the $x = 0$ sink becomes unstable, it bifurcates into the symmetric $T = 2$ sink in which $x_{n+1} = -x_n$. Following the results of [16], let's assume that this property persists (asymptotically) for $\alpha \in (0, 1)$. Equation (70) can be written as

$$x_{n+1} = x_n - \frac{K}{\Gamma(\alpha)}\Big\{W_\alpha(1)\sin(x_n) + \sum_{k=0}^{n-1} \sin(x_k)[W_\alpha(n-k+1) - W_\alpha(n-k)]\Big\}. \tag{73}$$

Because $W_\alpha(n-k+1) - W_\alpha(n-k) \to 0$ when $n \to \infty$, substituting $j = n-k$, asymptotically for large n Eq. (73) can be written as

$$x_n = \frac{K}{2\Gamma(\alpha)}\Big\{W_\alpha(1) + \sum_{j=1}^{\infty}(-1)^j[W_\alpha(j+1) - W_\alpha(j)]\Big\}\sin(x_n), \tag{74}$$

where the alternating series on the right side converges because its terms converge to 0 monotonically. This equation has real non-trivial solutions when

$$K > K_{s1} = \frac{2\Gamma(\alpha)}{W_\alpha(1) + \sum_{j=1}^{\infty}(-1)^j[W_\alpha(j+1) - W_\alpha(j)]}. \tag{75}$$

This expression for K_{s1} is used to plot the parts $0 < \alpha \le 1$ of the bottom thin and bold lines in Fig. 5a.

When the symmetric $T = 2$ sink becomes unstable it gives birth to the π-shift $T = 2$ sink in which $|x_{n+1} - x_n| = \pi$. The asymptotic analysis similar to the above performed for the symmetric $T = 2$ sink yields the following equation to define the asymptotic values for this sink

$$\pm\pi = \frac{K}{\Gamma(\alpha)}\Big\{W_\alpha(1) + \sum_{j=1}^{\infty}(-1)^j[W_\alpha(j+1) - W_\alpha(j)]\Big\}\sin(x_n), \quad (76)$$

which has solutions when

$$K > K_{s2} = \pi K_{s1}/2. \quad (77)$$

This expression for K_{s2} is used to plot the part $0 < \alpha \le 1$ of the middle thin line in Fig. 5a. It also could be used to calculate the middle bold line, but in this paper we use the results of the direct numerical calculations [16] instead. The difference between the results of the direct numerical simulations and the calculations using Eq. (77) is evident when $\alpha < 0.15$. This difference is due to the slow, as $n^{-\alpha}$, convergence of trajectories.

3.1.4 Fractional Difference Standard α-families of Maps for $1 < \alpha < 2$

When $1 < \alpha < 2$ the fractional difference Caputo standard α-FM Eq. (35) can be written as a two-dimensional map [16]

$$p_n = p_1 - \frac{K}{\Gamma(\alpha-1)} \times \sum_{s=2}^{n} \frac{\Gamma(n-s+\alpha-1)}{\Gamma(n-s+1)}\sin(x_{s-1}), \pmod{2\pi}, \quad (78)$$

$$x_n = x_{n-1} + p_n, \pmod{2\pi}, \quad n \ge 1, \quad (79)$$

where $p_n = \Delta x_{n-1} = x_n - x_{n-1}$.

As in the fractional standard map, in the fractional difference standard map, when the $(0, 0)$ fixed point becomes unstable, it bifurcates into the $T = 2$ antisymmetric sink $x_{n+1} = -x_n$, $p_{n+1} = -p_n$, which later, at K for which $x_n = \pi/2$, turns into two π-shift $T = 2$ sinks [16]. For the anti-symmetric sink, in the limit $n \to \infty$ Eq. (79) yields $p_n = 2x_n$ and Eq. (78)

$$p_n = \frac{K}{2\Gamma(\alpha-1)}\Big\{W_{\alpha-1}(1) + \sum_{j=1}^{\infty}(-1)^j[W_{\alpha-1}(j+1) - W_{\alpha-1}(j)]\Big\}\sin(x_n), \quad (80)$$

where, as in Eq. (70), $W_\alpha(s) = \Gamma(s+\alpha-1)/\Gamma(s)$. The equations for the antisymmetric $T = 2$ sink (x_n, p_n) are

$$x_n = \frac{K}{4\Gamma(\alpha-1)}\Big\{W_{\alpha-1}(1) + \sum_{j=1}^{\infty}(-1)^j[W_{\alpha-1}(j+1) - W_{\alpha-1}(j)]\Big\}\sin(x_n), \tag{81}$$

$$p_n = 2x_n. \tag{82}$$

They have solutions for

$$K > K_{s1}(\alpha) = 2K_{s1}(\alpha-1), \tag{83}$$

where $K_{c1d}(\alpha-1)$ is defined by Eq. (75). This result is used to plot the part $1 < \alpha \le 2$ of the bottom bold line in Fig. 5a.

Equations defining the π-shift $T = 2$ sink can be written as

$$\pm\pi = \frac{K}{2\Gamma(\alpha-1)}\Big\{W_{\alpha-1}(1) + \sum_{j=1}^{\infty}(-1)^j[W_{\alpha-1}(j+1) - W_{\alpha-1}(j)]\Big\}\sin(x_n), \tag{84}$$

$$p_n = \pm\pi. \tag{85}$$

π-shift sink exists when

$$K > K_{s2}(\alpha) = \frac{\pi}{2}K_{s1}(\alpha). \tag{86}$$

This result is used to plot the part $1 < \alpha \le 2$ of the middle bold line in Fig. 5a.

The fixed point in the origin is stable for $K < K_{s1}$ and the convergence of trajectories to the fixed point follows the power law $x_n \sim n^{1-\alpha}$ and $p_n \sim n^{-\alpha}$.

3.2 $T = 2^n$ Sinks

Investigation of the $T = 2^n$-sinks' stability with $n > 2$ by analytic methods is complicated. In papers [11–20] this is done by numerical simulations on individual trajectories with various values of parameters (K and α) and initial conditions. As in the case of the fixed point and $T = 2$-sink, stability of the high order sinks is asymptotic. Trajectories, which converge to $T = 2^n$-sinks (nth order sinks) stable in the limit $n \to \infty$, may first converge to low order sinks and then, through cascades of period doubling bifurcations, converge to the sinks of the nth order. Cascade of bifurcations type trajectories are the fundamental features of the discrete fractional systems. Their presence makes drawing of various kinds of fractional bifurcation diagrams (the subject of the next section) difficult. Fractional bifurcation diagrams strongly depend on the number of iterations and initial conditions of individual trajectories used in the analysis. The larger the number of iterations used in the calculations, the closer the calculated values of the sinks to their limiting values.

4 Fractional Bifurcation Diagrams

Fractional maps demonstrate the universal scenario of transition to chaos through the period doubling cascade of bifurcations with the change in a nonlinearity parameter, similar to the one described in [35]. This is illustrated in Fig. 3, in which the bifurcation diagrams (x vs. K) for various considered families of maps with $\alpha = 0.8$ are presented. Compared to the integer case $\alpha = 1$, bifurcation diagrams for the fractional maps are stretched along the K axis while bifurcation diagrams for the fractional difference maps are contracted. The existence of self-similarity, corresponding constants (analogs of the Feigenbaum constants), and their dependence on α in fractional maps are not investigated. The dependence of the bifurcation diagrams on the number of iterations is demonstrated in Fig. 4. Bifurcation diagrams obtained after five thousand iteration look much nicer than those obtained after two hundred. But, as it follows from Figs. 1 and 2, even 5000 iterations are not enough for computation of the asymptotic bifurcation diagrams.

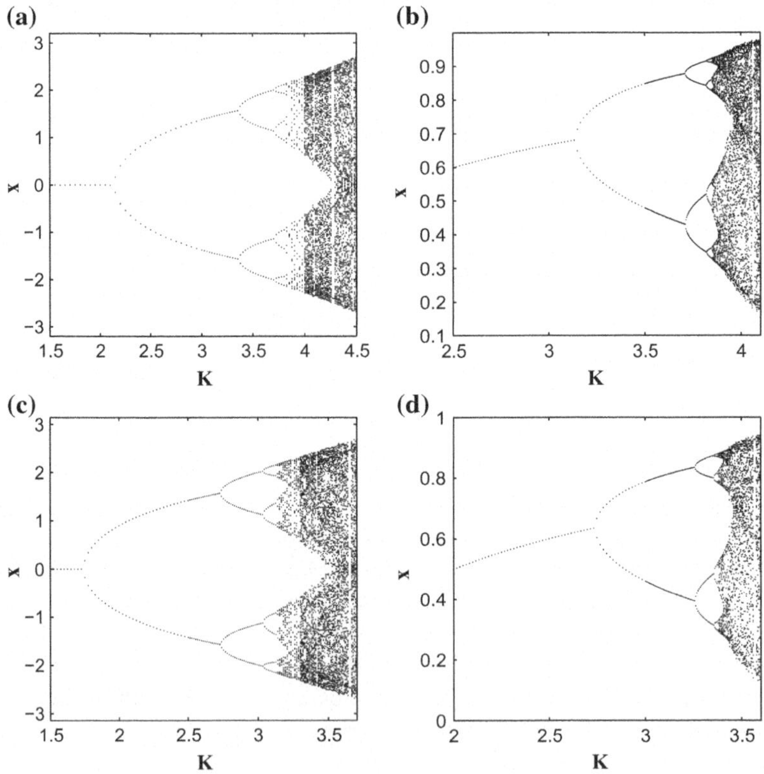

Fig. 3 The bifurcation diagrams for fractional Caputo standard (**a**) and logistic (**b**) maps and for fractional difference Caputo standard (**c**) and logistic (**d**) maps with $\alpha = 0.8$

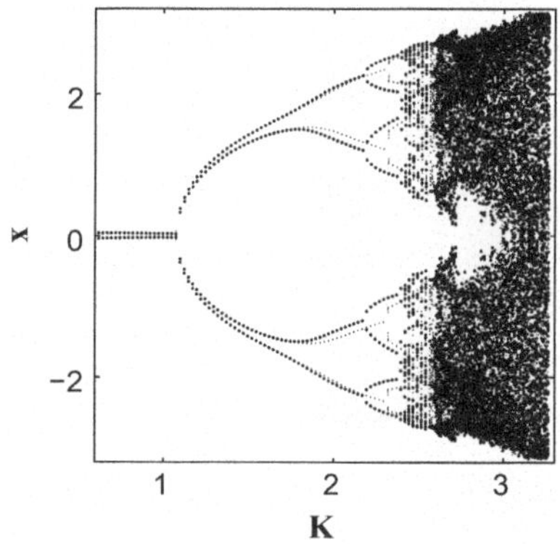

Fig. 4 Two bifurcation diagrams for the fractional difference Caputo Standard αFM with $\alpha = 0.1$ and $x_0 = 0.1$ calculated after 200 iterations (regular points) and 5000 iterations (bold points). This figure is reprinted from [16], with the permission of AIP Publishing

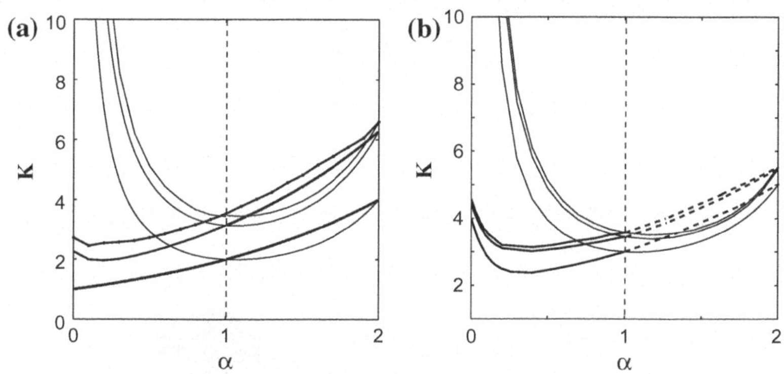

Fig. 5 2D bifurcation diagrams for fractional (thing lines) and fractional difference (bold lines) Caputo standard (**a**) and logistic (**b**) maps. First bifurcation, transition from the stable fixed point to the stable period two ($T = 2$) sink, occurs on the bottom curves. $T = 2$ sink (in the case of the standard α-families of maps, antisymmetric $T = 2$ sink with $x_{n+1} = -x_n$) is stable between the bottom and the middle curves. Transition to chaos occurs on the top curves. Period doubling bifurcations leading to chaos occur in the narrow band between the middle and the top curves. The bottom curves in (**a**) are obtained using Eqs. (52), (75), and (83). The thin middle curve in (a) is obtained using Eqs. (57) and (77). The $1 < \alpha \leq 2$ part of the middle bold line in (**a**) is obtained using Eq. (86). The $1 < \alpha \leq 2$ part of the bottom thin line in (**b**) is obtained using Eq. (69). The remaining curves, except the bold dashed curves in (**b**), are results of the direct numerical simulations. The bold dashed curves in (**b**) are obtained by interpolation

The two-dimensional bifurcation diagrams Fig. 5 are obtained by combining the results of the computations of the bifurcation (x vs. K) diagrams after 5000 iterations for fixed values of α. Because for small α convergence of trajectories to their asymptotic values is very slow, the results in the diagrams for $\alpha < 0.15$ do not represent well the asymptotic values and can be improved in future.

Looking at the 2D bifurcation diagrams one may note that systems with power-law memory should demonstrate bifurcations with changes in the memory parameter

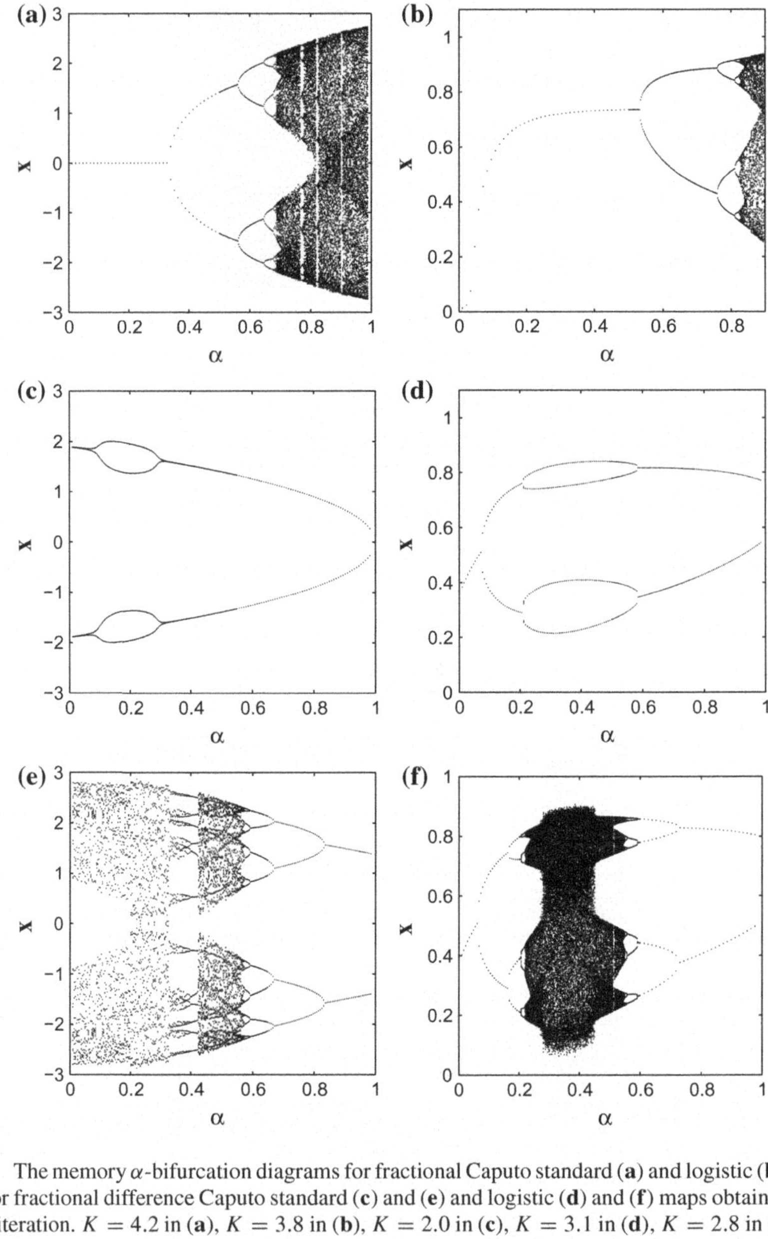

Fig. 6 The memory α-bifurcation diagrams for fractional Caputo standard (**a**) and logistic (**b**) maps and for fractional difference Caputo standard (**c**) and (**e**) and logistic (**d**) and (**f**) maps obtained after 5000 iteration. $K = 4.2$ in (**a**), $K = 3.8$ in (**b**), $K = 2.0$ in (**c**), $K = 3.1$ in (**d**), $K = 2.8$ in (**e**), and $K = 3.2$ in (7)

α when the nonlinearity parameter K stays constant. This property of systems with power-law memory is demonstrated in Fig. 6 and it may explain how changes/failures in live biological species can be caused by changes in their memory and nervous system. This also may explain how some diseases may be treated by treating the nervous system.

5 Conclusion

The following citation from Wikipedia, "universality is the observation that there are properties for a large class of systems that are independent of the dynamical details of the system", defines the notion of the universality in dynamical systems. The universality in systems with power-law memory goes beyond the period doubling with changes in nonlinearity and memory parameters and the universal scenario of transition to chaos. Individual trajectories of such systems also demonstrate cascade of bifurcations type behavior. In regular dynamics the universality has a mathematical expression in the form of the Feigenbaum function and constants. This is only the beginning of the research on fractional universality and most of the results are obtained by numerical simulations. Those results introduce more questions than answers. Some of those questions are:

- What is the nature and the corresponding analytic description of the bifurcations on a single trajectory of a fractional system?
- What kind of self-similarity can be found in CBTT?
- How to describe a self-similar behavior corresponding to the bifurcation diagrams of fractional systems? Can constants, similar to the Feigenbaum constants be found?
- Can cascade of bifurcations type trajectories be found in continuous systems?

Behavior of fractional systems at low values of α $(0 < \alpha < 0.15)$ is very important in biological applications but is not well established and requires an additional investigation.

As mentioned in the introduction, there is a possibility for multiple applications of the fractional universality in biology. The human body is a system with power-law memory, which implies the possibility of medical applications. Figure 2a suggests that, assuming some distribution (e.g., uniform) of the initial conditions of an asymptotically $T = 2$ system with power-law memory, it is possible to calculate the probability distribution of times before the stable fixed point behavior of the system bifurcates. Comparison of probability distributions for various values of K and α to the statistics of the times before sudden changes (e.g., deaths) after serious surgeries (e.g., heart transplants) may help to understand the state of a human body after the surgery and suggest some remedies.

The intermittent cascade of bifurcations type behavior is typical for systems with power-law memory (see Fig. 7). May this intermittency explain the intermittent behavior, transitions from stability to chaos and back to stability, in various

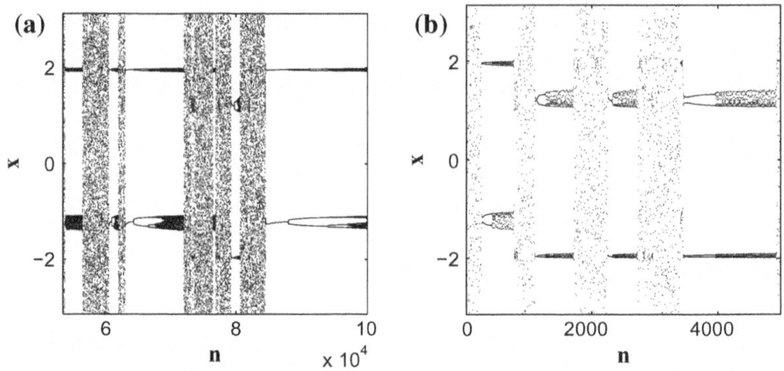

Fig. 7 Intermittent cascade of bifurcations type trajectories for the Caputo (a) and fractional difference Caputo (b) Standard αFMs. In (a) $\alpha = 1.65$, $K = 4.5$, $x_0 = 0.$, and $p_0 = 0.3$. In (b) $\alpha = 1.5$, $K = 4.82$, $x_0 = 0.$, and $p_0 = 0.01$. This figure is reprinted from [17], with the permission of L&H Scientific Publishing

socio-economic systems, which are systems with power-law memory? May the history of human society, with repeating periods of dictatorship, democracy, and chaos, be modeled by the equations with power-law memory? There are many questions related to the topic of this review which motivate research on systems with power-law memory.

Acknowledgements The author expresses his gratitude to R. Cole and R. V. Kohn, for the opportunity to complete this work at the Courant Institute. The author is grateful to the organizers of the 6th International Conference on Nonlinear Science and Complexity in Sao Jose dos Campos, Brazil, for financial support. The author acknowledges continuing support from Yeshiva University.

References

1. Aguila-Camacho, N., Duarte-Mermoud, M.A., Gallegos, J.A.: Lyapunov functions for fractional order systems. Commun. Nonlin. Sci. Numer. Simul. **19**, 29512957 (2014)
2. Anderson, J.R.: Learning and Memory: An Integrated Approach. Wiley, New York 1(995)
3. Anastassiou, G.A.: Nabla discrete fractional calculus and nabla inequalities. Math. Comput. Modelling **51**, 562–571 (2010)
4. Atici, F., Eloe, P.: Initial value problems in discrete fractional calculus. Proc. Am. Math. Soc. **137**, 981–989 (2009)
5. Atici, F., Eloe P.: Discrete fractional calculus with the nabla operator. Electron. J. Qual. Theory Differ. Equ. Spec. Ed. **I3**, 1–12 (2009)
6. Baleanu, D., Wu, G.-C., Bai, Y.-R., Chen, F.-L.: Stability analysis of Caputo-like discrete fractional systems. Commun. Nonlin. Sci. Numer. Simul. **48**, 520–530 (2017)
7. Bastos, N.R.O., Ferreira, R.A.C., Torres, D.F.M.: Discrete-time fractional variational problems. Signal Process. **91**, 513–524 (2011)
8. Bastos, N.R.O., Ferreira, R.A.C., Torres, D.F.M.: Necessary optimality conditions for fractional difference problems of the calculus of variations. Discrete Contin. Dyn. Syst. **29**, 417–437 (2011)

9. Chen, F., Luo, X., and Zhou, Y.: Existence results for nonlinear fractional difference equation. Adv. Differ. Eq. **2011**, 713201, (2011)
10. Cvitanovic, P.: Universality in Chaos. Adam Hilger, Bristol and New York (1989)
11. Edelman, M.: Fractional standard map: Riemann-Liouville vs. Caputo. Commun. Nonlin. Sci. Numer. Simul. **16**, 4573–4580 (2011)
12. Edelman, M.: Fractional maps and fractional attractors. Part I: α-families of maps. Discontinuity Nonlinearity Complex. **1**, 305–324 (2013)
13. Edelman, M.: Universal fractional map and cascade of bifurcations type attractors. Chaos **23**, 033127 (2013)
14. Edelman, M.: Universality in fractional dynamics. In: International Conference on Fractional Differentiation and Its Applications (ICFDA), 2014, pp. 1–6 (2014). https://doi.org/10.1109/ICFDA.2014.6967376
15. Edelman, M.: Fractional maps as maps with power-law memory. In: Afraimovich, A., Luo, A.C.J., Fu, X. (eds.) Nonlinear Dynamics and Complexity; Series: Nonlinear Systems and Complexity, pp. 79–120, Springer, New York (2014)
16. Edelman, M.: Caputo standard α-family of maps: fractional difference vs. fractional. Chaos **24**, 023137 (2014)
17. Edelman, M.: Fractional maps and fractional attractors. Part II: Fractional Difference α-Families of Maps. Discontinuity Nonlinearity Complex **4**, 391–402 (2015)
18. Edelman, M.: On the fractional Eulerian numbers and equivalence of maps with long term power-law memory (integral Volterra equations of the second kind) to Grünvald-Letnikov fractional difference (differential) equations. Chaos **25**, 073103 (2015)
19. Edelman, M., Tarasov, V.E.: Fractional standard map. Phys. Lett. A **374**, 279–285 (2009)
20. Edelman, M., Taieb, L.A.: New types of solutions of non-linear fractional differential equations. In: Almeida, A., Castro, L., Speck F.-O. (eds.) Advances in Harmonic Analysis and Operator Theory; Series: Operator Theory: Advances and Applications, vol. 229, pp. 139–155. Springer, Basel (2013)
21. Fairhall, A.L., Lewen, G.D., Bialek, W., de Ruyter van Steveninck R.R.: Efficiency and ambiguity in an adaptive neural code. Nature 787–792 (2001)
22. Ferreira, R.A.C., Torres, D.F.M.: Fractional h-difference equations arising from the calculus of variations. Appl. Anal. Discrete Math. **5**, 110–121 (2011)
23. Frederico, G.S.F., Torres, D.F.M.: A formulation of Noether's theorem for fractional problems of the calculus of variations. J. Math. Appl. Anal. Appl. **334**, 834–846 (2007)
24. Gray, H.L., Zhang, N.-F.: On a new definition of the fractional difference. Math. Comput. **50**, 513–529 (1988)
25. Kahana, M.J.: Foundations of human memory. Oxford University Press, New York (2012)
26. Kilbas, A.A., Bonilla, B., Trujillo, J.J.: Nonlinear differential equations of fractional order is space of integrable functions. Dokl. Math. **62**, 222–226 (2000)
27. Kilbas, A.A., Bonilla, B., Trujillo, J.J.: Existence and uniqueness theorems for nonlinear fractional differential equations. Demonstratio Math. **33**, 583–602 (2000)
28. Kilbas, A.A., Srivastava, H.M., Trujillo, J.J.: Theory and Application of Fractional Differential Equations. Elsevier, Amsterdam (2006)
29. Leopold, D.A., Murayama, Y., Logothetis, N.K.: Very slow activity fluctuations in monkey visual cortex: implications for functional brain imaging. Cerebr. Cortex **413**, 422–433 (2003)
30. Li, Y., Chen, Y.Q., and Podlubny, I.: Stability of fractional-order nonlinear dynamic systems: lyapunov direct method and generalized Mittag-Leffler stability. Comput. Math. Appl. **59**, 181021 (2010)
31. Lundstrom, B.N., Fairhall, A.L., Maravall, M.: Multiple time scale encoding of slowly varying whisker stimulus envelope incortical and thalamic neurons in vivo. J. Neurosci **30**, 5071–5077 (2010)
32. Lundstrom, B.N., Higgs, M.H., Spain, W.J., Fairhall, A.L.: Fractional differentiation by neocortical pyramidal neurons. Nat. Neurosci **11**, 1335–1342 (2008)
33. Machado, J.A.T, Pinto, C.M.A., Lopes, A.M.: A review on the characterization of signals and systems by power law distributions. Signal Process. **107**, 246–253 (2015)

34. Matignon, D.: Stability properties for generalized fractional differential systems. ESAIM Proc. **5**, 145–58 (1998)
35. May, R.M.: Simple mathematical models with very complicated dynamics. Nature **261**, 459–467 (1976)
36. Miller, K.S., Ross, B.: Fractional difference calculus. In: Srivastava, H.M., Owa, S. (eds.) Univalent Functions. Fractional Calculus, and Their Applications, pp. 139–151. Ellis Howard, Chichester (1989)
37. Mozyrska, D., Girejko, E.: Overview of the fractional h-difference operators. In: Almeida, A., Castro, L., Speck F.-O. (eds.) Advances in Harmonic Analysis and Operator Theory; Series: Operator Theory: Advances and Applications, vol. 229, pp. 253–267. Springer, Basel (2013)
38. Mozyrska, D., Girejko, E., Wirwas, M.: Fractional nonlinear systems with sequential operators. Cent. Eur. J. Phys. **11**, 1295–1303 (2013)
39. Mozyrska, D., Pawluszewicz, E.: Local controllability of nonlinear discrete-time fractional order systems. Bull. Pol. Acad. Sci. Techn. Sci. **61**, 251–256 (2013)
40. Mozyrska, D., Pawluszewicz, E., Girejko, E.: Stability of nonlinear h-difference systems with N fractional orders. Kibernetica **51**, 112–136 (2015)
41. Petras, I.: Fractional-Order Nonlinear Systems. Springer, Berlin (2011)
42. Podlubny, I.: Fractional Differential Equations. Academic Press, San Diego (1999)
43. Pozzorini, C., Naud, R., Mensi, S., Gerstner, W.: Temporal whitening by power-law adaptation in neocortical neurons. Nat.Neurosci. **16**, 942–948 (2013)
44. Rivero, M., Rogozin, S.V., Machado, J.A.T., Trujilo, J.J.: Stability of fractional order systems. Math. Probl. Eng. **2013**, 356215 (2013)
45. Rubin, D.C., Wenzel, A.E.: One hundred years of forgetting: a quantitative description of retention. Psychol. Rev. **103**, 743–760 (1996)
46. Samko, S.G., Kilbas, A.A., Marichev, O.I.: Fractional Integrals and Derivatives Theory and Applications. Gordon and Breach, New York (1993)
47. Stanislavsky, A.A: Long-term memory contribution as applied to the motion of discrete dynamical system. Chaos **16**, 043105 (2006)
48. Tarasov, V.E.: Differential equations with fractional derivative and universal map with memory. J. Phys. A **42**, 465102 (2009)
49. Tarasov, V.E.: Discrete map with memory from fractional differential equation of arbitrary positive order. J. Math. Phys. **50**, 122703 (2009)
50. Tarasov, V.E.: Fractional dynamics: application of fractional calculus to dynamics of particles. In: Fields and Media. HEP, Springer, Heidelberg (2011)
51. Tarasov, V.E., Zaslavsky, G.M.: Fractional equations of kicked systems and discrete maps. J. Phys. A **41**, 435101 (2008)
52. Toib, A., Lyakhov, V., Marom, S.: Interaction between duration of activity and recovery from slow inactivation in mammalian brain Na+ channels. J. Neurosci. **18**, 1893–1903 (1998)
53. Ulanovsky, N., Las, L., Farkas, D., Nelken, I.: Multiple time scales of adaptation in auditory cortex neurons. J Neurosci. **24**, 10440–10453 (2004)
54. Wixted, J.T.: Analyzing the empirical course of forgetting. J. Exp. Psychol. Learn. Mem. Cognit. **16**, 927–935 (1990)
55. Wixted, J.T., Ebbesen, E.: On the form of forgetting. Psychol. Sci. **2**, 409–415 (1991)
56. Wixted, J.T., Ebbesen, E.: Genuine power curves in forgetting. Mem. Cognit. **25**, 731–739 (1997)
57. Wu, G.-C., Baleanu, D.: Discrete fractional logistic map and its chaos. Nonlin. Dyn. **75**, 283–287 (2014)
58. Wu, G.-C., Baleanu, D., Zeng, S.-D.: Discrete chaos in fractional sine and standard maps. Phys. Lett. A **378**, 484–487 (2014)
59. Wyrwas, M., Pawluszewicz, E., Girejko, E.: Stability of nonlinear h-difference systems with N fractional orders. Kybernetika **15**, 112–136 (2015)
60. Zaslavsky, G.M.: Hamiltonian Chaos and Fractional Dynamics. Oxford University Press, Oxford (2005)

61. Zaslavsky, G.M., Stanislavsky, A.A.., Edelman, M: Chaotic and pseudochaotic attractors of perturbed fractional oscillator. Chaos **16**, 013102 (2006)
62. Zilany, M.S., Bruce, I.C., Nelson, P.C., Carney, L.H.: A phenomenological model of the synapse between the inner hair cell and auditory nerve: long-term adaptation with power-law dynamics. J. Acoust. Soc. Am. **126**, 2390–2412 (2009)

Fractional Deterministic Factor Analysis of Economic Processes with Memory and Nonlocality

Valentina V. Tarasova and Vasily E. Tarasov

Mathematics Subject Classification: 26A33 Fractional derivatives and integrals · 91B02 Fundamental topics (basic mathematics, applicable to economics in general) · 91B55 Economic dynamics

1 Introduction

The economic analysis actively uses methods of the deterministic factor analysis, which gives exact characteristics of influence of factors on change of the effective indicator. The main methods of the deterministic factor analysis are the method of differential calculus and the integral method. These methods are based on the mathematical theory of derivatives and integrals of integer orders. In modern mathematics, derivatives and integrals of non-integer (fractional) orders are well known [3, 8, 13, 14]. This mathematical tool allows us to describe various types of processes, which are characterized by memory and non-locality. This tool is actively used in the natural sciences [10, 17, 40]. In this paper we propose generalizations of the method of differential calculus and the integral method of factor analysis. We analyze a possibility to apply this mathematical tool to quantitative description of the influence of individual factors on the change in the resultant economic indicator.

Let us briefly describe the standard method of factor analysis, which is based on the differential calculus. The mathematical basis of this method is the derivatives (differentiation) of non-integer orders and the Taylor series. Let $z = f(x, y)$ be a function of two independent real variables x and y. We will use the notation

V.V. Tarasova (✉)
Lomonosov Moscow State University Business School, Lomonosov Moscow State University, Leninskie Gory, 1 (52A), Moscow 119991, Russia
e-mail: v.v.tarasova@mail.ru

V.E. Tarasov
Skobeltsyn Institute of Nuclear Physics, Lomonosov Moscow State University, Leninskie Gory, 1 (2), Moscow 119991, Russia
e-mail: v.e.tarasov@bk.ru; tarasov@theory.sinp.msu.ru

M. Edelman et al. (eds.), *Chaotic, Fractional, and Complex Dynamics: New Insights and Perspectives*, Understanding Complex Systems,
https://doi.org/10.1007/978-3-319-68109-2_9

$$f_x^{(1)}(x, y) := \frac{\partial f(x, y)}{\partial x}, f_y^{(1)}(x, y) := \frac{\partial f(x, y)}{\partial y} \tag{1}$$

to denote the first-order partial derivatives with respect to x and y. The Taylor formula for the function $z = f(x, y)$ can be written in the form

$$f(x, y) = f(x_0, y_0) + f_x^{(1)}(x_0, y_0) \cdot \Delta x + f_y^{(1)}(x_0, y_0) \cdot \Delta y + R_2(x, y), \tag{2}$$

where $f_x^{(1)}(x_0, y_0)$ and $f_y^{(1)}(x_0, y_0)$ are the values of the partial derivatives of the first order with respect to x and y at the point (x_0, y_0); Δx and Δy are factor increments of the corresponding variables ($\Delta x := x - x_0$, $\Delta y := y - y_0$); $R_2(x, y)$ is the remainder term, which is of an infinitesimal value in the neighborhood of the point (x_0, y_0).

Neglecting the remainder term $R_2(x, y)$, the total increment of the function $\Delta z := f(x_1, y_1) - f(x_0, y_0)$ can be written by the increment of the factors Δx and Δy in the form

$$\Delta z \approx f_x^{(1)}(x_0, y_0) \cdot \Delta x + f_y^{(1)}(x_0, y_0) \cdot \Delta y. \tag{3}$$

The influence of the factor x on the generalizing indicator z is calculated by the equation

$$\Delta z_x = f_x^{(1)}(x_0, y_0) \cdot \Delta x, \tag{4}$$

and the influence of the factor y is described by the equation

$$\Delta z_y = f_y^{(1)}(x_0, y_0) \cdot \Delta y, \tag{5}$$

where x_0, y_0 are the basic (planned) values of the factors x and y that have influence on effective indicator; x_1, y_1 are the actual values of these factors; $\Delta x := x_1 - x_0$ and $\Delta y := y_1 - y_0$ are the absolute changes (deviations) of the factors x and y.

In the differential calculus method, it is assumed that the total increment of the functions Δz is decomposed into the terms Δz_x and Δz_y. The value of each of these terms is calculated as the product of the corresponding partial derivative and the increment of the variable (factor). In this method, an indecomposable remainder (the remainder term $R_2(x, y)$), which is neglected, is interpreted as a logical error of the differential calculus method. The neglecting of an indecomposable residue is one of the disadvantages of this method, since for economic calculations the exact balance of the change of the effective indicator and the algebraic sum of the influence of all factors are often required.

In the standard approach to factor analysis, the differential calculus method uses the derivatives of integer orders. It is known that the derivatives of integer orders are determined by the properties of the differentiable function only in an infinitesimal neighborhood of the considered point. As a result, the differential equations with derivatives of integer orders with respect to time cannot describe processes with dynamic memory. In fact, these equations describe only such economic processes, in which agents actually have a total amnesia. In other words, economic models, which use derivatives of integer orders, can be applied, when economic agents forget the history of changes of economic indicators and factors during an infinitesimal period of time. Obviously, the assumption of lack of the memory in economic agents is a strong restriction.

The standard differential method of the factor analysis uses the derivatives of integer (first) orders. In modern mathematics, there are concepts of the derivatives of non-integer (fractional, arbitrary) orders [3, 8, 13, 14]. Derivatives and integrals of non-integer orders are actively used to describe various physical processes characterized by memory and non-locality [10, 17, 40]. Derivatives and integrals of non-integer order have recently been used to describe various financial and economic processes in articles [6, 9, 11, 15, 37–39] and our papers [22–29, 31–36].

There are different types of fractional derivatives, which have been proposed by Riemann, Liouville, Sonin, Letnikov, Marsho, Riesz, Weil, Hadamard, and Caputo [8, 13, 14]. To construct factor analysis methods, which are based on the fractional integro-differentiation of arbitrary (non-integer) orders, it is convenient to use the Caputo fractional derivatives [[8], pp. 90–99]. The main distinctive feature of these derivatives is that their effect on the constant function gives zero. This leads us to the zero effect of a constant factor on the indicator. Let us give the definitions of the Caputo derivative [[8], p. 92].

Definition The left-sided and right-sided Caputo derivatives of order $\alpha \geq 0$ on the interval $[a, b]$ are defined by the equations

$$\left(D^{\alpha}_{a+} f\right)(x) := \frac{1}{\Gamma(n-\alpha)} \int_a^x \frac{f^{(n)}(\chi)\, d\chi}{(x-\chi)^{\alpha-n+1}}, \tag{6}$$

$$\left(D^{\alpha}_{b-} f\right)(x) := \frac{(-1)^n}{\Gamma(n-\alpha)} \int_x^b \frac{f^{(n)}(\chi)\, d\chi}{(\chi-x)^{\alpha-n+1}}, \tag{7}$$

where $\Gamma(\alpha)$ is the gamma function, $a< x <b$, and $f^{(n)}(\chi)$ is the derivative of the integer order $n=[\alpha]+1$ of the function $f(\chi)$ with respect to the variable χ. It is assumed that the function $f(x)$ has integer derivatives up to $(n\text{-}1)$th order, which are absolutely continuous functions on the interval $[a, b]$.

From Eqs. (6) and (7), we can see that the Caputo derivative of non-integer order is actually an integro-differential operator with a power-law kernel with non-integer $\alpha > 0$. For positive integer values $\alpha = n$, the Caputo derivatives coincide [[8], p. 92] with the standard derivative of the integer order n:

$$\left(D^{n}_{a+} f\right)(x) = f^{(n)}(x), \quad \left(D^{n}_{b-} f\right)(x) = (-1)^n f^{(n)}(x). \tag{8}$$

Using property (8), the Caputo derivatives are considered as generalizations of standard derivatives of positive integer orders. Moreover, the standard derivatives are special cases of derivatives of fractional (arbitrary) order.

In general, the indicator of processes with memory cannot be represented as a single-valued function of factors [30]. One can only consider indicators and factors as single-valued functions of time. If the parametric dependence of indicator z(t) on factors x(t) and y(t) cannot be represented by a single-valued function, then we should use the parametric derivatives and integrals. An important role in factor analysis of processes with power-law memory can be played by the parametric fractional derivatives and integrals. The Riemann-Liouville parametric fractional derivative has been considered in Sect. 18.2 of [[14], pp. 325–329] and Sect. 2.5 of [8]. The Caputo parametric fractional derivatives and its properties have been described in [1].

Definition The Caputo parametric fractional derivatives of order $\alpha > 0$ are defined by the equations

$$\left(D_{a+}^{\alpha,x} f\right)(t) := \frac{1}{\Gamma(n-\alpha)} \int_a^t d\tau \frac{x^{(1)}(\tau)}{(x(t)-x(\tau))^{\alpha+1-n}} \cdot \left(\frac{1}{x^{(1)}(\tau)} \cdot \frac{d}{d\tau}\right)^n f(\tau), \tag{9}$$

$$\left(D_{b-}^{\alpha,x} f\right)(t) := \frac{1}{\Gamma(n-\alpha)} \int_t^b d\tau \frac{x^{(1)}(\tau)}{(x(\tau)-x(t))^{\alpha+1-n}} \cdot \left(-\frac{1}{x^{(1)}(\tau)} \cdot \frac{d}{d\tau}\right)^n f(\tau), \tag{10}$$

where $a < t < b$, $n-1 \leq \alpha \leq n$, and $x(\tau)$ is a monotone function having a continuous derivative $x^{(1)}(\tau) = dx(\tau)/d\tau$. The derivatives (9) and (10) are also called the Caputo fractional derivatives of function f(t) by a function x(t) of the order $\alpha > 0$.

An application of fractional derivatives of non-integer orders in the natural sciences allows us to describe processes and systems with nonlocality and memory [10, 17, 40]. It allows us to expect that this mathematical tool can be expanded to analyzing the financial and economic processes. The nonlocal economic processes in the state space are considered in [38, 39]. The concept of dynamic memory for economics is considered in [22, 25]. In recent works [23, 24, 26–29, 31–36] the Caputo derivatives were applied to describe the economic processes with power-law memory.

In this paper we propose methods of deterministic factor analysis that can be used in the analysis of economic indicators (endogenous variables), which are represented by power functions of factors (exogenous variables) of processes with nonlocality and memory. The power functions are widely used in economic analysis, including the well-known Cobb-Douglas functions [2] and other functions of the power laws described in the reviews [4, 5]. To describe the non-locality and fading memory of power-law types, we use equations with integrals and derivatives of non-integer orders [3, 8, 13].

2 Method of Differential Calculus of Arbitrary (non-Integer) Order

The standard differential method of factor analysis is based on the Taylor formula. Therefore we need a generalization of the Taylor formula for the Caputo fractional derivative to generalize this method by using the derivatives of non-integer orders. Let us consider the generalization of the Taylor series, which was proposed in [1, 12].

The function f(x) with $x \geq a$ ($x \in [a, b]$) can be expanded [[12, p. 289] by using the generalized Taylor series with the left-sided Caputo derivatives of order $0 < \alpha \leq 1$ in the form

$$f(x) = \sum_{k=0}^{N-1} \frac{\left(\left(D_{a+}^{\alpha}\right)^k f\right)(a)}{\Gamma(k\alpha + 1)} \cdot (x - a)^{k\alpha} + R_{N\alpha}(x, a+), \tag{11}$$

where $R_{N\alpha}(x, a+)$ is the remainder term, which can be represented in the form

$$R_{N\alpha}(x, a+) = \frac{\left(\left(D_{a+}^{\alpha}\right)^N f\right)(\chi_+)}{\Gamma(N\alpha + 1)} \cdot (x - a)^{N\alpha}, \tag{12}$$

where $a \leq \chi_+ \leq x$. The generalized Taylor series (11), is applicable to the functions f(x) that satisfy the condition $\left(\left(D_{a+}^{\alpha}\right)^k f\right)(x) \in C[a, b]$, for $k = 0, 1, \ldots, N$ [[12], p.289].

For $x \leq a$, formula (11) cannot be applied if the parameter α is not an integer. This problem can be solved by using the right-sided Caputo derivative of the order $\alpha > 0$. The function f(x), which is defined on the domain $x \leq a$ ($x \in [c, a]$) can be represented [[8], p. 95] as the generalized Taylor series of order $0 < \alpha \leq 1$ with the right-sided Caputo derivatives

$$f(x) = \sum_{k=0}^{N-1} \frac{\left(\left(D_{a-}^{\alpha}\right)^k f\right)(a)}{\Gamma(k\alpha + 1)} \cdot (a - x)^{k\alpha} + R_{N\alpha}(x, a-), \tag{13}$$

where $R_{N\alpha}(x, a-)$ is the remainder term, which can be represented in the form

$$R_{N\alpha}(x, a-) = \frac{\left(\left(D_{a-}^{\alpha}\right)^N f\right)(\chi_-)}{\Gamma(N\alpha + 1)} \cdot (a - x)^{N\alpha}, \tag{14}$$

and $x \leq \chi_- \leq a$.

For example, series (11) and (13) with $N=2$ have the form

$$f(x) = f(a) + \frac{\left(D^{\alpha}_{a+}f\right)(a)}{\Gamma(\alpha+1)} \cdot (x-a)^{\alpha} + R_{2\alpha}(x, a+), \tag{15}$$

$$f(x) = f(a) + \frac{\left(D^{\alpha}_{a-}f\right)(a)}{\Gamma(\alpha+1)} \cdot (a-x)^{\alpha} + R_{2\alpha}(x, a-), \tag{16}$$

where $x \in [a, b]$ for (15) and $x \in [c, a]$ for (16).

The factor analysis of processes with memory [22, 24, 25, 27] should take into account that the generalizing indicators cannot be represented as single-valued functions of factors in general [30]. These indicators and factors can be described by single-valued functions of time. If the parametric dependence of indicator z(t) on factors x(t) and y(t) cannot be represented by a single-valued function, then we should use the fractional Taylor formula with the parametric fractional derivatives. Using Theorem 18 of [[1], p. 474], we can write the fractional Taylor formula with the Caputo parametric fractional derivative as

$$f(x(t)) = \sum_{k=0}^{N-1} \frac{\left(\left(D^{\alpha,x}_{a+}\right)^k f\right)(a)}{\Gamma(k\alpha+1)} \cdot (x(t)-x(a))^{k\alpha} + R_{N\alpha}(x(t), a+), \tag{17}$$

$$f(x(t)) = \sum_{k=0}^{N-1} \frac{\left(\left(D^{\alpha,x}_{b-}\right)^k f\right)(b)}{\Gamma(k\alpha+1)} \cdot (x(b)-x(t))^{k\alpha} + R_{N\alpha}(x(t), b-), \tag{18}$$

where $R_{N\alpha}(x(t), a+)$ and $R_{N\alpha}(x(t), b-)$ are the remainder terms, which can be represented in the form

$$R_{N\alpha}(x(t), a+) = \frac{\left(\left(D^{\alpha,x}_{a+}\right)^N f\right)(\chi_+)}{\Gamma(N\alpha+1)} \cdot (x(t)-x(a))^{N\alpha}, \tag{19}$$

$$R_{N\alpha}(x(t), b-) = \frac{\left(\left(D^{\alpha,x}_{b-}\right)^N f\right)(\chi_-)}{\Gamma(N\alpha+1)} \cdot (x(b)-x(t))^{N\alpha}, \tag{20}$$

where we can consider $b=a$.

For Eqs. (11), (13), (15)–(18) there is an additional problem, which is caused by the coincidence of the initial and final values in the derivatives $\left(D^{\alpha}_{a+}f\right)(a)$ and $\left(D^{\alpha}_{a-}f\right)(a)$. It is important to have the generalized Taylor series at an arbitrary point x_0, which does not coincide with the initial point of fractional derivative D^{α}_{a+}, i.e., $x_0 \neq a$. In this case we can avoid some restrictions on a possible application of Eqs. (11), (13), (15)–(18) to calculating the influence of factors on the change in the effective indicator. To solve this problem, we propose a generalization of the Taylor formulas (11) to the case when the Caputo derivative is considered at an arbitrary point $x_0 \geq a$. Consideration of the Taylor series (15) for the function f(x) and for the Caputo derivatives $\left(D^{\alpha}_{a+}f\right)(x)$ at the point $x = x_0$ allows us to obtain the equation

$$f(x) = f(x_0) + \frac{\left(D^{\alpha}_{a+}f\right)(x_0)}{\Gamma(\alpha+1)} \cdot \Delta_{\alpha}x + R_{2\alpha}(x, x_0, a+), \tag{21}$$

where $x_0 \geq a$, $x \geq a$, and the remainder term $R_{2\alpha}(x, x_0, a+)$ is equal to the difference between $R_{2\alpha}(x, a-)$ and $R_{2\alpha}(x_0, a-)$. In Eq. (21) we use the notations

$$\Delta_{\alpha}x := (x-a)^{\alpha} - (x_0-a)^{\alpha}. \tag{22}$$

For the Eqs. (17)–(20) with the parametric fractional derivatives, we should use

$$\Delta_{\alpha}x := (x(t)-x(a))^{\alpha} - (x(t_0)-x(a))^{\alpha}. \tag{23}$$

Analogous formulas can also be obtained for the generalized Taylor series for the right-sided Caputo fractional derivative. For simplicity, we consider only the case of the left-sided derivative with initial point $a=0$, since many factors can be described by positive real numbers. In this case, equation (21) can be written in the form

$$f(x) = f(x_0) + \frac{\left(D^{\alpha}_{0+}f\right)(x_0)}{\Gamma(\alpha+1)} \cdot \Delta_{\alpha}x + R_{2\alpha}(x, x_0, 0+), \tag{24}$$

where $x_0 \geq 0$, $x \geq 0$, and $\Delta_{\alpha}x := x^{\alpha} - x_0^{\alpha}$.

For a function of two variables $z=f(x, y)$, the generalized Taylor formula has the form

$$f(x, y) = f(x_0, y_0) + \frac{\left(D^{\alpha}_{0+;x}f\right)(x_0, y_0)}{\Gamma(\alpha+1)} \cdot \Delta_{\alpha}x + \frac{\left(D^{\beta}_{0+;y}f\right)(x_0, y_0)}{\Gamma(\beta+1)} \cdot \Delta_{\beta}y + R_{2\alpha,2\beta}(x, y, 0+), \tag{25}$$

where $R_{2\alpha,2\beta}(x, y, 0+)$ is the remainder term. For $\alpha = \beta = 1$, the Taylor formula (25) gives the standard formula (2).

As a result, neglecting the remainder term $R_{2\alpha,2\beta}(x, y)$, the total increment of the function $\Delta z := f(x_1, y_1) - f(x_0, y_0)$ is written in terms of the increments $\Delta_{\alpha}x$ and $\Delta_{\beta}y$ of the factors x and y in the form

$$\Delta z \approx \frac{\left(D^{\alpha}_{0+;x}f\right)(x_0, y_0)}{\Gamma(\alpha+1)} \cdot \Delta_{\alpha}x + \frac{\left(D^{\beta}_{0+;y}f\right)(x_0, y_0)}{\Gamma(\beta+1)} \cdot \Delta_{\beta}y, \tag{26}$$

where x_0, y_0 are the basic (planned) values of the factors x and y that have influence on effective indicator z; the values x_1, y_1 describe the actual values of these factors, $\Delta_\alpha x = x_1^\alpha - x_0^\alpha$ and $\Delta_\beta y = y_1^\alpha - y_0^\beta$ are the generalized absolute changes (deviations) of the factors x and y.

As a result, the influence of the factor x on the indicator z will be calculated by the formula

$$\Delta z_{x,\alpha} = \frac{\left(D_{0+;x}^\alpha f\right)(x_0, y_0)}{\Gamma(\alpha+1)} \cdot \Delta_\alpha x, \tag{27}$$

and the influence of the factor y can be calculated by the equation

$$\Delta z_{y,\beta} = \frac{\left(D_{0+;y}^\beta f\right)(x_0, y_0)}{\Gamma(\beta+1)} \cdot \Delta_\beta y. \tag{28}$$

For $\alpha = \beta = 1$, Eqs. (27) and (28), which describe the influence of the factors x and y, take standard form (4) and (5), respectively. It is known that the Caputo derivative of order $\alpha > 0$ (and $\beta > 0$) of a constant function is equal to zero. As a result, for the case $z = f(x, y) = \text{const}$, we get a zero effect of both factors $\Delta z_{x,\alpha} = \Delta z_{y,\beta} = 0$. In addition, for the indicator, which does not depend on one of the factors, the effect of this factor will be zero.

3 Comparison with the Standard Method of Differential Calculus

The standard Taylor's formulas (2) without remainder term cannot give exact results for power functions with non-integer exponents. The Taylor formula (25) with derivatives of non-integer order even without remainder term is a more accurate tool for approximating the power functions. In the beginning, to illustrate this statement, we consider the nonlinear power function $f(x) = c_\alpha x^\alpha + c_0$, where α is non-integer positive number. If we use the standard Taylor formula for the expansion of this function in a neighborhood of the point $x_0 > 0$, then we get an infinite power series. For this function, the expression of the generalized Taylor formula (24) with the Caputo derivatives can be obtained using the formulas [[8], p. 95] for the left-sided Caputo derivative of the power function

$$D_{a+}^\alpha (x-a)^\beta = \frac{\Gamma(\beta+1)}{\Gamma(\beta-\alpha+1)}(x-a)^{\beta-\alpha}, \tag{29}$$

where $n - 1 < \alpha < n$, $\beta > n - 1$, and

$$D_{a+}^\alpha (x-a)^k = 0, \ (k = 0, 1, ..., n-1). \tag{30}$$

In particular, we have $D^{\alpha}_{a+}1 = 0$ and $D^{\alpha}_{a+}(x-a)^{\alpha} = \Gamma(\alpha+1)$. Using these equations, we get $\left(D^{\alpha}_{0+}f\right)(x_0) = c_{\alpha}\Gamma(\alpha+1)$. In this case, the higher-order Caputo derivatives will be zero, that is, $\left(\left(D^{\alpha}_{0+}\right)^{k} f\right)(x) = 0$ for k=1, 2, …. Substituting $\left(D^{\alpha}_{0+}f\right)(x_0) = c_{\alpha}\Gamma(\alpha+1)$ and $f(x_0) = c_{\alpha}x_0^{\alpha} + c_0$ into equation (24), we get

$$f(x) = c_{\alpha}x_0^{\alpha} + c_0 + c_{\alpha}\cdot\Delta_{\alpha}x = c_{\alpha}x^{\alpha} + c_0. \tag{31}$$

As a result, the Taylor series (24) gives the function $f(x) = c_{\alpha}x^{\alpha} + c_0$ exactly. Consequently, the generalized Taylor formula of such a power function gives an exact result (the function itself), and not an approximate expression. Therefore it can be concluded that the use of fractional derivatives in the deterministic factor analysis can yield more accurate results than standard methods.

If the indicators are given as parametric function of factors, then we can use the parametric left-sided Caputo derivative. For example, we can use the equation [[1], p. 464] in the form

$$D^{\alpha,x}_{a+}(x(t)-x(a))^{\beta} = \frac{\Gamma(\beta+1)}{\Gamma(\beta-\alpha+1)}(x(t)-x(a))^{\beta-\alpha}, \tag{32}$$

where $\alpha > 0$ and $\beta > -1$,

An example of a power function of two real variables is the Cobb-Douglas production function P = P(L, C). This function describes the dependence of the volume of production P on the labor costs L and capital costs C. In this case the production is the indicator (z = P) and the variables L and C are the factors (x = L, y = C). The Cobb-Douglas function has the form $P(L,C) := A\cdot L^{a}\cdot C^{b}$, where A is the aggregate productivity of the factors, and a and b are interpreted as capital and labor elasticities, respectively. The values of the constant $0 \le a < 1$ and $0 \le b < 1$ are determined by the available technologies. The Cobb-Douglas function was first proposed in [[2], p. 151] for the US industry for the period 1899–1922 in the form $P(L,C) := 1.01\cdot L^{3/4}\cdot C^{1/4}$, that is, A 1.01, a = 0.75 and b = 0.25. Using equation (29) for the Cobb-Douglas function $P(L, C) := A\cdot L^{a}\cdot C^{b}$, we obtain

$$\left(D^{\alpha}_{0+;L}P\right)(L,C) := \frac{A\cdot\Gamma(a+1)}{\Gamma(a-\alpha+1)}\cdot L^{a-\alpha}\cdot C^{b}, \tag{33}$$

$$\left(D^{\beta}_{0+;C}P\right)(L,C) := \frac{A\cdot\Gamma(b+1)}{\Gamma(b-\beta+1)}\cdot L^{a}\cdot C^{b-\beta}. \tag{34}$$

If the orders of the Caputo derivatives are taken equal to the exponents of the Cobb-Douglas function ($\alpha = a$, $\beta = b$), then Eqs. (33) and (34) give the expressions $\left(D^{\alpha}_{0+;L}P\right)(L,C) := A\cdot\Gamma(\alpha+1)\cdot C^{\beta}$ and $\left(D^{\beta}_{0+;C}P\right)(L,C) := A\cdot L^{\alpha}\cdot\Gamma(\beta+1)$. If the orders of the Caputo derivatives are equal to one ($\alpha = \beta = 1$), then Eqs. (33) and (34) give the standard expressions.

As a result of the influence of the factors L and C on the indicator P, which is calculated by formulas (27) and (28), have the form

$$\Delta P_{L,\alpha} = \frac{\left(D^{\alpha}_{0+;L}P\right)(L_0, C_0)}{\Gamma(\alpha+1)} \cdot \Delta_{\alpha}L =$$

$$\frac{A \cdot \Gamma(a+1)}{\Gamma(a-\alpha+1) \cdot \Gamma(\alpha+1)} \cdot L_0^{a-\alpha} \cdot C_0^{b} \cdot \Delta_{\alpha}L, \tag{35}$$

$$\Delta P_{C,\beta} = \frac{\left(D^{\beta}_{0+;C}P\right)(L_0, C_0)}{\Gamma(\beta+1)} \cdot \Delta_{\beta}C =$$

$$\frac{A \cdot \Gamma(b+1)}{\Gamma(b-\beta+1) \cdot \Gamma(\beta+1)} \cdot L_0^{a} \cdot C_0^{b-\beta} \cdot \Delta_{\beta}C, \tag{36}$$

where $\Delta_{\alpha}L = L_1^{\alpha} - L_0^{\alpha}$ and $\Delta_{\beta}C = C_1^{\alpha} - C_0^{\beta}$.

If we consider the orders of the derivatives $\alpha = a$ and $\beta = b$, then formulas (35) and (36) can be written in the form $\Delta P_{L,a} = A \cdot C_0^{b} \cdot \Delta_a L$, and $\Delta P_{C,b} = A \cdot L_0^{a} \cdot \Delta_b C$. As a numerical example, we consider the production function in the form $P(L,C) := 1.01 \cdot L^{\frac{3}{4}} \cdot C^{\frac{1}{4}}$, that is, $A = 1.01$, $a = 0.75$ and $b = 0.25$. For simplicity, the basis (planned) and actual values of the factors are chosen in the form of quaternary powers: $L_0 = 6.1^4, L_1 = 6.2^4$, and $C_0 = 0.2^4, C_1 = 0.5^4$. In this case, formulas (35) and (36) lead to the following values

$$\Delta P_{L,\alpha=a} = A \cdot C_0^{b} \cdot \Delta_a L = 2.2920940 \approx 2.29, \tag{37}$$

$$\Delta P_{C,\beta=b} = A \cdot L_0^{a} \cdot \Delta_b C = 68.775243 \approx 68.8. \tag{38}$$

The total influence of the factors L and C is

$$\Delta P_{L,\alpha} + \Delta P_{C,\beta} = 71.067337 \approx 71.1. \tag{39}$$

In this case, the increment $\Delta P = P(L_1, C_1) - P(L_0, C_0)$ of the effective indicator P, which is given by the production function $P(L,C) := A \cdot L^{a} \cdot C^{b}$, has the form

$$\Delta P = A \cdot L_1^{a} \cdot C_1^{b} - A \cdot L_0^{a} \cdot C_0^{b} = 74.505478 \approx 74.5. \tag{40}$$

The total influence of factors, calculated by the method of differentiation of fractional order, differs from the actual change in the effective index ΔP approximately by 4.6%. The additional increase of the effective indicator from the interaction of factors is determined by the expression

$$\delta_{\alpha,\beta} := \Delta P - \left(\Delta P_{L,\alpha} + \Delta P_{C,\beta}\right) = 3.438141 \approx 3.44. \tag{41}$$

Let us now compare the results with the standard differential method. If the orders of the Caputo derivatives in Eqs. (35) and (36) are equal to one ($\alpha = \beta = 1$), then we obtain standard expressions of the standard differential method given by Eqs. (4) and (5). For the standard approach with the Cobb-Douglas production function $P(L, C) := 1.01 \cdot L^{\frac{3}{4}} \cdot C^{\frac{1}{4}}$, the influence of the factors L and C is defined by the values

$$\Delta P_{L,1} = A \cdot a \cdot L_0^{a-1} \cdot C_0^{b} \cdot \Delta L = 2.310983484 \approx 2.31, \tag{42}$$

$$\Delta P_{C,1} = A \cdot b \cdot L_0^{a} \cdot C_0^{b-1} \cdot \Delta C = 436.2929478 \approx 436. \tag{43}$$

The corresponding total influence of the factors L and C is

$$\Delta P_{L,1} + \Delta P_{C,1} = 438.6039313 \approx 439. \tag{44}$$

The total influence of factors calculated by the standard differential method differs from the real change in the effective index ΔP approximately by 4.9 times, i.e., by 390%. We emphasize that for the proposed method the difference is 4.6%. The standard expression of the additional change in the effective index from the interaction of factors is given by

$$\delta = \delta_{1,1} := \Delta P - \left(\Delta P_{L,1} + \Delta P_{C,1}\right) = \quad 364.0984533 \approx -364. \tag{45}$$

Note that all the first numerical values of Eqs. (37)–(45) are absolutely exact values like $5^4 = 625$. Approximate values of these numbers are given for simplification.

From a comparison of the values of $\delta_{1,1}$, and the expressions $\delta_{\alpha,\beta}$, it can be seen that the error of the standard method is greater than the error of the new methods by more than 100 times ($|\delta_{1,1}/\delta_{\alpha,\beta}| \approx 106$).

4 Integral Method of Arbitrary (non-Integer) Order

The mathematical basis of the standard integral method is the first-order integration, the fundamental theorem of mathematical analysis, and the classical Newton-Leibniz formula

$$\int_a^b f_x^{(1)}(x)\, dx = f(b) - f(a). \tag{46}$$

The integral method is one of the most common methods of factor analysis, which allows us to decompose the overall increase in the effective index by factor increments. We give an example of standard formulas that describe the relationship between the increment of a function and the increment of factor characteristics. For simplicity, we consider the function of two real variables $Z = f(x, y)$. The formula of the integral method, which makes it possible to calculate the influence of the factor x on the resultant indicator, has the form

$$\Delta Z_x = \int_{x_0}^{x_1} f_x^{(1)}(x, y)dx. \quad (47)$$

The influence of factor y is calculated by the formula

$$\Delta Z_y = \int_{y_0}^{y_1} f_y^{(1)}(x, y)dx. \quad (48)$$

It is obvious that as integration variables we can consider $x = \chi$ and $y = \eta$. The integral method allows us to obtain accurate estimations of factor influences and does not imply the separation of factors into quantitative and qualitative ones.

In this section, we investigate the possibility of applying the calculus of integrals of non-integer orders in the economic factor analysis. As was shown in [18] and [[17], p. 241–264], the inverse operation for the Caputo derivative is the Riemann-Liouville integration of the same order. Let us give the definition of this integration [[8], pp. 69–70].

Definition The left-sided and right-sided Riemann-Liouville integrals of order $\alpha \geq 0$ on the interval [a, b] are defined by the equations

$$\left(I_{a+}^{\alpha} f\right)(x) := \frac{1}{\Gamma(\alpha)} \int_a^x \frac{f(\chi)\, d\chi}{(x-\chi)^{1-\alpha}}, \quad (49)$$

$$\left(I_{b-}^{\alpha} f\right)(x) := \frac{1}{\Gamma(\alpha)} \int_x^b \frac{f(\chi)\, d\chi}{(\chi-x)^{1-\alpha}}, \quad (50)$$

where $\Gamma(\alpha)$ is the gamma function, $a<x<b$, and the function $f(\chi)$ is assumed measurable on the interval (a,b) and satisfies the condition $\int_a^b |I(\chi)|\, d\chi < \infty$.

The Riemann-Liouville integrations (49) and (50) are generalizations of the standard n-fold integration [8, 13, 14]. We note that the Riemann-Liouville integrals (49) and (50) for the order equal to one ($\alpha=1$) are equal to the standard integral of the first order

$$\left(I_{a+}^{1} f\right)(x) = \int_a^x f(\chi)\, d\chi, \; \left(I_{b-}^{1} f\right)(x) = \int_x^b f(\chi)\, d\chi. \quad (51)$$

Thus, the integration, which is used in the basic formulas of the standard integral method of factor analysis, can be considered as a special case of the integration of fractional order.

If the parametric dependence of indicator z(t) on factors x(t) and y(t) cannot be represented by a single-valued function, then we should use the parametric derivatives and integrals. The parametric fractional integrals are defines [[8], pp. 99–100] be the following expression.

Definition The Riemann-Liouville parametric fractional integrals of a function f with respect to another function x of the order $\alpha > 0$ on [a, b] are defined by the equations

$$(I^{\alpha,x}_{a+}f)\,(t) := \frac{1}{\Gamma(\alpha)} \int_a^t d\tau \, \frac{x^{(1)}(\tau)}{(x(t) - x(\tau))^{1-\alpha}} f(\tau), \tag{52}$$

$$(I^{\alpha,x}_{b-}f)\,(t) := \frac{1}{\Gamma(\alpha)} \int_t^b d\tau \, \frac{x^{(1)}(\tau)}{(x(t) - x(\tau))^{1-\alpha}} f(\tau), \tag{53}$$

where $a < t < b$, $\alpha > 0$, and $x(\tau)$ is a monotone function that has a continuous derivative $x^{(1)}(\tau) = dx(\tau)/d\tau$.

Since the integral method is based on the Newton-Leibniz formula and the fundamental theorem of mathematical analysis, we need to generalize this theorem and the Newton-Leibniz formula to the case of fractional (non-integer) operators. The fundamental theorem of the theory of integro-differentiation of a fractional (non-integer) order was formulated in [18] and book [[17], pp. 247–248]. Some additional aspects of this theorem are discussed in [7]. We give a generalization of the Newton-Leibniz formula to the case of integrals and derivatives of non-integer order. For the left-sided operators, the following generalized Newton-Leibniz formula holds

$$(I^{\alpha}_{a+}D^{\alpha}_{a+}f)\,(b) = f(b) - f(a) - \sum_{k=1}^{n-1} \frac{f^{(k)}(a)}{k!}(b-a)^k, \tag{54}$$

where $n\text{-}1 \leq \alpha < n$, and for the right-sided operators the formula has the form

$$\left(I^{\alpha}_{b-}D^{\alpha}_{b-}f\right)(a) = f\,(a) - f\,(b) - \sum_{k=1}^{n-1} \frac{(-1)^k f^{(k)}(b)}{k!}(b-a)^k. \tag{55}$$

Using the expressions

$$\left(I^{1}_{b-}D^{1}_{b-}f\right)(a) = -\int_a^b f^{(1)}_x(x)\,dx,$$

$$\left(I^{1}_{a+}D^{1}_{a+}f\right)(b) = \int_a^b f^{(1)}_x(x)\,dx, \tag{56}$$

we find that Eqs. (54) and (55) for $\alpha = 1$ give the standard Newton-Leibniz formula (46).

To apply the method of fractional integration of non-integer order in a factor analysis, we need formulas for the left-sided and right-sided Riemann-Liouville integral of order $\alpha \geq 0$ for the power function [[8], p. 71] that has the form

$$I^{\alpha}_{a+}(x-a)^{\beta} = \frac{\Gamma(\beta+1)}{\Gamma(\beta+\alpha+1)}(x-a)^{\beta+\alpha},\ (\beta > 0), \tag{57}$$

$$I^{\alpha}_{b-}(b-x)^{\beta} = \frac{\Gamma(\beta+1)}{\Gamma(\beta+\alpha+1)}(b-x)^{\beta+\alpha}, \ (\beta > 0)\,. \tag{58}$$

The use of the fractional integro-differentiation methods makes it possible to obtain more accurate results for the influence of factors in comparison with the method of differential calculus of fractional order. This is due to the fact that the additional increase in the resultant indicator, which arises from the interaction of factors, is distributed among them in equal proportions. As an example, we give formulas for a power function of two variables, which is analogous to the Cobb-Douglas function considered above. For the function $f(x,y) = A \cdot x^{\alpha} \cdot y^{\beta}$ with $\alpha > 0$ and $\beta > 0$, similarly to the standard deterministic factor analysis [16], we have

$$\delta_{\alpha,\beta} := \Delta Z - \Delta z_{x,\alpha} - \Delta z_{y,\beta} = A \cdot \Delta_{\alpha} x \cdot \Delta_{\beta} y, \tag{59}$$

where $\Delta Z = f(x_1, y_1) - f(x_0, y_0)$, and $\Delta z_{x,\alpha}$, $\Delta z_{y,\beta}$ are defined by Eqs. (27) and (28). Then using $\Delta Z = \Delta Z_{x,\alpha} + \Delta Z_{y,\beta}$ and

$$\Delta Z_{x,\alpha} = \frac{\left(D^{\alpha}_{0+;x} f\right)(x_0, y_0)}{\Gamma(\alpha+1)} \cdot \Delta_{\alpha} x + \frac{1}{2}\delta_{\alpha,\beta}, \tag{60}$$

$$\Delta Z_{y,\beta} = \frac{\left(D^{\beta}_{0+;y} f\right)(x_0, y_0)}{\Gamma(\beta+1)} \cdot \Delta_{\beta} y + \frac{1}{2}\delta_{\alpha,\beta}, \tag{61}$$

we obtain the expressions

$$\Delta Z_{x,\alpha} = A \cdot y_0^{\beta} \cdot \Delta_{\alpha} x + \frac{A}{2} \cdot \Delta_{\alpha} x \cdot \Delta_{\beta} y = \frac{A}{2} \cdot (y_1^{\beta} + y_0^{\beta}) \cdot \Delta_{\alpha} x, \tag{62}$$

$$\Delta Z_{y,\beta} = A \cdot x_0^{\alpha} \cdot \Delta_{\beta} y + \frac{A}{2} \cdot \Delta_{\alpha} x \cdot \Delta_{\beta} y = \frac{A}{2} \cdot (x_1^{\alpha} + x_0^{\alpha}) \cdot \Delta_{\beta} y. \tag{63}$$

For $\alpha = \beta = 1$, Eqs. (59)–(63) give the standard equations for the integral calculus method for a multiplicative model with the function $f(x,y) = A \cdot x \cdot y$ [[16], p. 67, 323].

For the production function $P(L,C) := 1.01 \cdot L^{\frac{3}{4}} \cdot C^{\frac{1}{4}}$, the planned and actual values of the factors $L_0 = 6.1^4$, $L_1 = 6.2^4$, $C_0 = 0.2^4$, $C_1 = 0.5^4$, and Eqs. (62)–(63) lead to the values

$$\Delta P_{L,\alpha=a} = A \cdot C_0^{b} \cdot \Delta_a L + \frac{A}{2} \cdot \Delta_{\alpha} L \cdot \Delta_{\beta} C = 4.0111645 \approx 4.0, \tag{64}$$

$$\Delta P_{C,\beta=b} = A \cdot L_0^{a} \cdot \Delta_b C + \frac{A}{2} \cdot \Delta_{\alpha} L \cdot \Delta_{\beta} C = 70.4943135 \approx 70. \tag{65}$$

As a result, we get $\Delta P_{L,a} + \Delta P_{C,b} = \Delta P = 74.5054780 \approx 74$.

In addition to (57) and (58), there is a set of integration formulas of a non-integer order that are given in Tables 9.1–9.2 of book [[14], pp. 173–174]. To apply the proposed generalization of the integral method, it is required to use these tables of formulas that allow us to develop final working formulas for the most common types of factor dependencies and make this method more accessible.

5 Conclusion

The proposed methods of integro-differentiation of a non-integer order expand the possibilities of deterministic factor economic analysis. The method of differential calculus of non-integer order can give more accurate results in comparison with the standard method (the method of differential calculus used the derivatives of an integer order) for a wide class of functions including the power functions. In addition, the methods of integro-differentiation of a non-integer order make it possible to take into account the effects of memory and nonlocality in economic processes.

The proposed fractional differential and integral methods of the deterministic factor analysis can be used in the analysis of economic or financial processes, in which indicators (endogenous variables) are power functions of factors (exogenous variables). These methods can be used to study the processes described by power law, including the processes described by the Cobb-Douglas production function. The suggested methods allow us to more accurately describe the total influence of factors compared with the standard methods.

The proposed methods give significant advantages over the standard methods for the processes with dynamic memory and nonlocality in the factor space. The advantage of the proposed methods is especially clear for processes described by power functions and functions, which are representable by series with non-integer powers.

In application of suggested methods of the deterministic factor analysis, it should be taken into account that derivatives of non-integer order have a number of non-standard properties, including violation of the standard product differentiation rule (for example, see [[14], pp. 280–284], [[3], p. 59], [19, 20]) and the violation of the standard rule for differentiating a composite function (for example, see [[13], pp. 91–92], [[3], pp. 59–60], [21]). For the purpose of practical application of integro-differentiation methods of non-integer order for the economic factor analysis, it is necessary to develop final working formulas on the basis of the table of integrals [[14], pp. 173–174] for various kinds of factor dependencies, so that these methods become available to a wide range of analysts.

References

1. Almeida, R.A.: Caputo fractional derivative of a function with respect to another function. Commun. Nonlin. Sci. Numer. Simul. **44** 460–481. https://doi.org/10.1016/j.cnsns.2016.09.006. (arXiv:1609.04775) (2017)
2. Cobb, C.W., Douglas, P.H.: A theory of production. Am. Econom. Rev. **18**(Supplement), 139–165 (1928)
3. Diethelm, K.: The Analysis of Fractional Differential Equations: An Application-Oriented Exposition Using Differential Operators of Caputo Type. Berlin: Springer. p. 247. https://doi.org/10.1007/978-3-642-14574-2 (2010)
4. Gabaix, X.: Power laws in economics and finance. Ann. Rev. Econom. **1**(1), 255–293 (2009). 1941-1383/09/0904-0255
5. Gabaix, X.: Power laws in economics: an introduction. J. Econom. Perspect. **30**(1), 185–206 (2016). https://doi.org/10.1257/jep.30.1.185
6. Gorenflo, R., Mainardi, F., Scalas, E., Raberto, M.: Fractional calculus and continuous-time finance III: the diffusion limit. In: Kohlmann, M., Tang, S. (Eds.) Mathematical Finance. Trends in Mathematics. Basel: Birkhauser. pp. 171–180 (2001). https://doi.org/10.1007/978-3-0348-8291-0_17
7. Grigoletto, E.C., De Oliveira, E.C.: Fractional versions of the fundamental theorem of calculus. Appl. Math. **4**, 23–33 (2013). https://doi.org/10.4236/am.2013.47A006
8. Kilbas, A.A., Srivastava, H.M., Trujillo, J.J.: Theory and Applications of Fractional Differential Equations, p. 540. Elsevier, Amsterdam (2006)
9. Laskin, N.: Fractional market dynamics. Physica A. **287**(3), 482–492 (2000). https://doi.org/10.1016/S0378-4371(00)00387-3
10. Mainardi, F.: Fractional Calculus and Waves Linear Viscoelasticity: An Introduction to Mathematical Models. London: Imperial College Press. p. 368 (2010). https://doi.org/10.1142/P614
11. Mainardi F., Raberto M., Gorenflo R., Scalas E. Fractional calculus and continuous-time finance II: the waiting-time distribution. Physica A. **287** 3–4. 468–481 (2000). https://doi.org/10.1016/S0378-4371(00)00386-1
12. Odibat, Z.M., Shawagfeh, N.T.: Generalized Taylor's formula. Appl. Math. Comput. **186**(1), 286–293 (2007). https://doi.org/10.1016/j.amc.2006.07.102
13. Podlubny, I.: Fractional Differential Equations, p. 340. Academic Press, San Diego (1998)
14. Samko, S.G., Kilbas, A.A., Marichev, O.I.: Fractional Integrals and Derivatives Theory and Applications, p. 1006. Gordon and Breach, New York (1993)
15. Scalas, E., Gorenflo, R., Mainardi, F.: Fractional calculus and continuous-time finance. Physica A. **284**(1–4), 376–384 (2000). https://doi.org/10.1016/S0378-4371(00)00255-7
16. Sheremet, A.D.: Theory of Economic Analysis, 2nd edn, p. 366. Infra-M, Moscow (2005)
17. Tarasov, V.E.: Fractional Dynamics: Applications of Fractional Calculus to Dynamics of Particles, Fields and Media. New York: Springer p. 505 (2010). https://doi.org/10.1007/978-3-642-14003-7
18. Tarasov, V.E.: Fractional vector calculus and fractional Maxwell's equations. Ann. Phys. **323**(11), 2756–2778 (2008). https://doi.org/10.1016/j.aop.2008.04.005
19. Tarasov, V.E.: Leibniz rule and fractional derivatives of power functions. J. Comput. Nonlin. Dynam. **11**(3), 031014 (2016). https://doi.org/10.1115/1.4031364
20. Tarasov, V.E.: No violation of the Leibniz rule. No fractional derivative. Commun. Nonlin. Sci. Numeric. Simul. **18**(11), 2945–2948 (2013). https://doi.org/10.1016/j.cnsns.2013.04.001
21. Tarasov, V.E.: On chain rule for fractional derivatives. Commun. Nonlin. Sci. Numeric. Simul. **30**(1–3), 1–4 (2016). https://doi.org/10.1016/j.cnsns.2015.06.007
22. Tarasov, V.E., Tarasova, V.V.: Long and short memory in economics: fractional-order difference and differentiation. IRA-Int. J. Manag. Soc. Sci. **5**(2), 327–334 (2016). https://doi.org/10.21013/jmss.v5.n2.p10
23. Tarasov, V.E., Tarasova, V.V.: Time-dependent fractional dynamics with memory in quantum and economic physics. Annal. Phys. (2017). https://doi.org/10.1016/j.aop.2017.05.017

24. Tarasova, V.V., Tarasov, V.E.: A generalization of concepts of accelerator and multiplier to take into account memory effects in macroeconomics. J. Econom. Entrepreneur. [Ekonomika i Predprinimatelstvo]. **10–3** (75–3). 1121–1129 [in Russian] (2016)
25. Tarasova, V.V., Tarasov, V.E.: Concept of dynamic memory in economics. Commun. Nonlin. Sci. Numer. Simul. https://doi.org/10.1016/j.cnsns.2017.06.032 Accepted for publication
26. Tarasova, V.V., Tarasov, V.E.: Dynamic intersectoral models with power-law memory // Commun. Nonlin. Sci. Numer. Simul. **54**, 100–117 (2018). https://doi.org/10.1016/j.cnsns.2017.05.015
27. Tarasova, V.V., Tarasov, V.E.: Economic accelerator with memory: discrete time approach. Prob. Modern Sci. Educat. [Problemy Sovremennoj Nauki i Obrazovaniya]. **36** (78). 37–42 (2016). https://doi.org/10.20861/2304-2338-2016-78-002
28. Tarasova, V.V., Tarasov, V.E.: Economic growth model with constant pace and dynamic memory. Prob. Modern Sci. Educat. **2**(84), 40–45 (2017). https://doi.org/10.20861/2304-2338-2017-84-001
29. Tarasova, V.V., Tarasov, V.E. Economic indicator that generalizes average and marginal values. J. Econom. Entrepreneur. [Ekonomika i Predprinimatelstvo]. **11–1** (76-1) 817–823 (2016) [in Russian]
30. Tarasova, V.V., Tarasov, V.E.: Economic interpretation of fractional derivatives. Prog. Fraction. Differ. Appl. **3**(1), 1–7 (2017). https://doi.org/10.18576/pfda/030101
31. Tarasova, V.V., Tarasov, V.E.: Elasticity for economic processes with memory: fractional differential calculus approach. Fract. Differ Calcul. **6**(2), 219–232 (2016). https://doi.org/10.7153/fdc-06-14
32. Tarasova, V.V., Tarasov, V.E.: Fractional dynamics of natural growth and memory effect in economics. Eur. Res. **12**(23), 30–37 (2016). https://doi.org/10.20861/2410-2873-2016-23-004
33. Tarasova, V.V., Tarasov, V.E.: Logistic map with memory from economic model. Chaos, Solitons Fract. **95**, 84–91 (2017). https://doi.org/10.1016/j.chaos.2016.12.012
34. Tarasova, V.V., Tarasov, V.E.: Macroeconomic models with dynamic memory. J. Econom. Entrepreneur. [Ekonomika i Predprinimatelstvo] **3–2**(80–2) 26–35 [in Russian] (2017)
35. Tarasova, V.V., Tarasov V.E.: Marginal utility for economic processes with memory. Almanac of Modern Science and Education [Almanah Sovremennoj Nauki i Obrazovaniya] **7** (109) 108–113 [in Russian] (2016)
36. Tarasova, V.V., Tarasov, V.E.: Marginal values of non-integer order in economic analysis // Azimuth Scientific Research: Economics and Management [Azimut Nauchnih Issledovanii: Ekonomika i Upravlenie] **3**(16) 197–201 [in Russian] (2016)
37. Tenreiro Machado, J.A., Duarte, F.B., Duarte, G.M.: Fractional dynamics in financial indeces. Int. J. Bifur. Chaos. **22**(10). ar1250249 (2012) https://doi.org/10.1142/S0218127412502495
38. Tenreiro Machado, J.A., Mata, M.E.: Pseudo phase plane and fractional calculus modeling of western global economic downturn. Communications in Nonlinear Science and Numerical Simulation. **22**(1–3), 396–406 (2015). https://doi.org/10.1016/j.cnsns.2014.08.032
39. Tenreiro Machado, J.A., Mata, M.E., Lopes, A.M.: Fractional state space analysis of economic systems. Entropy. **17**(8), 5402–5421 (2015). https://doi.org/10.3390/e17085402
40. Uchaikin, V.V.: Fractional Derivatives for Physicists and Engineers. Vol. II. Applications. Berlin: Springer, p. 466 (2013)

Fractional-Order Model of Wine

António M. Lopes, J.A. Tenreiro Machado and Elisa Ramalho

1 Introduction

Electrical Impedance Spectroscopy (EIS) measures the electrical impedance of a specimen within a given bandwidth [27, 38, 39]. This technique has the advantages of being non-destructive and avoiding time consuming experiments. EIS has been used for studying vegetable [3, 10, 33, 58, 61] and animal [1, 16, 21, 32, 63] tissues, materials [19, 26, 62], and devices [2, 20, 31].

This chapter uses EIS to characterize wine. Wine is a beverage produced by fermented grapes, and its most important characteristics are the sweetness, acidity, tannins, balance, and body. However, the mathematical description of wine poses a challenging problem.

Zheng et al. [65] used EIS for studying the dielectric properties of ethanol and organic acids of grape musts. They found correlations between the concentrations of the pure solutions and the dielectric parameters. Riul et al. [51] proposed a taste sensor based on EIS. They processed the EIS data using principal component analysis and artificial neural networks, concluding that the system could distinguish between certain types of red wines.

A.M. Lopes (✉)
Faculty of Engineering, UISPA–LAETA/INEGI, University of Porto,
Rua Dr. Roberto Frias, 4200-465 Porto, Portugal
e-mail: aml@fe.up.pt

J.A.T. Machado
Department of Electrical Engineering, Institute of Engineering,
Polytechnic of Porto, R. Dr. António Bernardino de Almeida, 431,
4249-015 Porto, Portugal
e-mail: jtm@isep.ipp.pt

E. Ramalho
Chemical Engineering Department, CIETI/DEQ/ISEP, Centre of Innovation
on Engineering and Industrial Technology, Institute of Engineering,
Polytechnic of Porto, R. Dr. António Bernardino de Almeida, 431,
4249-015 Porto, Portugal
e-mail: err@isep.ipp.pt

M. Edelman et al. (eds.), *Chaotic, Fractional, and Complex Dynamics: New Insights and Perspectives*, Understanding Complex Systems,
https://doi.org/10.1007/978-3-319-68109-2_10

The EIS requires exciting a specimen with frequency-variable electric sinusoidal signals and registering the response. The voltage $v(t)$ and current $i(t)$ across the specimen at steady-state are sinusoidal functions of time given by:

$$\begin{cases} v(t) = V \cos(\omega t + \theta_V) \\ i(t) = I \cos(\omega t + \theta_I) \end{cases}, \tag{1}$$

where $\{V, I\}$ are the amplitudes of the voltage and current, $\{\theta_V, \theta_I\}$ denote their phase shifts, and ω represents the angular frequency.

The variables $v(t)$ and $i(t)$ can be represented in the frequency-domain:

$$\begin{cases} \mathbf{V}(j\omega) = V \cdot e^{j\theta_V} \\ \mathbf{I}(j\omega) = I \cdot e^{j\theta_I} \end{cases}, \tag{2}$$

where $j = \sqrt{-1}$. The complex impedance $\mathbf{Z}(j\omega)$ is defined as the ratio of phasors:

$$\mathbf{Z}(j\omega) = \frac{\mathbf{V}(j\omega)}{\mathbf{I}(j\omega)} = \frac{V}{I} \cdot e^{j(\theta_V - \theta_I)} = |\mathbf{Z}(j\omega)| \cdot e^{j \arg[\mathbf{Z}(j\omega)]}. \tag{3}$$

For characterizing wine, we start by measuring its electrical impedance, and then we apply fractional calculus (FC) [4, 41, 42, 45] to design an appropriate heuristic model. We adopt hierarchical clustering (HC) to visualize the relationships between the measurements. The methodology is illustrated with a set of red and white wines. The impedance parameters are correlated with chemical data, showing that the technique is able to characterize wine samples.

Having these ideas in mind, this chapter is organized as follows. Section 2 introduces the empirical FC models. Section 3 models the electrical impedance by means of FC tools and correlates the impedance parameters with chemical data. Section 4 uses HC for visualizing the results. Finally, Sect. 5 draws the main conclusions.

2 Empirical Fractional-Order Models

Given an experimental impedance spectrum, we need to find a description, or "model", that fits the data, and has a limited number of parameters [35, 46, 47]. Different empirical models in the scope of the dielectric relaxation phenomenon were proposed [15].

Many materials are characterized by dynamic processes that occur at different scales in space-time. The Debye (D) model [8] is insufficient to describe the behavior of such materials, since it does not consider the interactions among distinct relaxing effects and long-memory phenomena [13]. Fractional-order (FO) formulations of the D model are capable of dealing with the phenomena taking place within a material [7, 17, 59, 64]. Non-Debye relaxation in condensed matter and dielectrics are described in [28, 29, 29, 30].

The Cole-Cole (CC), Cole-Davidson (CD) and Havriliak-Negami (HN) expressions generalize the D model, describing anomalous relaxation processes [5, 6, 11, 23, 36, 54, 59]. These models are empirical and do not use fractional derivatives or integrals, explicitly, but they represent pioneer approaches to the application of FC [4, 9, 34, 44], as discussed by several researchers [18, 24, 48–50, 52, 53]. FO formulations of the non-Debye relaxation and models dealing with phenomena taking place within condensed matter are studied in [55–57]. Nevertheless, just a few studies discuss the fractional dynamics and complex properties of wine [12, 40, 60].

In the Laplace domain, the D, CC, CD and HN models are [52]:

$$\tilde{\varepsilon}_D(s) = \frac{\varepsilon^*(s) - \varepsilon_\infty}{\varepsilon_0 - \varepsilon_\infty} = \frac{1}{1 + s\tau}, \tag{4}$$

$$\tilde{\varepsilon}_{CC}(s) = \frac{1}{1 + (s\tau)^\alpha}, \tag{5}$$

$$\tilde{\varepsilon}_{CD}(s) = \frac{1}{(1 + s\tau)^\beta}, \tag{6}$$

$$\tilde{\varepsilon}_{HN}(s) = \frac{1}{[1 + (s\tau)^\alpha]^\beta}, \tag{7}$$

where $0 < \alpha \leq 1, 0 < \beta \leq 1$, $\tilde{\varepsilon}$ denotes the complex susceptibility, $\{\varepsilon_0, \varepsilon_\infty\}$ represent the low and high-frequency limits of the complex dielectric permittivity, ε^*, τ is the relaxation time and $f = \omega/2\pi$ is the frequency.

3 EIS Analysis of Wine

The impedance spectra for $N = 16$ Portuguese wines are determined by means of EIS. The set includes samples from distinct wine regions [14, 40], involving a mixed of ripe and green, both red and white, styles (Table 1). Given the characteristics of $\mathbf{Z}(j\omega)$, FO models are considered for fitting the experimental data with a minimum of parameters.

The experimental set-up used for EIS is depicted in Fig. 1 [38, 39]. An $l \times w \times h = 120 \times 100 \times 55$ mm parallelepipedic container is filled with 200 ml of wine, at room temperature (23 °C). Two 0.5 mm diameter, 15 mm length, copper electrodes

Table 1 The set of wine samples analysed

i	Tag	Wine region	Wine style
1	W_1	Alentejo	White
2	W_2	Alentejo	White
3	W_3	Alentejo	White
4	W_4	Península de Setúbal	White
5	W_5	Tejo	white
6	W_6	Douro	White
7	W_7	Vinhos Verdes	Green white
8	W_8	Vinhos Verdes	Green white
9	R_1	Alentejo	Red
10	R_2	Alentejo	Red
11	R_3	Península de Setúbal	Red
12	R_4	Península de Setúbal	Red
13	R_5	Bairrada	Red
14	R_6	Douro	Red
15	R_7	Vinhos Verdes	Green red
16	R_8	Vinhos Verdes	Green red

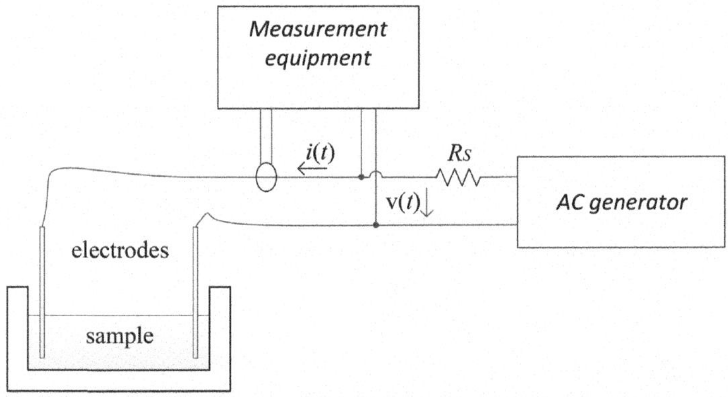

Fig. 1 Experimental set-up for measuring impedance

are used to connect the samples to the excitation and measurement equipment. The electrodes are immersed at 5 mm distance from the bottom of the container, and are placed diametrically opposed to each other. Electrodes with different sizes and shapes were tested, but their influence on the results was found to be negligible. An adaptation resistance, $R_s = 15$ kΩ, is used in series with the specimen under analysis, and $\mathbf{Z}(j\omega)$ is measured in the bandwidth $BW_1: \quad 3 \leq f \leq 10^6$ Hz, at $L = 33$ logarithmically spaced points.

The FO models are adjusted to the data minimizing the Canberra-like distance, J, between the experimental, $\mathbf{Z}_e$, and model, $\mathbf{Z}_m$, impedances:

$$J = \frac{1}{L}\sum_{k=1}^{L} \frac{|\text{Re}[\mathbf{Z}_e(j\omega_k)] - \text{Re}[\mathbf{Z}_m(j\omega_k)]|}{|\text{Re}[\mathbf{Z}_e(j\omega_k)]| + |\text{Re}[\mathbf{Z}_m(j\omega_k)]|} + \frac{|\text{Im}[\mathbf{Z}_e(j\omega_k)] - \text{Im}[\mathbf{Z}_m(j\omega_k)]|}{|\text{Im}[\mathbf{Z}_e(j\omega_k)]| + |\text{Im}[\mathbf{Z}_m(j\omega_k)]|}, \tag{8}$$

where $\text{Re}(\cdot)$ and $\text{Im}(\cdot)$ represent the real and imaginary parts.

Expression (8) captures the relative error of the adjustment and avoids "saturation" effects caused by the simultaneous presence of large and small values.

A good fit between approximation and real data in the bandwidth BW_1 is found for the 6-parameter model:

$$\mathbf{Z}_m(j\omega) = K \cdot \frac{\left(1 + \frac{j\omega}{z_1}\right)^{\alpha_1} \cdot \left(1 + \frac{j\omega}{z_2}\right)^{\alpha_2}}{(j\omega)^{\beta}}, \tag{9}$$

with $\{K, \alpha_1, z_1, \alpha_2, z_2, \beta\} > 0$. This heuristic expression represents a good compromise between complexity and fitting quality.

Table 2 summarizes the optimal impedance, or electrical, parameters obtained for the $N = 16$ samples analyzed (to be denoted in the follow-up by the $N \times S = 16 \times 6$ matrix $\mathscr{E} = [\mathbf{e}_1, \ldots, \mathbf{e}_6]$).

The polar diagrams of the experimental data and model (9) for sample W_2 and R_2 are depicted in Fig. 2 for BW_1, where the detail figures correspond to a zoom within the bandwidth $BW_2: \ 3 \leq f \leq 500$ Hz. For the remaining wine samples the results are similar.

Table 2 also comprises chemical data for all samples. The values were obtained by standard measurement procedures and correspond to: reducing substances (sug)—ref: OIV-MA-AS311-01A, alcoholic strength by volume (alc)—ref: OIV-MA-AS312-01A, total acidity (aci)—ref: OIV-MA-AS313-01, density (dens) at 20 °C (by pycnometry)—ref: OIV-MA-AS2-01A [25], and tannins (tan): ref: AWRI (Australian Wine Research Institute) with methyl cellulose precipitable tannin assay (to be denoted in the follow-up by the $N \times T = 16 \times 5$ matrix $\mathscr{Q} = [\mathbf{q}_1, \ldots, \mathbf{q}_5]$).

To unveil the correlations between the descriptors, that is, the electrical parameters and the chemical results, we define the $16 \times (6 + 5)$ dimensional matrix $\mathscr{P} = [\mathscr{E}|\mathscr{Q}] = [\mathbf{p}_1, \ldots, \mathbf{p}_{11}]$ and we calculate the Pearson correlation, r_{ij}, $i, j = 1, \ldots, 11$, $j > i$, between the vectors:

$$r_{ij} = \frac{(\mathbf{p}_i - \bar{p}_i)^T \cdot (\mathbf{p}_j - \bar{p}_j)}{||\mathbf{p}_i - \bar{p}_i|| \cdot ||\mathbf{p}_j - \bar{p}_j||}, \tag{10}$$

where $\bar{p}_i$ and $\bar{p}_j$ denote the average values of columns i and j of matrix $\mathscr{P}$.

Table 2 Impedance parameters and chemical data of the wine samples

i	Tag	Impedance parameters, $\mathscr{E}$							Chemical data, $\mathscr{Q}$				
		K ($\times 10^3$)	z_1 ($\times 10^3$)	α_1 ($\times 10^4$)	z_2	α_2	β	J	Alc (% v/v)	Aci (g tartaric acid/L)	Sug (g/L)	Den (g/L)	Tan (g epicatechin eq/L)
1	W_1	6.7	1.0	0.33	24	0.88	0.29	0.23	12.7	5.47	3.0	988.9	–
2	W_2	5.0	1.0	0.32	19	0.87	0.32	0.242	12.7	5.35	1.4	989.0	–
3	W_3	5.8	1.3	0.40	22	0.88	0.31	0.27	13.0	5.50	1.6	988.2	–
4	W_4	7.0	1.0	0.35	22	0.86	0.31	0.26	12.6	4.31	1.0	988.6	–
5	W_5	5.5	1.0	0.32	19	0.87	0.33	0.29	13.1	6.21	1.4	988.8	–
6	W_6	6.0	0.9	0.32	20	0.88	0.29	0.29	13.1	6.63	1.0	989.0	–
7	W_7	7.5	1.6	0.40	15	0.75	0.36	0.27	9.1	8.01	6.4	1001.0	–
8	W_8	7.5	1.6	0.40	17	0.75	0.36	0.23	10.0	6.90	6.4	996.0	–
9	R_1	5.5	0.95	0.33	19	0.87	0.32	0.29	12.8	5.83	1.4	991.8	1.51
10	R_2	6.8	1.1	0.33	23	0.92	0.33	0.29	13.5	4.93	1.9	991.0	1.50
11	R_3	5.0	1.0	0.33	23	0.86	0.31	0.27	13.4	5.67	2.8	992.8	1.61
12	R_4	6.0	1.0	0.34	22	0.88	0.33	0.29	13.4	5.68	2.2	992.9	1.65
13	R_5	7.5	1.6	0.39	15	0.75	0.35	0.26	12.6	6.77	8.0	994.8	1.12
14	R_6	5.0	1.0	0.32	20	0.88	0.32	0.30	12.6	6.06	1.5	994.4	1.59
15	R_7	6.5	1.1	0.30	20	0.89	0.35	0.37	10.3	8.62	2.3	996.0	1.63
16	R_8	7.2	1.7	0.41	19	0.88	0.38	0.30	10.4	8.11	4.3	995.7	0.92

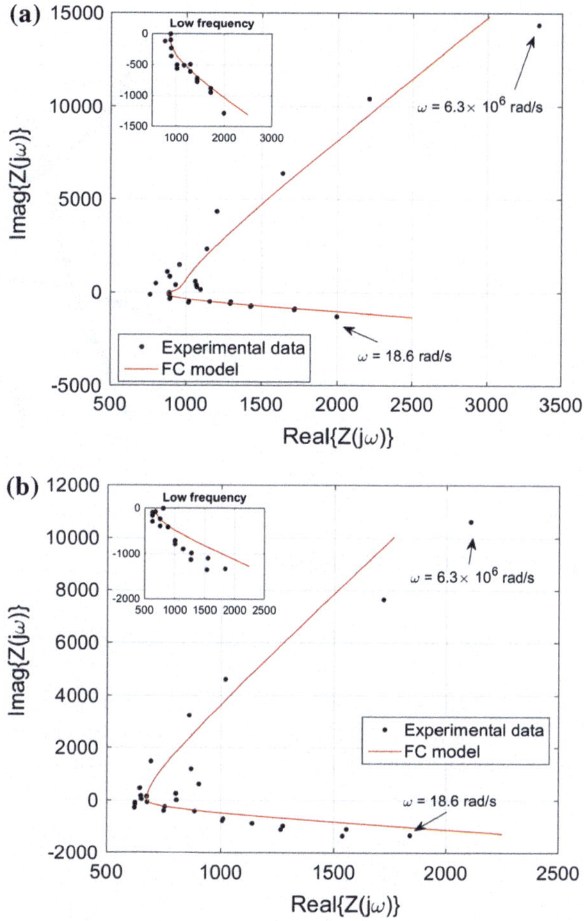

Fig. 2 Polar diagrams of $\mathbf{Z}(j\omega)$ for the wine sample W_2 and R_2 and bandwidth BW_1. The detail figures correspond to a zoom within the bandwidth BW_2

Figure 3 depicts the correlations, where all pairs of descriptors with $|r_{ij}| \geq 0.79$ are connected by a double arrow. We observe in Fig. 3 intra- and inter-model correlations, particularly between sugars and $\{\alpha_1, z_1, \alpha_2\}$.

In conclusion, the results demonstrate that model (9) yields a convincing description and reliable characterization of wine. Given the correlations between electrical and chemical results, EIS represents a straightforward procedure to implement in the winery industry.

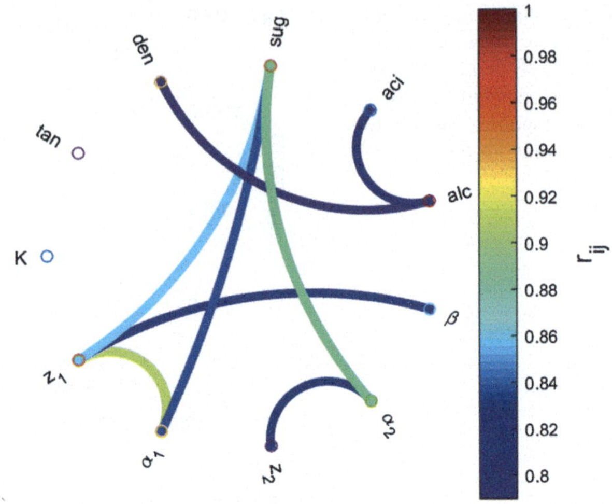

Fig. 3 Pearson correlation between the descriptors that characterize wine. Arcs connect pairs of descriptors with $|r_{ij}| \geq 0.79$, and colors represents the magnitude of the correlation

4 HC and Visualizing

HC is a technique for analyzing and visualizing relationships embedded in data. The algorithm generates groups of objects that are similar to each other in some sense, and the results are usually depicted in a dendrogram or a tree diagram [22].

We adopt the HC for visualizing the relationships between 16 wine samples. The HC processes matrix $\mathbf{M} = [c_{ij}]$, based on the distance:

$$c_{ij} = \frac{1}{L} \sum_{k=1}^{L} \frac{\left|\mathrm{Re}[\mathbf{Z}_i(j\omega_k)] - \mathrm{Re}[\mathbf{Z}_j(j\omega_k)]\right|}{\left|\mathrm{Re}[\mathbf{Z}_i(j\omega_k)]\right| + \left|\mathrm{Re}[\mathbf{Z}_j(j\omega_k)]\right|} + \frac{\left|\mathrm{Im}[\mathbf{Z}_i(j\omega_k)] - \mathrm{Im}[\mathbf{Z}_j(j\omega_k)]\right|}{\left|\mathrm{Im}[\mathbf{Z}_i(j\omega_k)]\right| + \left|\mathrm{Im}[\mathbf{Z}_j(j\omega_k)]\right|}, \tag{11}$$

where the indices $i, j = 1, \ldots, 16$, denote 2 wine samples, and $\mathbf{Z}$ represents the impedances measured by means of EIS.

Figure 4 depicts the tree generated by applying the successive (agglomerative) and average-linkage methods [22, 37, 43] for experimental data in the bandwidth BW_1. Three main clusters emerge: $\mathscr{C}_1 = \{W_2, W_3\}$, $\mathscr{C}_2 = \{W_1, W_4, W_5, W_6, W_7\}$ and $\mathscr{C}_3 = \{R_1, R_2, R_3, R_4, R_5, R_6, R_7, R_8, W_8\}$. That is, clusters $\{\mathscr{C}_1, \mathscr{C}_2\}$ include white styles, while $\mathscr{C}_3$ comprises all red wines plus the green white W_8. Therefore, we verify that the information in $\mathbf{M}$ together with the HC technique is able to distinguish the types of wine.

Figure 5 represents the tree generated by the HC and $\mathbf{M}$ when focusing on the low frequency bandwidth $BW_2: \ 3 \leq f \leq 500$ Hz, of $\mathbf{Z}$. We verify the emergence of clusters $\mathscr{D}_1 = \{W_1, W_4, W_5, W_6\}$, $\mathscr{D}_2 = \{W_2, W_3, W_8, R_3\}$ and $\mathscr{D}_3 = \{R_1, R_2, R_4, R_5, R_6, R_7, R_8, W_7\}$. Therefore, the ability to distinguish between white and red

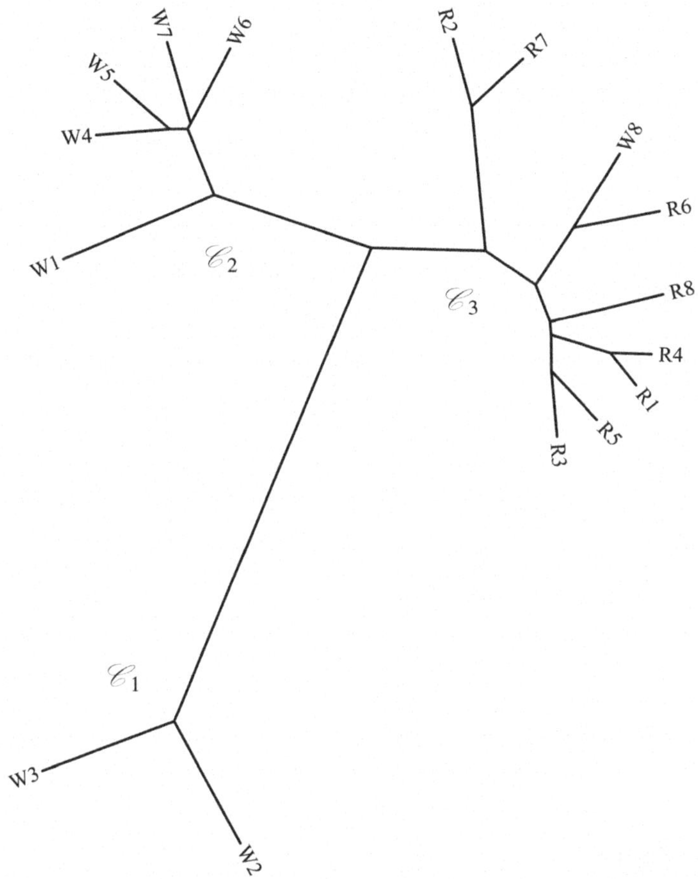

Fig. 4 The hierarchical tree for the $N = 16$ wine samples and matrix $\mathbf{M}$ (experimental data in the bandwidth BW_1)

styles has some similarities to the one obtained considering the full bandwidth BW_1, and we conclude that a considerable part of the EIS characteristics of wine emerges at low frequencies.

In conclusion, the EIS proved to be a valuable technique leading to results comparable with standard chemical analysis, while allowing an easier industrial implementation. Future developments using EIS in winery processes show that the adoption of more sensitive equipment seems important to have a better description of distinct wine samples.

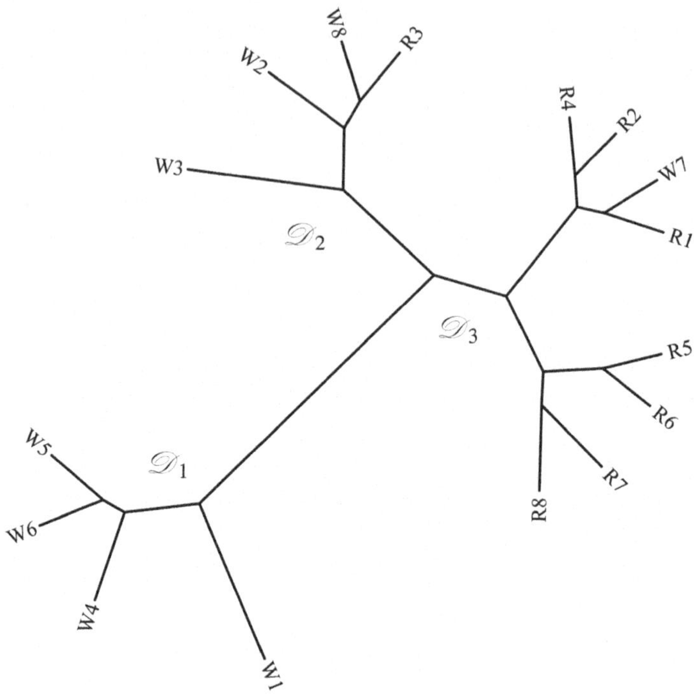

Fig. 5 The hierarchical tree for the $N = 16$ wine samples, matrix $\mathbf{M}$ (experimental data in the bandwidth BW_2)

5 Conclusions

EIS was used to determine the electrical impedance spectra of 16 wines. FO transfer functions requiring 6 parameters describe adequately the data. Based on the correlations between electrical and chemical measurements, we conclude that impedance obtained from EIS leads to an assertive wine characterization. Due to its easy implementation and short time needed to obtain results, the EIS has high potential in the winery industry, both to replace, or to complement, traditional chemical analysis.

References

1. Åberg, P., Birgersson, U., Elsner, P., Mohr, P., Ollmar, S.: Electrical impedance spectroscopy and the diagnostic accuracy for malignant melanoma. Exp. Dermatol. **20**(8), 648–652 (2011)
2. Adachi, M., Sakamoto, M., Jiu, J., Ogata, Y., Isoda, S.: Determination of parameters of electron transport in dye-sensitized solar cells using electrochemical impedance spectroscopy. J. Phys. Chem. B **110**(28), 13872–13880 (2006)

3. Ando, Y., Maeda, Y., Mizutani, K., Wakatsuki, N., Hagiwara, S., Nabetani, H.: Effect of air-dehydration pretreatment before freezing on the electrical impedance characteristics and texture of carrots. J. Food Eng. **169**, 114–121 (2016)
4. Baleanu, D., Diethelm, K., Scalas, E., Trujillo, J.J.: Fractional Calculus: Models and Numerical Methods, vol. 3. World Scientific (2012)
5. Cole, K.S., Cole, R.H.: Dispersion and absorption in dielectrics I. Alternating current characteristics. J. Chem. Phys. **9**(4), 341–351 (1941)
6. Davidson, D., Cole, R.: Dielectric relaxation in glycerol, propylene glycol, and n-propanol. J. Chem. Phys. **19**(12), 1484–1490 (1951)
7. Debye, P.: Interferenz von Röntgenstrahlen und Wärmebewegung. Annalen der Physik **348**(1), 49–92 (1913)
8. Debye, P.J.W.: Polar Molecules. Chemical Catalog Company, Incorporated (1929)
9. Diethelm, K., Ford, N.J.: Analysis of fractional differential equations. J. Math. Anal. Appl. **265**(2), 229–248 (2002)
10. El Khaled, D., Castellano, N., Gazquez, J., Salvador, R.G., Manzano-Agugliaro, F.: Cleaner quality control system using bioimpedance methods: a review for fruits and vegetables. J. Cleaner Prod. (2015)
11. Emmert, S., Wolf, M., Gulich, R., Krohns, S., Kastner, S., Lunkenheimer, P., Loidl, A.: Electrode polarization effects in broadband dielectric spectroscopy. Eur. Phys. J. B **83**(2), 157–165 (2011)
12. Eremenko, Z., Skresanov, V., Shubnyi, A., Anikina, N., Gerzhikova, V., Zhilyakova, T.: Complex permittivity measurement of high loss liquids and its application to wine analysis. In: Electromagnetic Waves. InTech (2011)
13. Feldman, Y., Puzenko, A., Ryabov, Y.: Non-Debye dielectric relaxation in complex materials. Chem. Phys. **284**(1), 139–168 (2002)
14. Fraga, H., Malheiro, A., Moutinho-Pereira, J., Jones, G., Alves, F., Pinto, J.G., Santos, J.: Very high resolution bioclimatic zoning of Portuguese wine regions: present and future scenarios. Reg. Environ. Change **14**(1), 295–306 (2014)
15. Freeborn, T.J.: A survey of fractional-order circuit models for biology and biomedicine. IEEE J. Emerg. Sel. Top. Circ. Syst. **3**(3), 416–424 (2013)
16. Freeborn, T.J., Elwakil, A.S., Maundy, B.: Compact wide frequency range fractional-order models of human body impedance against contact currents. Math. Prob. Eng. **2016** (2016)
17. Fröhlich, H.: Theory of Dielectrics: Dielectric Constant and Dielectric Loss. Clarendon Press (1958)
18. Garra, R., Giusti, A., Mainardi, F., Pagnini, G.: Fractional relaxation with time-varying coefficient. Fract. Calc. Appl. Anal. **17**(2), 424–439 (2014)
19. Ghasemi, S., Darestani, M.T., Abdollahi, Z., Gomes, V.G.: Online monitoring of emulsion polymerization using electrical impedance spectroscopy. Polym. Int. **64**(1), 66–75 (2015)
20. Glatthaar, M., Riede, M., Keegan, N., Sylvester-Hvid, K., Zimmermann, B., Niggemann, M., Hinsch, A., Gombert, A.: Efficiency limiting factors of organic bulk heterojunction solar cells identified by electrical impedance spectroscopy. Solar Energy Mater. Solar Cells **91**(5), 390–393 (2007)
21. Groeber, F., Engelhardt, L., Egger, S., Werthmann, H., Monaghan, M., Walles, H., Hansmann, J.: Impedance spectroscopy for the non-destructive evaluation of in vitro epidermal models. Pharm. Res. **32**(5), 1845–1854 (2015)
22. Hartigan, J.A.: Clustering Algorithms. Wiley (1975)
23. Havriliak, S., Negami, S.: A complex plane analysis of α-dispersions in some polymer systems. J. Polym. Sci. Part C: Polym. Symp. **14**, 99–117 (1966). Wiley Online Library
24. Hilfer, R.: Analytical representations for relaxation functions of glasses. J. Non-Cryst. Solids **305**(1), 122–126 (2002)
25. International Organisation of Vine and Wine: Compendium of International Methods of Wines and Musts Analysis, vol. 1. International Organisation of Vine and Wine, Paris (2016)
26. Irvine, J.T., Sinclair, D.C., West, A.R.: Electroceramics: characterization by impedance spectroscopy. Adv. Mater. **2**(3), 132–138 (1990)

27. Jesus, I.S., Machado, J.T., Cunha, J.B.: Fractional electrical impedances in botanical elements. J. Vib. Control **14**(9–10), 1389–1402 (2008)
28. Jonscher, A.: Low-frequency dispersion in carrier-dominated dielectrics. Philos. Mag. B **38**(6), 587–601 (1978)
29. Jonscher, A.: Dielectric Relaxation in Solids. Chelsea Dielectrics Press, London (1983)
30. Jonscher, A.K.: The 'universal' dielectric response. Nature **267**(5613), 673–679 (1977)
31. Kern, R., Sastrawan, R., Ferber, J., Stangl, R., Luther, J.: Modeling and interpretation of electrical impedance spectra of dye solar cells operated under open-circuit conditions. Electrochim. Acta **47**(26), 4213–4225 (2002)
32. Kerner, T.E., Paulsen, K.D., Hartov, A., Soho, S.K., Poplack, S.P.: Electrical impedance spectroscopy of the breast: clinical imaging results in 26 subjects. IEEE Trans. Med. Imaging **21**(6), 638–645 (2002)
33. Kertész, Á., Hlaváčová, Z., Vozáry, E., Staroňová, L.: Relationship between moisture content and electrical impedance of carrot slices during drying. Int. Agrophys. **29**(1), 61–66 (2015)
34. Kiryakova, V.S.: Generalized Fractional Calculus and Applications. Longman Scientific & Technical, Harlow (1994)
35. Kuang, W., Nelson, S.: Dielectric relaxation characteristics of fresh fruits and vegetables from 3 to 20 GHz. J. Microw. Power Electromagn. Energy **32**(2), 115–123 (1997)
36. Laufer, S., Ivorra, A., Reuter, V.E., Rubinsky, B., Solomon, S.B.: Electrical impedance characterization of normal and cancerous human hepatic tissue. Physiol. Meas. **31**(7), 995 (2010)
37. Lopes, A.M., Machado, J.T.: Analysis of temperature time-series: embedding dynamics into the MDS method. Commun. Nonlinear Sci. Numer. Simul. **19**(4), 851–871 (2014)
38. Lopes, A.M., Machado, J.T.: Fractional order models of leaves. J. Vib. Control **20**(7), 998–1008 (2014)
39. Lopes, A.M., Machado, J.T.: Modeling vegetable fractals by means of fractional-order equations. J. Vib. Control **22**(8), 2100–2108 (2016)
40. Lopes, A.M., Machado, J.T., Ramalho, E.: On the fractional-order modeling of wine. European Food Research and Technology, pp. 1–9 (2016)
41. Machado, J., Lopes, A., Duarte, F., Ortigueira, M., Rato, R.: Rhapsody in fractional. Fract. Calc. Appl. Anal. **17**(4), 1188–1214 (2014)
42. Machado, J., Mata, M.E., Lopes, A.M.: Fractional state space analysis of economic systems. Entropy **17**(8), 5402–5421 (2015)
43. Machado, J.A.T., Lopes, A.M.: Analysis and visualization of seismic data using mutual information. Entropy **15**(9), 3892–3909 (2013)
44. Machado, J.T.: Matrix fractional systems. Commun. Nonlinear Sci. Numer. Simul. **25**(1), 10–18 (2015)
45. Magin, R.L.: Fractional Calculus in Bioengineering. Begell House Redding (2006)
46. Maundy, B., Elwakil, A., Allagui, A.: Extracting the parameters of the single-dispersion Cole bioimpedance model using a magnitude-only method. Comput. Electron. Agric. **119**, 153–157 (2015)
47. Nigmatullin, R., Nelson, S.: Recognition of the "fractional" kinetics in complex systems: dielectric properties of fresh fruits and vegetables from 0.01 to 1.8 GHz. Signal Process. **86**(10), 2744–2759 (2006)
48. Nigmatullin, R., Ryabov, Y.E.: Cole-Davidson dielectric relaxation as a self-similar relaxation process. Phys. Solid State **39**(1), 87–90 (1997)
49. Novikov, V., Wojciechowski, K., Komkova, O., Thiel, T.: Anomalous relaxation in dielectrics. Equations with fractional derivatives. Mater. Sci.-Pol. **23**(4), 977 (2005)
50. de Oliveira, E.C., Mainardi, F., Vaz Jr., J.: Models based on Mittag-Leffler functions for anomalous relaxation in dielectrics. Eur. Phys. J. Spec. Top. **193**(1), 161–171 (2011)
51. Riul, A., de Sousa, H.C., Malmegrim, R.R., dos Santos, D.S., Carvalho, A.C., Fonseca, F.J., Oliveira, O.N., Mattoso, L.H.: Wine classification by taste sensors made from ultra-thin films and using neural networks. Sens. Actuators B: Chem. **98**(1), 77–82 (2004)
52. Rosa, E.C., de Oliveira, E.C.: Relaxation Equations: Fractional Models. arXiv:1510.01681 (2015)

53. Sibatov, R.T., Uchaikin, D.V.: Fractional relaxation and wave equations for dielectrics characterized by the Havriliak-Negami response function. arXiv:1008.3972 (2010)
54. Stanislavsky, A., Weron, K., Trzmiel, J.: Subordination model of anomalous diffusion leading to the two-power-law relaxation responses. EPL (Europhys. Lett.) **91**(4), 40,003 (2010)
55. Tarasov, V.E.: Fractional equations of Curie–von Schweidler and Gauss laws. J. Phys. Condens. Matter **20**(14), 145,212 (2008)
56. Tarasov, V.E.: Universal electromagnetic waves in dielectric. J. Phys. Condens. Matter **20**(17), 175,223 (2008)
57. Tarasov, V.E.: Fractional integro-differential equations for electromagnetic waves in dielectric media. Theoret. Math. Phys. **158**(3), 355–359 (2009)
58. Tomkiewicz, D., Piskier, T.: A plant based sensing method for nutrition stress monitoring. Precis. Agric. **13**(3), 370–383 (2012)
59. Vosika, Z., Lazarević, M., Simic-Krstić, J., Koruga, D.: Modeling of bioimpedance for human skin based on fractional distributed-order modified Cole model. FME Trans. **42**(1), 74–81 (2014)
60. Watanabe, K., Taka, Y., Fujiwara, O.: Cole-Cole measurement of dispersion properties for quality evaluation of red wine. Meas. Sci. Rev. **9**(5), 113–116 (2009)
61. Watanabe, T., Orikasa, T., Shono, H., Koide, S., Ando, Y., Shiina, T., Tagawa, A.: The influence of inhibit avoid water defect responses by heat pretreatment on hot air drying rate of spinach. J. Food Eng. **168**, 113–118 (2016)
62. West, A.R., Sinclair, D.C., Hirose, N.: Characterization of electrical materials, especially ferroelectrics, by impedance spectroscopy. J. Electroceram. **1**(1), 65–71 (1997)
63. Zhang, L., Shen, H., Luo, Y.: A nondestructive method for estimating freshness of freshwater fish. Eur. Food Res. Technol. **232**(6), 979–984 (2011)
64. Zhang, M., Repo, T., Willison, J., Sutinen, S.: Electrical impedance analysis in plant tissues: on the biological meaning of Cole-Cole α in Scots pine needles. Eur. Biophys. J. **24**(2), 99–106 (1995)
65. Zheng, S., Fang, Q., Cosic, I.: An investigation on dielectric properties of major constituents of grape must using electrochemical impedance spectroscopy. Eur. Food Res. Technol. **229**(6), 887–897 (2009)

Dynamics of Particles and Bubbles Under the Action of Acoustic Radiation Force

Lev A. Ostrovsky and Yury A. Stepanyants

1 Introduction

The acoustic radiation force (ARF) is the period-average force created by a sound wave. The studies of ARF have a rich history associated with such names as Lord Rayleigh, Carl and Vilhelm Bjerkneses, Leon Brillouin, Paul Langevin, and others. The story commenced from the concept of *radiation pressure* which was known for electromagnetic waves and was extended in early 1900s by Rayleigh to any vibrations, including acoustic waves. Later Langevin suggested a somewhat different expression for the radiation pressure; both formulae derived by Rayleigh and Langevin are being used now for the corresponding conditions (see, e.g., [7]). A historical outline of this early development is given by Beyer [6].

In 1920s Wood and Loomis experimentally demonstrated that the acoustic force is capable to support glass objects in an oil bath [47, 48]. They were also apparently the first to predict possible biological and chemical effects of ultrasonics (see, e.g., [28] and references therein).

Another line of research refers to the interaction of bubbles with acoustic field. As shown by Carl Bjerknes in the end of XIX century, sound wave exerts a force on a small bubble, and two pulsating bubbles can attract or repel each other, depending on the phases of their oscillations [32].

The progress in the study of particle dynamics under the action of ARF is related with the seminal works of the frontier of 50s–60s by Yosioka and Kavasima [50] and Gor'kov [17] who derived an expression for the ARF exerting on a small particle of an arbitrary density based on the summing up of the incident and scattered acoustic fields.

Another line of research refers to the interaction of bubbles with an acoustic field. As shown by Carl Bjerknes in the end of XIX century, a sound wave exerts a

L.A. Ostrovsky (✉)
University of Colorado, Boulder, USA
e-mail: lev.ostrovsky@gmail.com

Y.A. Stepanyants
University of Southern Queensland, Toowoomba, Australia
e-mail: Yury.Stepanyants@usq.edu.au

M. Edelman et al. (eds.), *Chaotic, Fractional, and Complex Dynamics: New Insights and Perspectives*, Understanding Complex Systems,
https://doi.org/10.1007/978-3-319-68109-2_11

force on a small bubble, and two pulsating bubbles can attract or repel each other, depending on the phases of their oscillations [32]. In 1970s–1980s a more complex dynamics of bubble ensembles interacting with acoustic field was considered [20]. A self-consistent system including the kinetic equation for the bubble distribution and the acoustic field affected by the bubbles was derived (a similar but different kinetic equation is well-known for plasma in the electromagnetic field). New acoustic effects such as the self-focusing [5] and "self-transparency" [21] of sound waves in the fluid with gaseous bubbles were found and then, were confirmed experimentally. A mathematical model for the many-bubbles interaction (without their coalescence) based on the Lagrangian approach was developed in Ref. [14]. Various optical-acoustical analogies such as stimulated scattering and wave front reversal were discussed in 1980s (see, e.g., [11]); we do not consider these effects in details, because they are not directly related to the ARF action. For an outline of nonlinear acoustical phenomena associated with the bubbles we refer to the book [26].

In the subsequent decades, the ARF-related effects have found many important applications in medical diagnostics (see [22, 37, 45] and references therein), and recently also in food quality control as briefly described below.

Here we outline a class of effects related to the ARF action on particles and gaseous bubbles in liquids. This is one of the most promising areas of applications in biology and chemistry (such as concentration, separation, and stirring of small solid particles and gaseous bubbles in ultrasonic resonators). Also, in the context of this book, complex motion of particles and bubbles in acoustic fields is a challenging problem of nonlinear dynamics.

In what follows we consider some recent theoretical and experimental results in this area. Note that some important theoretical problems remained unresolved until very recently. In particular, the theory of these processes is mainly limited to the motion of a single particle due to the ARF. However, to evaluate the result of the action of ARF on an ensemble of many particles, their collective behavior should be analyzed as well, including the last stages of concentration when the particles begin effectively interact. It is also of interest to extend the range of particle parameters to include inorganic micro- and nanoparticles which can be used in modern technology, e.g., such as ultrasonic cleaning and delivery of drugs to specific organs in vivo.

Finally, in some cases, a commonly used model describing a balance between the ARF and viscous Stokes drag force can be insufficient. In such cases, the equation of particle motion should be supplemented by the inertial force with the added mass effect, as well as the history integral drag (HID) force reducible to the Boussinesq–Basset drag (BBD) force for solid particles.

The aim of this Chapter is threefold. First, to overview the studies of particle and bubble motions under the action of ARF. Second, to describe recent experiments in acoustic resonators of a plane and cylindrical configurations. Third, to present some new models and ideas of potential ARF applications in technology.

2 Dynamics of Particles in an Acoustic Field

According to Gor'kov (1962), the force acting in a fluid on a small spherical particle of a radius a in an acoustic field can be represented as

$$\mathbf{F} = -\nabla U, \quad \text{where} \quad U = 2\pi\rho_f a^3 \left(\frac{\langle p^2 \rangle}{3\rho_f^2 c_f^2} f_1 - \frac{\langle u^2 \rangle}{2} f_2 \right). \tag{1}$$

Here and further the angular brackets $\langle \ldots \rangle$ denote a period averaged quantities, in particular, the incident sound wave pressure p and velocity u, and

$$f_1 = 1 - \frac{1}{qs^2}, \qquad f_2 = 2\frac{q-1}{2q+1}, \tag{2}$$

where $q = \rho_p/\rho_f$, $s = c_p/c_f$, and subscripts f and p pertain to the density ρ and sound speed c in the ambient fluid and particle material respectively.

Equation (1) is based on the dipole scattering of a sound wave on a spherical particle. If the particle is non-spherical, only the factor in front of brackets in Eq. (1) for the potential, U would change.

The particle velocity $\mathbf{u}$ in the quiescent fluid can be found from the equation of motion:

$$\left(\rho_p + m\rho_f\right) V \frac{d\mathbf{u}}{dt} = -6\pi\rho_f \nu a \, \mathbf{u} - I - \mathbf{F}, \tag{3}$$

where V is the particle volume, m is the geometric factor of a particle which accounts for the added mass effect (e.g., for a spherical particle $m = 1$), ν is the kinematic viscosity of an ambient fluid, and $\mathbf{I}$ is the integral Boussinesq–Basset drag (BBD) force (it will be specified below).

For small particles the inertial and BBD forces in Eq. (3) can be neglected in most cases (see, however, some special cases below when these forces can be important), so that the velocity can be found from the balance between the radiation force and Stokes drag force:

$$\mathbf{u} = -\frac{\mathbf{F}}{6\pi\nu\rho_f a}. \tag{4}$$

Since in most cases Eq. (4) works well enough for small particles, we begin with the consideration of this simplified case.

2.1 A Particle in a Standing Wave

In many experiments, particles are operated in acoustic resonators where a standing wave produces minimums and maximums of average force potential. In a plane standing wave defined by the potential

$$\varphi(x,t) = -\frac{u_0}{k}\cos\omega t\cos kx \tag{5}$$

the radiation force has the form

$$F(x) = \frac{VkP_0^2}{2\rho_f c_f^2}\Phi\sin 2kx, \qquad \Phi = \frac{q+(2/3)(q-1)}{2q+1} - \frac{1}{2qs^2}, \tag{6}$$

where $P_0 = \rho_f c_f u_0$ is the maximal pressure amplitude in a standing wave.

Using Eq. (6), we obtain from Eq. (4) for the particle velocity

$$u(x) = -\frac{F}{6\pi\rho_f \nu a} = -\frac{2\pi a^2 f\, P_0^2\Phi}{9\nu\rho_f^2 c_f^3}\sin\frac{4\pi f\, x}{c_f}, \tag{7}$$

where f is the sound frequency.

Hence, depending on the sign of parameter σ, the particle is at rest in the nodes or antinodes of the acoustic field. Considering (7) as a differential equation for the coordinate x (i.e., substituting $x = du/dt$), it is easy to obtain a solution for a particle motion:

$$x = k^{-1}\arctan\left[\tan kx_0\exp\left(-2k\Phi t\right)\right], \tag{8}$$

where x_0 is the initial position of a particle at $t = 0$. Figure 1 illustrates return of a hard particle coordinate from the initial distance $x_0 = 0.15\lambda$, where λ is the wavelength of a sound, to the closest equilibrium position in a standing wave.

The trend of the particles towards the equilibrium can be used for concentration of them in certain areas of a resonator. Moreover, as mentioned above, the particles with a small and large densities can gather in different regions of a resonator (nodes and antinodes of ultrasonic pressure amplitude). Such effects were considered in the

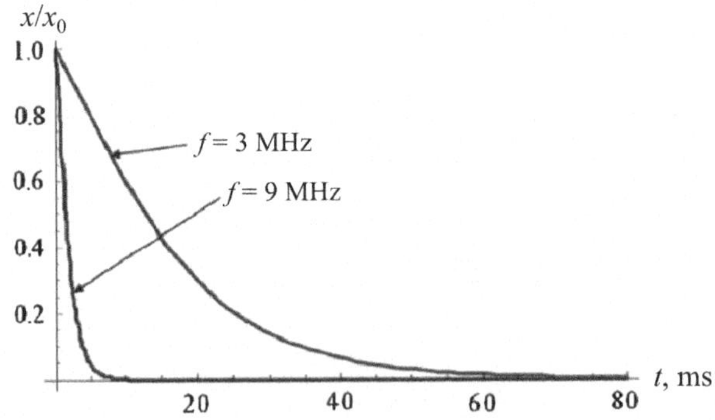

Fig. 1 Variation of particle coordinate from a distance $x_0 = 0.15\lambda$ to the equilibrium at the acoustic wave frequencies of 3 and 9 MHz (from [36]). Particle radius $a = 10\,\mu$m, pressure amplitude $P_0 = 200$ kPa

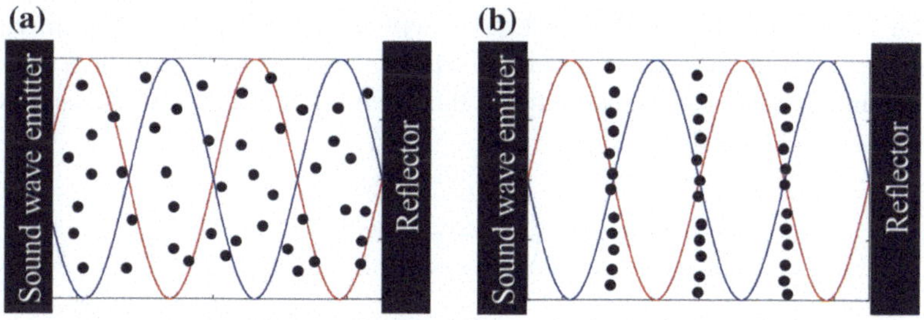

Fig. 2 Schematic illustration of particles gathering in the nodes of a standing acoustic wave in a plane resonator with initially randomly distributed particles (from panel **a** to panel **b**). Lines show a standing acoustic wave in two instances of time when the wave phase differs on π

review papers by Wiklund and Hertz [45] and Kuznetsova and Coackley [22]. One illustration is shown in Fig. 2.

2.2 *Particle Control by Means of Switching Modes*

As mentioned, a particle under the influence of acoustic radiative force moves to a stable equilibrium state which can be a node or antinode of a standing wave of pressure depending on the particle property. If the frequency of the acoustic wave in a resonator is changed, then the configuration of the standing field changes according to the mode structure and therefore, the particle position can be switched to a new equilibrium. Sarvazyan and Ostrovsky [36] considered switching between two different modes (3rd and 4th) of a resonator by changing the excitation frequency as a way of stirring and mixing of fluids in small volumes. In this particular case, the dimensionless radiation force has the form

$$F_a = \left[\sin\left(n_3 K \xi\right) e^{-2b\gamma\tau} + \sin\left(n_4 \frac{f_4}{f_3} K \xi\right) \left(1 - e^{-b\gamma\tau}\right)^2 \right] \Delta H, \tag{9}$$

where $\xi = 2\pi x/KL, \tau = t/\gamma, \gamma = a^2/9\nu$, $K \frac{2\pi^2 a^4 P_0^2 f}{81 \rho_f^2 \nu^2 c_f^3 L} \Phi(q, s), n_3 = 3$ and $n_4 = 4$ are the mode numbers, f_3 and f_4 are the mode frequencies, b is the damping rate of the mode which is caused by the dissipation of acoustic field in the fluid and energy losses through the resonator walls (for simplicity it is assumed that b does not depend on frequency), $\Delta H = H(\pi) - H(\pi - \pi_s)$, $H(\pi)$ is the Heaviside unit function, and π_s is the time interval during which the third mode decays and the fourth mode increases. In the next time interval π_s we assume that due to the mode switching the fourth mode decays, and the third mode increases. Then the switching repeats periodically with the same time interval π_s. As the initial condition, we assume that at $t = 0$ a particle starts moving from the rest being at the initial position $x = x_0$.

When the parameter $b \neq 0$, then Eq. (3) with F_a from Eq. (9) is non-integrable even in the quasi-static approximation. Only the approximate analytical solutions can be obtained in this case separately for the decaying mode 3 and increasing mode 4 (or vice versa) [18, 36]:

$$\xi_d = \frac{2}{Kn_3} \mathrm{atan} \left[\tan\left(\frac{Kn_3}{2}\xi_0\right) \exp\left(\frac{Kn_3}{2b\gamma} e^{-2b\gamma\tau}\right) \right], \tag{10}$$

$$\xi_i = \frac{2f_3}{Kf_4 n_4} \mathrm{atan} \left\{ \tan\left(\frac{Kf_4 n_4}{2f_3}\xi_0\right) \exp\left[\frac{Kf_4 n_4}{f_3}\left(\tau + \frac{4e^{-b\gamma\tau} - 3 - e^{-2b\gamma\tau}}{2b\gamma}\right)\right]\right\}. \tag{11}$$

According to solution (10), a particle being initially at the position $\xi = \xi_0$ moves under the action of the acoustic field towards the equilibrium state at $\xi = 0$. If the switching between the modes occurs at $\tau = \tau_s$, then the particle starts moving to another equilibrium state $\xi_{eq} = \pi f_3/(Kf_4 n_4)$. The latter is illustrated by Fig. 3, where the switching time in the dimensional variables was chosen at $t_s =$ 180 ms ($\tan_s = 1.62 \times 10^4$).

In the frame (a) one can see a behaviour of a relatively heavy gold particle with $\sigma = 2.375$, and in the frame (b)—relatively light ice particle with $\sigma = 0.823$. Odd lines $1, 3$, etc., pertain to particle motion under the action of the third acoustic mode and even lines 2, 4, etc.—to particle motion under the action of the fourth acoustic mode. Dashed horizontal lines in Fig. 3 show the equilibrium position ξ_{eq} which a particle would attain if only the fourth mode acts on it.

2.3 Acoustic Levitation and Tweezers

A balance of different forces acting on a particle can be used for a sophisticated manipulation of particle position and motion. One of the interesting particular effects known for a relatively long time is the possibility of a particle or a liquid droplet to levitate in the gravity field (see, e.g., [29] and references therein). The last-mentioned can happen when the action of gravity is balanced by the ARF. By placing a reflector some distance away from a transducer, one can create a standing acoustic field in the vertical direction. Depending on the particle position, the acoustic wave can exert a constant downward or upward pressure. At some places below the standing field nodes where the pressure directed upward and the pressure gradient is strong enough to balance the pull of gravity, the particle or droplet can levitate as shown in the photo of Fig. 4.

Note that there are other, non-acoustic mechanisms which can also cause particle levitation. In particular, in Ref. [2] the authors have theoretically shown that particles can be dynamically levitated and lifted by interacting vortices even when their densities are of few orders of magnitude greater than the fluid density.

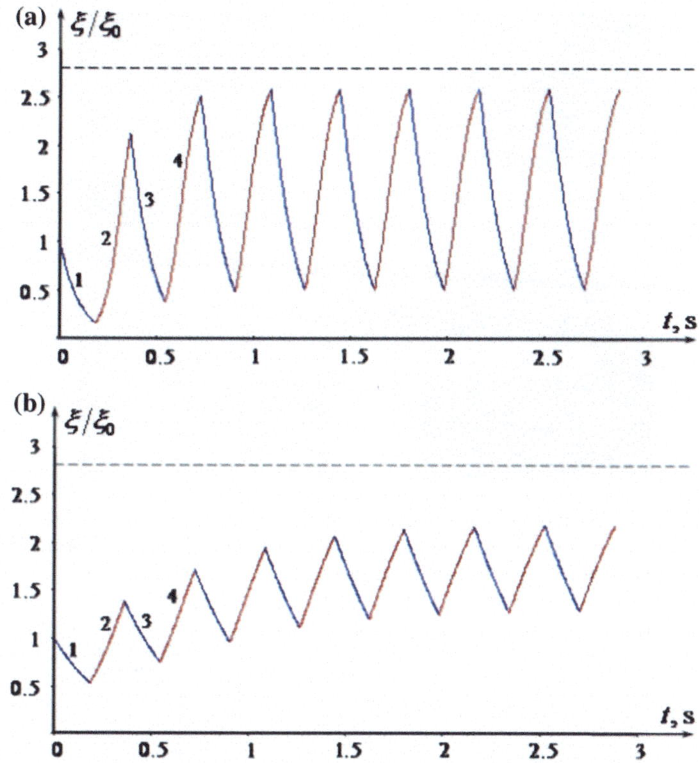

Fig. 3 Time dependences of a particle position when the acoustic field periodically switches from mode 3 to mode 4 and back. Panel **a**—for a gold particle, panel **b**—for an ice particle. From [18]

In other studies, particles were manipulated in a precise way by controlling the balance between ARF and friction as considered above. This technique is used in the so-called acoustic tweezers (see [1] and numerous references therein). Recently a similar process was realized by Ding et al. [12, 13] for surface waves in a lithium niobate piezoelectric over a fluid layer. Upon switching between excitation frequencies, the particles (10 μm fluorescent polystyrene beads) moved at the velocity ∼1.6 mm/s which corresponds to ARF of 150 pN. They also experimented with two-dimensional standing waves which allowed to manipulate particles and placing them in the prescribed positions. Figure 5 illustrates this possibility.

In the simplest implementation, the theoretical basis for tweezers' operation can be constructed as the two-dimensional generalisation of the model described above in this section. Creating two-dimensional standing acoustic wave by means of two sets of transmitter and reflector operating in the perpendicular directions, one can undertake the particle control on a plane by periodic switching between two modes in each direction. Then, the particle can drift in each direction and eventually oscillate, around the equilibrium position as shown in Fig. 3 for one-dimensional case. Obvi-

Fig. 4 Levitation of a small droplet in the standing acoustic field. The transducer below the droplet generates the acoustic wave which forms a standing acoustic field reflecting from the metallic cylinder shown on the top. From [46] (see also [35])

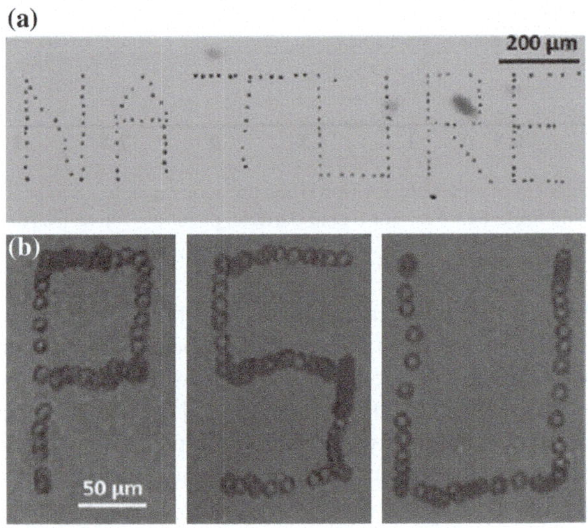

Fig. 5 **a** Stacked images demonstrating the particle motion control in x and y directions using 7 μm fluorescent polystyrene beads to write the word "NATURE". **b** Stacked images showing dynamic control of a bovine red blood cells to trace the letters "PSU" (Penn-State University). From [12]

ously, this arrangement can be further generalised for the three-dimensional case. The examples of numerous applications of acoustic tweezers can be found in [1].

In the two-dimensional case, a standing acoustic wave creates a cellular field, so that in Eq. (5) the force becomes a function of two variables:

$$F(x, y) = \frac{VP_0^2}{2\rho_f c_f^2}\Phi k_x k_y \sin(2k_x x)\sin\left(2k_y y\right). \tag{12}$$

As was shown in [15, 16] for hydrodynamic flows, the particle motion in a such 2D field can be fairly complex and even chaotic, especially if an initial particle position is close to the boundaries of cells. The particle can move within one cell, and at a certain condition after a while, it can jump to another cell and spend some time on it, then jump to another, and so on. To undertake an effective control of particle motion, it is necessary to study the conditions for the onset of regular and chaotic particle motions in such arrangement.

3 Cylindrical Resonators

In a certain number of cases, cylindrical resonators have a practical advantage as compared to plane resonators, due to the axial energy concentration [31, 33, 34].

In the case of a symmetric, circular geometry acoustic potential in the lowest mode of such a system has the form $\varphi = A\mathrm{J}_0(kr)e^{-i\omega t}$, where J_0, as well as J_1 below, are the Bessel functions of the first kind, r is the cylindrical coordinate, and k is the radial wave number. Then the radial velocity and pressure are

$$u = -\varphi_r = Ak\mathrm{J}_1(kr)e^{-i\omega t}, \quad p = \rho_f\omega\phi = i\rho_f\omega A\mathrm{J}_0(kr)e^{-i\omega t}. \tag{13}$$

Substituting this into Eq. (1), we obtain the radiation force:

$$F = \frac{8}{3}\frac{\pi^2 a^3 P_0^2 f}{\rho_f c_f^2} J_1(kr)\left[\Phi J_0(ka) - 3\frac{q-1}{2q+1}J_2(ka)\right]. \tag{14}$$

Here $P_0 = i\rho_f\omega A$ is the pressure amplitude at the axis ($r = 0$) of the resonator.

In the quasi-static approximation, the velocity field can be found from Eq. (3) in terms of hypergeometric function, but Eq. (3) can be easily integrated numerically. Figure 6 shows the radial distribution of potential energy U and particle velocity u for the excitation frequency $f = 1.5$ MHz.

The solution in the cylindrical case is not periodic in space. Therefore the trajectory and the final position of a particle depend on the initial coordinate ξ_0 more complexly than in the plane standing wave field. The equilibrium states can be found from Eq. (3) by equating to zero the function in the right-hand side of this equation. Figure 7 illustrates typical dependences of particle positions on time for a very heavy ($q \gg 1$)

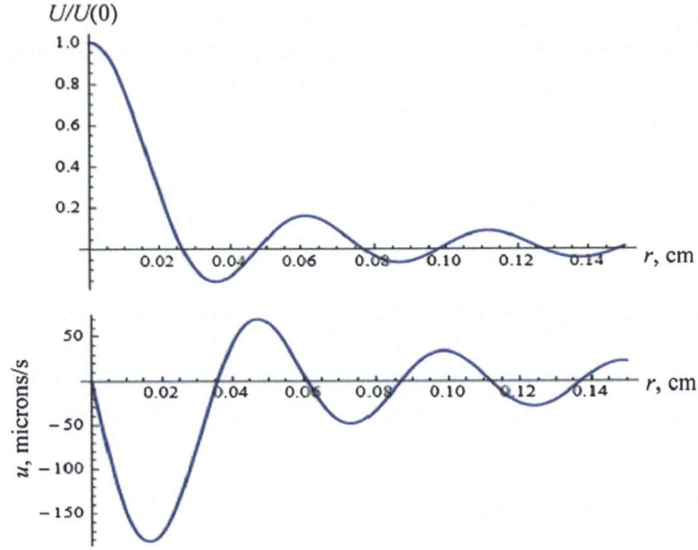

Fig. 6 Upper frame—radial distribution of ARF potential; lower frame—particle velocity distribution in a cylindrical resonator of radius 15 mm at $f = 1.5$ MHz. From [30]

and hard ($s \gg 1$) particle with $\sigma = 5/2$ in the field of the third radial mode of a cylindrical resonator.

Within the limits of this figure, the stable equilibrium states correspond to $\xi = 0$ and 2.36×10^4 (dashed line 2), and the unstable equilibrium states correspond to $\xi = 1.24 \times 10^4$ (dashed line 1) and 3.31×10^4 (dashed line 3). If the initial particle displacement is less than $\xi_1 = 1.24 \times 10^4$, then the particle moves to the center $\xi = 0$ (see three lines below the dashed line 1). If the initial particle displacement is between $\xi_1 = 1.24 \times 10^4$ and $\xi_3 = 3.31 \times 10^4$, then the particle moves to the next stable equilibrium state $\xi_2 = 2.36 \times 10^4$ (see three lines between the dashed lines 1 and 3), and so on.

One can easily show that the stable/unstable equilibrium states occur at the nodes of the acoustic force $F_a(r)$ as per Eq. (14) with the positive/negative gradient with respect to r. Thus, one can conclude that if there is an ensemble of particles randomly distributed in the cylindrical acoustic resonator, then under the action of a standing acoustic field they will be separated and bunched at certain nodes of the acoustic field. Similar results were obtained for other particles (gold, aluminium, silica, ice) with the finite values of q and s [18].

In the general case the acoustic pressure in a resonator can have a more complex structure in all three directions, radial, longitudinal, and angular:

$$p = PJ_0(k_r r)\cos(k_x x)e^{i(\omega t \pm m\phi)}, \qquad k = \frac{\omega_{res}}{c_0}, \quad m = 0,\ 1,\ 2,\ \ldots \tag{15}$$

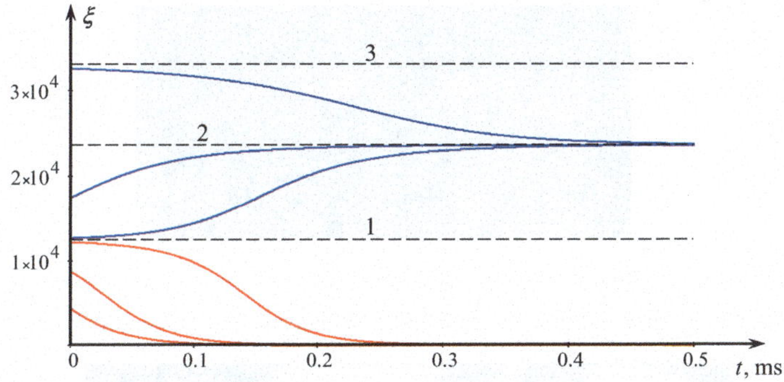

Fig. 7 Dependences of particle positions $\xi = x/x_0$ on time for different initial conditions in a cylindrical resonator. From [18]. See comments in the text

Fig. 8 Schematic of a standing wave beam in a cylindrical resonator. From [33]

To illustrate this, we show a few experimental pictures from [31, 33, 34]. The cylindrical resonators used there were fluid-filled tubes, typically of 5–10 mm long, which were excited at resonant frequencies of radial oscillations (sometimes by a focused transducer) (Fig. 8). One experimental result is shown in Fig. 9. The submicron particles (bacteria) were concentrated at the axis under the action of ARF, in agreement with the theory.

Figure 10 shows another result that refers to milk in a glass capillary excited at the first mode. Light fat is pushed to the walls whereas heavier somatic (protein) cells are gathered at the axis. This is presumably an efficient way to test liquid foods.

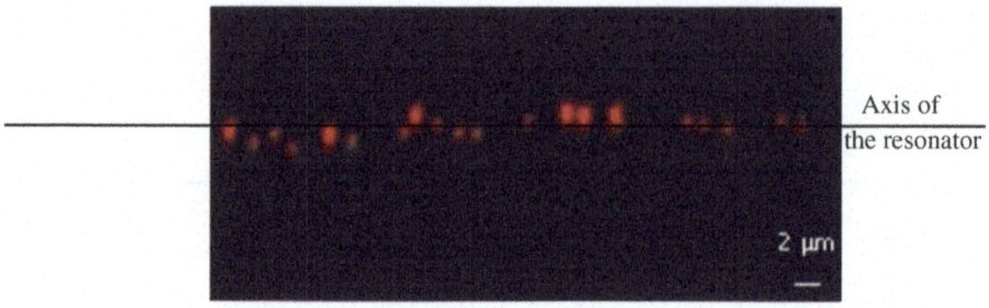

Fig. 9 Concentration of Salmonella bacteria in a cylindrical pipe resonator. Pipe radius is 9.85 mm, frequency $f = 2$ MHz. Scale bar is 2 microns. From [31]

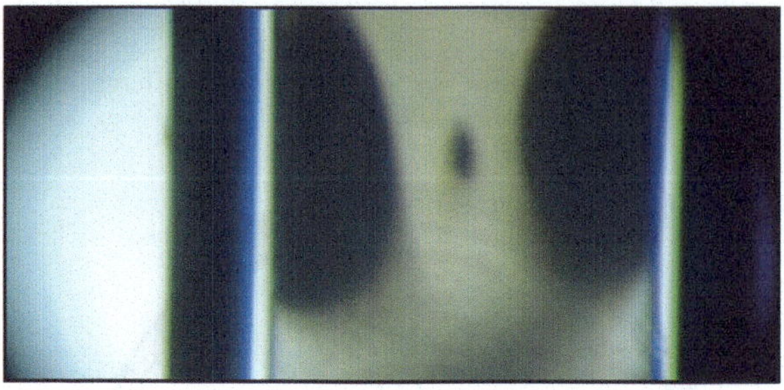

Fig. 10 Concentration of milk particles in a glass capillary of 0.4 mm diameter in which the first longitudinal mode is excited at frequency $f = 2$ MHz. Light fat is pushed to the wall whereas heavier somatic (protein) cells are concentrated on the axis. From [31]

4 Concentration Dynamics of Microparticles

In cases when the ARF is used to manipulate ensembles of particles, collective dynamics of particles is of a significant interest. Ostrovsky [30] considered the evolution of particle concentration in the plane and cylindrical standing waves. This allows one, in particular, to estimate the time necessary for a significant compression of the initially scattered particles. The basic equation for particle number density (concentration) C is

$$\frac{\partial C}{\partial t} + \nabla \cdot (\mathbf{u}C) = D\Delta C, \tag{16}$$

where $\mathbf{u}$ is the particle velocity vector as per Eq. (4) and D is the diffusion coefficient. Assuming that the distance between the particles is much larger than the particle radius a, one can use the expressions (1) and (4) obtained for a single particle.

One of the examples regarded a population of quasi-spherical particles (cells) which are initially almost uniformly distributed over the length of about 1.6 cm in a plane standing wave. The initial condition is prescribed in the form

$$\zeta(x, 0) = \frac{1}{2}\left[\tanh(8x + 6) - \tanh(8x - 6)\right], \tag{17}$$

where $\zeta = C/C_{max}$, where C_{max} is the peak value of the initial distribution.

To further specify the problem, consider particles (such as the biological cells) with the radius $a = 10\,\mu\text{m}$, density $\rho_s = 1.1\ \text{g/cm}^3$ and sound velocity $c_s = 1.575 \times 10^5$ cm/s. In this case the factor Φ in Eq. (6) is equal to 0.29. We assume also that the ultrasonic standing wave in a resonator has the frequency $f = 150$ kHz and pressure amplitude $P_0 = 200$ kPa. Then, from Eq. (6) we obtain $u = -0.0018\sin(4\pi x)$ cm/s. The diffusion coefficient is in this case $D = 2.2 \times 10^{-10}\,\text{cm}^2/\text{s}$. With these parameters Eq. (16) can be readily solved numerically. Figure 11 shows the distribution of concentration at three instants of time.

In this case, the amplitude of the particle velocity $u(x)$ defined by Eq. (7) and used for solving Eq. (1) is of the order of 10 μm/s; the time for strong concentration is of the order of tens of seconds which agrees with the available experimental data. For $f = 1.5$ MHz this time decreases to a few seconds.

For a cylindrical resonator Eq. (16) for the particle concentration in the axially symmetric case has the form

$$\frac{\partial C}{\partial t} + \frac{1}{r}\frac{\partial}{\partial r}(ruC) = D\frac{1}{r}\frac{\partial}{\partial r}\left(r\frac{\partial C}{\partial r}\right). \tag{18}$$

Defining the average velocity from Eqs. (4) and (14) and substituting into Eq. (18), we obtain a closed equation for the particle concentration C. One example in the form of the 3D plot of $\zeta(x, y)$ is shown in Fig. 12.

In Ref. [30] the distributions of a particle concentration were also calculated for the radial-spatial modes and for acoustic beams considered above. In all cases, the results are in a good agreement with the experimental data outlined above.

A more complex question concerns the maximal particle concentration achievable in the process. In particular, the Brownian diffusion coefficient D in Eq. (18) should establish a limit for this compression. However, in the considered cases the diffusion is significant only for the particles with a radius a less than one micron. Thus, in

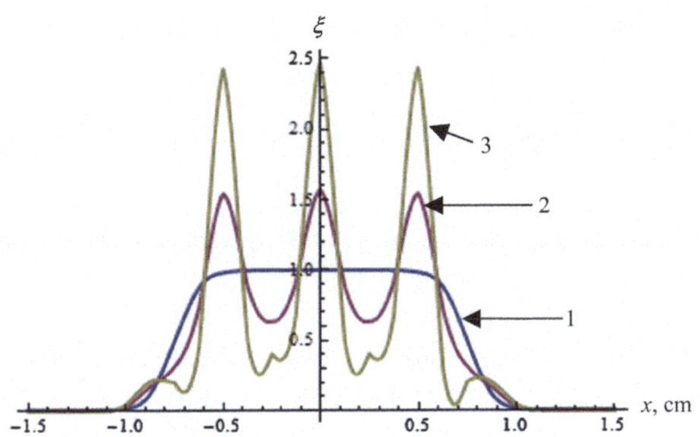

Fig. 11 Spatial distribution of particles defined by Eq. (14) at $t = 0$ (line 1), $t = 20$ s (line 2), and $t = 40$ s (line 3), for $f = 150$ kHz. From [30]

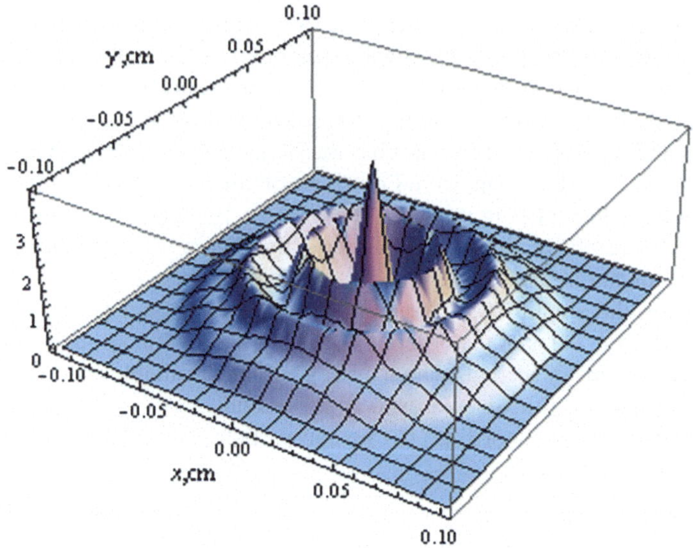

Fig. 12 The 3D representation of the particle density distribution $C(x, y)$ as per solution of Eq. (16) s for $f = 3$ MHz and $P_0 = 200$ kPa, at $t = 20$ ms

practice, the compression can be limited by just touching the particles or, at least, when particles approach each other at distances comparable with their diameters. The latter case was considered, e.g., in [40]. The authors have studied the acoustic interaction of two compressible spherical particles in an ideal fluid exposed to an external acoustic wave.

In most of the relevant experiments, the particles were eventually collected into a dense group as illustrated by the micro-photographs in Fig. 13. This figure shows the concentration of bacteria (submicron sizes) attached to latex beads (typical of immunoassays) by ultrasound.

5 Dynamics of Bubbles

The dynamics of microbubbles in the acoustic field is even more complex. There are two main reasons for that. Firstly, the bubble response to the oscillating field is strongly frequency-dependent and includes resonances. Secondly, whereas a solid particle in an acoustic field oscillates translationally, as a dipole, a small (compared to the wavelength) bubble most intensely pulsates radially, i.e., as a monopole. The average motion of a bubble under the action of radiation force is determined by the interaction between its monopole and dipole movements.

The classical Rayleigh–Plesset equation describes the monopole pulsations of a gas-filled bubble of a radius $a(t)$ (see, e.g., [10, 26]):

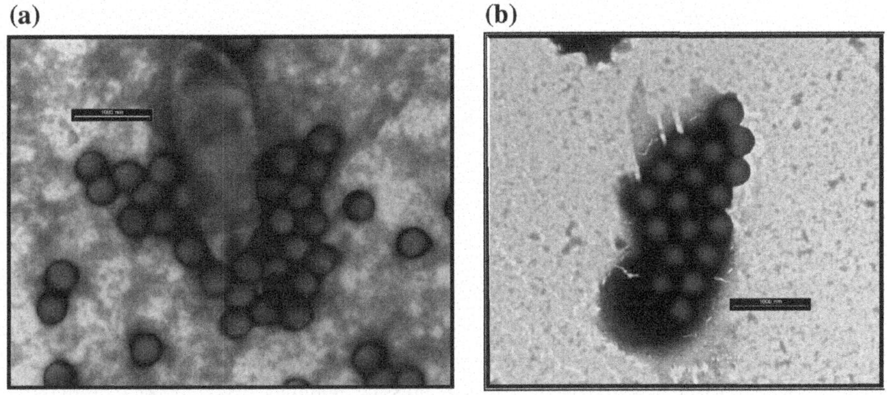

Fig. 13 Electron micrographs of USW-accelerated latex immunoassays of *Salmonella* (left) and *E. coli* (right). Ultrasound frequency $f = 2$ MHz, pressure $P = 100$ kPa, concentration time is 30–40 s (black horizontal stripes on both figures correspond to 1 μm length). See, e.g., [33]

$$a\ddot{a} + \frac{3}{2}\dot{a}^2 = \frac{1}{\rho_f}\left[p_0\left(\frac{a}{a_0}\right)^{3\gamma} - p_a\right]. \tag{19}$$

Here γ is the polytropic ratio of a gas inside the bubble, p_0 is static (e.g., atmospheric) pressure, a_0 is the equilibrium radius, and p_a is the external acoustic pressure in the absence of the bubble; dots on the top of the variable a stand for the time derivative.

Linearization of this equation yields that in the absence of an external field the bubble pulsates as a monopole, at the resonance frequency $\omega_0 = \sqrt{3\gamma p_0/\rho a_0^2}$ (the Minnaert formula [25], see also [26]). The bubble can also be involved in the translational (dipole-type) motion.

Interaction of these two types of motion, monopole and dipole, causes a significant specifics of the bubble case as regards the acoustic radiation force. In an acoustic field having a potential $\psi = \phi(\mathbf{r})e^{-i\omega t}$ the radiation force occurring as a result of this interaction [20] is $\langle \rho_f dV/dt \cdot \nabla\phi \rangle$, where V is the variable bubble volume. In the case of a small bubble it is again balanced by the viscous drag force:

$$\left\langle \frac{dV}{dt}\nabla\varphi \right\rangle = 4\pi\nu\mathbf{u}, \tag{20}$$

Note that the viscous force used here (the Hadamard–Rubzynski drag force—see, e.g., [24]) slightly differs from the Stokes drag force in Eq. (3) acting on a solid particle. This difference is caused by the different boundary conditions at the bubble surface in comparison with the solid surface; in the result, the coefficient 4 appears in Eq. (20) instead of 6 in Eq. (3).

From Eqs. (19) and (20) one can find the average bubble velocity under the action of the acoustic field:

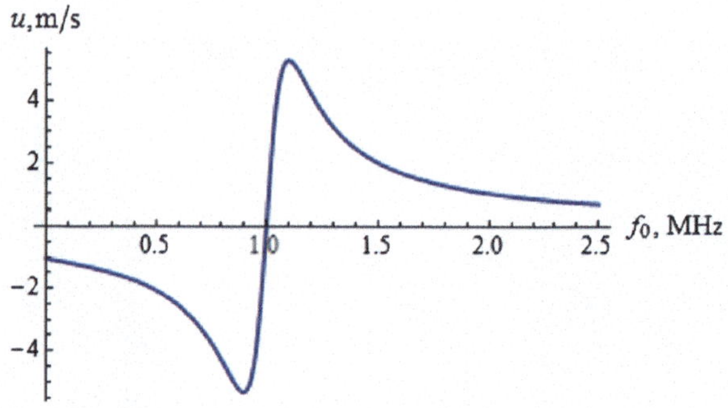

Fig. 14 The dependence of bubble velocity on the resonance frequency $f_0 = \omega_0/2\pi$. Here $Q = 10$, excitation frequency is $f = 1$ MHz and ultrasound pressure amplitude $P_0 = 200$ kPa. From [30]

$$u = \frac{1-\xi^2}{\left(1-\xi^2\right)^2 + Q^{-2}} \frac{\nabla |\phi|^2}{4\nu}, \tag{21}$$

where $\xi = \omega_0/\omega$, ω_0 is the bubble resonance frequency defined above, and Q is the quality factor of the bubble oscillating as a monopole. For example, in an acoustic beam, the bubble velocity is [20]:

$$u = \frac{1}{4\nu\left[\left(1-\xi^2\right)^2 + Q^{-2}\right]} \left[\left(\frac{kA^2}{Q}\right) n_x - \frac{1-\xi^2}{2} \frac{\partial A^2}{\partial r} n_\perp\right], \quad A = \frac{P_0}{k\rho c_f}. \tag{22}$$

Here $\mathbf{n}_x$ and $\mathbf{n}_\perp$ are the unit vectors directed along and across the beam axis respectively, and r is the radial distance from the beam axis. From here it follows that bubbles are pushed along the beam (x-axis). Simultaneously, bubbles smaller than the resonant ones (with $\omega_0 > \omega$) are drawn into the beam from the side, and vice versa.

In a 1D plane standing wave the period-averaged bubble velocity (21) is (cf. Eq. (7) for particles)

$$u = -\frac{P_0^2}{4\nu\omega_0\rho_f^2 c_f} \frac{\xi\left(1-\xi^2\right)}{\left(1-\xi^2\right)^2 + Q^{-2}} \sin 2kx. \tag{23}$$

Here the buoyant vertical motion is neglected. Evidently, the direction of the drift depends on both the bubble radius and the excitation frequency ω. Figure 14 shows that the bubble drift velocity can reach several m/s which by orders exceeds those of solid particles. Note that in this approximation, the exactly resonant bubbles with $f_0 = f$ do not drift at all.

Figure 15 illustrates a bubble motion towards the equilibrium from an initial non-equilibrium position. A comparison with Fig. 1 shows that even at a lower frequency (1 MHz vs. 3 MHz) a bubble returns to the equilibrium by at least two orders faster than a solid particle.

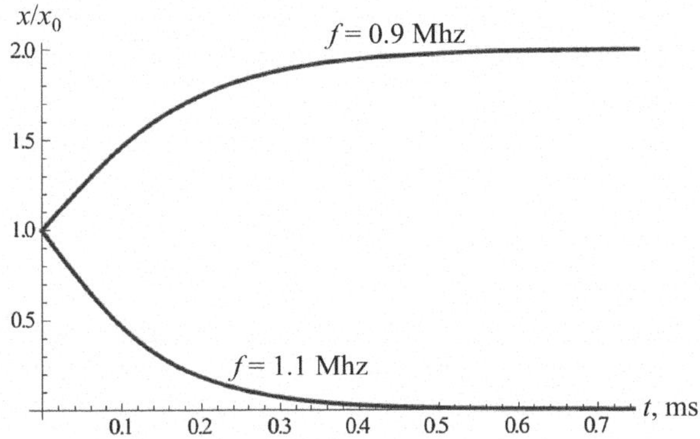

Fig. 15 The bubble coordinate variation in a standing wave from the initial deflection of $x_0 = 0.15\lambda$ from the nearest equilibrium for two ultrasonic frequencies, 0.9 and 1.1. MHz. The ultrasonic pressure amplitude is $P_0 = 100$ kPa, the bubble quality factor $Q = 10$. From [36]

Thus, if initially the bubble ensemble is randomly distributed in radii (as, for example, in the upper ocean layer), they will eventually be separated to the "small" and "large" groups under the action of an acoustic field.

5.1 *Interacting Bubbles*

At the final stage of concentration, when bubbles are close to each other (at distances of the order of their sizes) their motion radically changes since the oscillating bubbles begin to interact with each other via the secondary radiation force known as the Bjerknes force [8] (see also [23]), albeit the similar expressions were obtained independently in approximately the same time by Hicks and Pearson (see, e.g., [32] and references therein). For two pulsating bubbles (or other spherical particles) with volumes $V_1(t)$ and $V_2(t)$ this force is

$$F = \frac{\rho_f}{4\pi r_{12}^2}\left\langle \frac{dV_1}{dt}\frac{dV_2}{dt}\right\rangle. \tag{24}$$

Here $r_{12}(t)$ is the distance between the bubbles. The positive sign corresponds to the attractive force, and negative—to the repulsive.

As the result of such interaction, the bubbles can form clusters or even coalesce if they oscillate in phase. This effect was considered in [30] under a simplifying assumption that the bubbles are identical in radius, have almost isotropic distribution near the concentration point, and they are still well separated. At this stage, the bubble concentration varies as

$$C(t) = \frac{C(t_0)}{1 - 3\kappa C(t_0)t}, \quad \kappa = \frac{2R_0}{3\nu} \frac{|\phi|^2}{\left(1 - \xi^2\right)^2 + Q^{-2}}. \tag{25}$$

Here t_0 is the time when the Bjerknes force becomes dominating over the acoustically induced drift defined by Eq. (21). It should also be noted that, unlike the "rarefied" stage considered above, the interaction is the strongest for the resonant bubbles.

As one can see from Eq. (25), the bubble concentration becomes singular in a finite time, $t_{cr} = [3\kappa C(t_0)]^{-1}$. In reality, this is the time at which the bubbles either coalesce or, if the surface tension prevents it, form clusters of contacting bubbles. Both these possibilities were observed experimentally (see [21] and references therein).

As follows from (24), the bubbles can attract and repulse each other, depending on the relative phase shift between their pulsations. In particular, they can in principle interact being located in different half-waves of a standing wave. The expression (24) is valid in the quasi-static (near-field) approximation when a fluid disturbance produced by one oscillating particle impacts another particle without any time delay. This is correct if the distance between the particles is much less than the wavelength λ of acoustic wave in the surrounding fluid at the oscillation frequency f, i.e., $r_{1,2} \ll \lambda = c_f/f$.

If $r_{1,2} > \lambda$ the Bjerknes formula (24) requires modification, because the time delay occurs for the pressure perturbation produced by one particle to reach another one. Such modification has been done by Nemtsov [27] who derived the formula for the modified Bjerknes force. In the particular case of two bubbles of equal radii oscillating with the same amplitude $\varepsilon \ll a$ the formula is:

$$F_{1,2} = \frac{2\pi\rho_f \varepsilon^2 a^4 \omega^2}{r_{1,2}^2} \left[\cos\left(\Delta\phi - kr_{1,2}\right) - kr_{1,2} \sin\left(\Delta\phi - kr_{1,2}\right)\right], \tag{26}$$

where $k = 2\pi/\lambda$, $\Delta\phi$ is the relative phase of bubble oscillations.

The expression (26) reduces to the conventional formula for the Bjerknes force in the limit of $kr_{1,2} \to 0$. The modified formula (26) describes some new effects. Firstly, the interaction force between the bubbles decays slower with the distance: at large kr_{12} it is now proportional to r^{-1} rather than r^{-2} as in the quasi-static approximation. Secondly, the forces acting on two spheres are different, in general. Thirdly, the bubbles centre of mass may travel in the direction opposite to the radiation of a sound wave.

The interaction of two freely oscillating bubbles under the action of modified Bjerknes force has been studied in [41]. In that paper it was shown, in particular, that when the viscosity is neglected, two air bubbles in a water can (i) attract each other until they coalesce into one bubble, (ii) approach each other only up to a certain minimal distance and then go away from each other, or (iii) periodically approach to certain distance and move away. In the meantime, when the radiative effect is neglected, bubbles either monotonically draw together or move away from each other. This is illustrated by Fig. 16 where the phase planes are presented (the rate of change of a distance Δx between two bubbles as a function of Δx) for two bubbles oscillating in-phase and in antiphase with the conventional Bjerknes force (panels (a) and (b) respectively), and with the modified Bjerknes force, (panels (c) and (d)). As one can see, the portions of panels (c) and (d) are similar to (a) and (b) for small distances between the spheres $\Delta x < 100$.

In [20] a more general problem of mutual action of acoustic field and a bubble ensemble was considered. This is, however, beyond the scope of this review.

In conclusion to this section, we mention an interesting visual analogy between the modified Bjerknes force in fluids and in granular media that was observed by Shimada et al. [39]. The authors studied a model of periodic contraction and extension of two disks linked by a spring in a granular bed. They found optimal conditions that provide fluidization of sand beds and permit translational motion of the disks with a minimal resistance. Figure 17 schematically illustrates the disk motion and resembles the translational motion of pulsating spheres in a fluid under the action of modified Bjerknes force when a sound wave radiates in the opposite direction with respect to the translational motion. The authors hypothesize that this mechanism of "swimming" in a sand, is used by certain classes of reptiles.

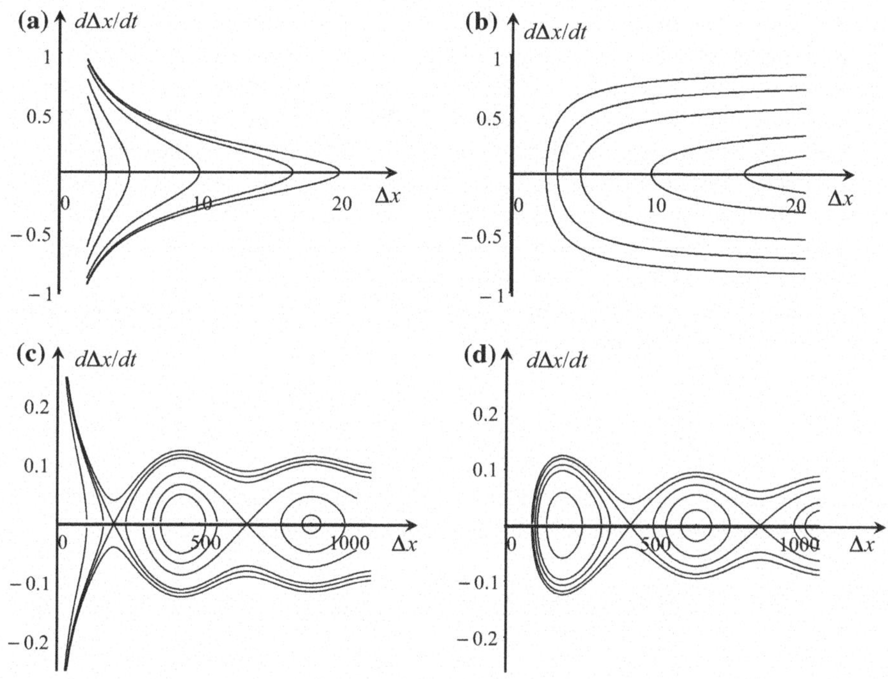

Fig. 16 A comparison of the phase planes for two bubbles oscillating in-phase (**a**) and (**c**) and in antiphase (**b**) and (**d**) with the conventional Bjerknes force (25) between them (**a**) and (**b**) and with radiation modified Bjerknes force (26) between them (**c**) and (**d**). From [41]

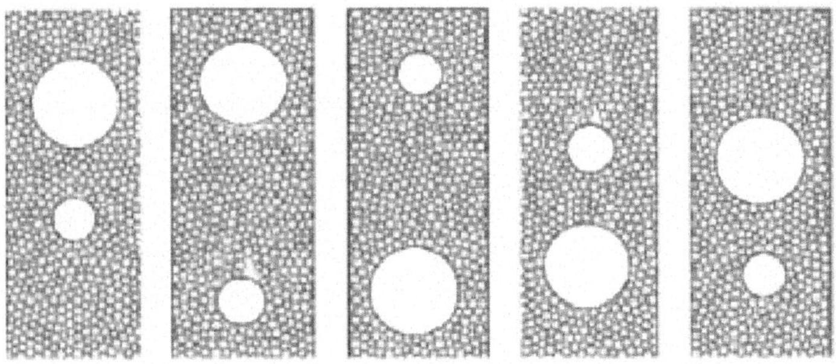

Fig. 17 The shapes of two disks at the end of each cycle of the motion tact. Movies are available on the web http://bopper.t.u-tokyo.ac.jp/~shimada/sandswim/movies.htm. The figure is taken from [39] and 90° rotated to save a space

6 The Effects of Memory and Inertia

As regards micro-particles, the above theory based on the balance between the ARF and viscous drag force is in most cases sufficient for the adequate quantitative description of the motion of particles and bubbles. However, in some special situations, additional dynamic factors can be important. This refers to the particle inertia and especially to the memory effect depending on viscosity, but also on the temporal rate of the process. The latter is the so-called Boussinesq–Basset drag (BBD) force. It was first considered by Boussinesq in 1885 [9] and Basset in 1888 [4] for small solid particles in a non-stationary flow with a small Reynolds number (the creeping flow regime). Here we take these effects into account.

The equation for a particle motion generalizing Eq. (3) and including the inertia (mass) and BBD integral term is

$$\left(\rho_p + \frac{\rho_f}{2}\right) V_p \frac{d^2\mathbf{r}}{dt^2} = -6\pi\rho_f \nu a \left(\frac{d\mathbf{r}}{dt} + \frac{a}{\sqrt{\pi\nu}} \int_{-\infty}^{t} \frac{d^2\mathbf{r}}{d\theta^2} \frac{d\theta}{\sqrt{t-\theta}}\right) + \mathbf{F}_a. \quad (27)$$

Substituting here the expression for the average acoustic force (6), produced by a plane standing wave, we obtain in the dimensionless variables:

$$(2r+1)\frac{d^2\xi}{d\tau^2} = -\frac{d\xi}{d\tau} - \frac{3}{\sqrt{\pi}} \int_{-\infty}^{\tau} \frac{d^2\xi}{d\vartheta^2} \frac{d\vartheta}{\sqrt{\tau-\vartheta}} - \sin(K\xi), \quad (28)$$

where $\xi = 4\pi x f / c_f$, $\tau = t/\gamma$, $\gamma = a^2/9\nu$, and $K = \frac{4\pi^2 a^4 P_0^2 f^2}{81 \rho_f^2 \nu^2 c_f^4} \Phi(r, s)$.

The examples of particle motion calculated on the basis of this equation are shown in Fig. 18 for different particles of the same radius $a = 10\,\mu$m in a resonator of the length $L = 2$ mm submerged in a water ($\rho_f = 10^3$ kg/m^3, $c_f = 1500$ m/s, $\nu = 10^{-6}$ m^2/s); the amplitude of acoustic pressure, $P_0 = 200$ KPa. The initial conditions were chosen the same as in Fig. 1 and as follows from Eq. (7), i.e., $x_0 = 0.15\lambda$ ($\xi_0 = 2.2 \times 10^3$ in dimensionless variables) and $u_0 \sim 1$ cm/s ($d\xi_0/d\tau = -0.1$).

It is seen that the quasi-static approximation is sufficient almost everywhere except for the final asymptotic, when ξ changes its sign because of the influence of the BBD force, i.e., the particle slightly passes the equilibrium position $\xi = 0$ and then returns to it. The qualitative difference in the effect of BBD force compared to the Stokes drag force, albeit it occurring only in a very small vicinity of equilibrium, is in the character of particle approach the equilibrium state. Under the action of Stokes drag force, only the particle approaches the equilibrium monotonically and exponentially quickly in time, whereas under the influence of BBD force it approaches the equilibrium non-monotonically and much slower, as $t^{-1/2}$ [42, 44].

Another difference occurs at the initial period of motion; this is depicted in Fig. 19. As one can see from this figure, the dependence of particle speed on the distance is well described in average by the approximate quasi-static formula (7). However, at the very beginning, when $\xi = \xi_0$, there is a significant difference in the character of particle motion (see panel b).

In general, as follows from the dimensionless equation (28), the first term in the left-hand side responsible for the Stokes drag force leads to the relatively quick relaxation of a motion to the "intermediate asymptotic" regime within the timescale of $\tau \sim 1$(or $t \sim \gamma = a^2/9\nu$). After that, the particle motion slowly relaxes to the final asymptotic state under the action of BBD force. This is underpinned by the exact analytical solution derived in Ref. [42] for the equation similar to (28), but with the gravity force instead of the acoustic radiative force in the right-hand side. Figure 20 from that paper illustrates the speed relaxation of a spherical aluminium

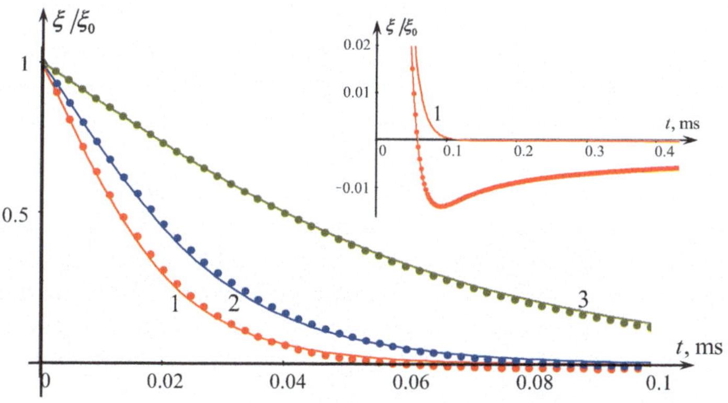

Fig. 18 The dependence of normalized particle position on the dimensional time as per approximate solution (7) (solid lines) for different particles. Line 1 pertains to the reference case of $\sigma = 5/2$ (very dense incompressible particle), line 2—to an aluminum particle ($\sigma = 1.777$), line 3—to an ice particle ($\sigma = 0.823$). Dots show the numerical data when all factors including the BBD and inertia forces were taken into account. The inset demonstrates the difference between the approximate (line 1) and numerical solutions (dots) for the reference case ($\sigma = 5/2$) when the particle approaches the equilibrium state. From [18]

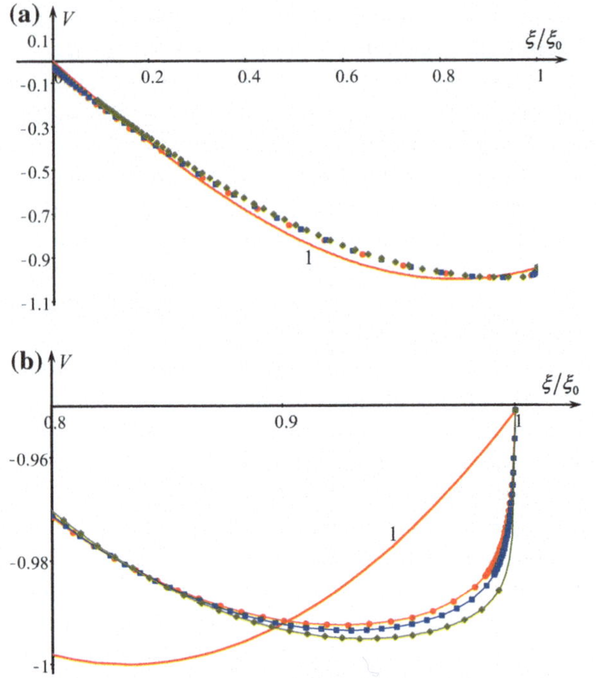

Fig. 19 The particle speed against the normalized distance as per quasi-static solution (7) (solid line 1). Symbols represent the numerical data as per Eq. (28) for different particles: dots—the reference case of a hard incompressible particle with $\Phi = 5/2$; squares pertain to the aluminium particle ($\Phi = 1.777$), and rhombuses relate to the ice particle ($\Phi = 0.823$). Panel **b** represents the magnified fragment of the panel (**a**) corresponding to the initial stage of particle motion. From [18]

particle of a radius $a = 44\,\mu$m and density $\rho_p = 2.7$ g/cm^3 falling down in a water of temperature $T = 45\,°$C and kinematic viscosity of $\nu_w = 6.05 \times 10^{-3}$ cm^2/s (such situation is typical to the pool vessel of a research nuclear reactor).

Lines 1–4 illustrate the dimensionless particle velocity versus dimensionless time for different initial conditions when both the Stokes and BBD forces were taken into account. Line 5 pertains to the case of exponential relaxation under the influence of Stokes drag force only. The dashed line 6 represents the "intermediate asymptotic" of particle motion $\upsilon(\tau) \sim \upsilon_t[1 - 3/(\pi\tau)^{1/2}]$, where υ_t is the terminal speed of particle motion in a water under the gravity force.

As one can see from this figure, in relatively short time a particle either attain the terminal speed, if the BBD force is neglected, or relaxes to the "intermediate asymptotic" (line 6), if the BBD force is taken into consideration. After that, the particle slowly approaches the terminal state and falls down in a viscous fluid (water) with the constant speed υ_t.

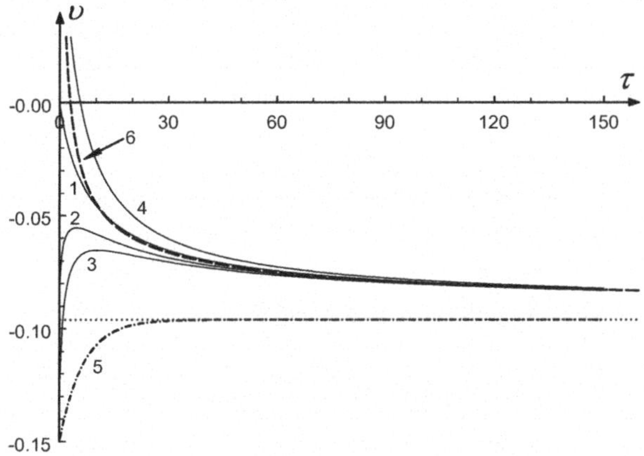

Fig. 20 Velocity versus time in dimensionless variables for a small aluminium particle. Line 1: $\upsilon_0 = 0$; line 2: $\upsilon_0 = \upsilon_t = -9.6 \times 10^{-2}$; line 3: $\upsilon_0 = -0.15$; line 4: $\upsilon_0 = 0.15$. Horizontal dotted line shows the terminal velocity. Lines 5 and 6 are explained in the text. From [42]

In conclusion to this section, we mention about works where the influence of intense sound waves on particle motion was studied (see, e.g., [3] and references therein). In particular, in Ref. [3] the authors investigated the effect of nonlinear steepening of an acoustic wave in a resonator on the dynamic of particles including the Boussinesq–Basset drag effect. It was also confirmed through the series of numerical calculations that the BBD force becomes important after a while when the relaxation to intermediate asymptotic regime occurs because of Stokes drag force. Apparently, further study of nonlinear effects on particle and bubble motion is required.

7 Conclusions

In this brief review, we intended to show that the period-average force exerted by acoustic waves on microparticles and bubbles can cause a complex dynamics resulting in a variety of non-trivial effects which have been harnessed to be used in biology, medicine, chemistry, food control, and others. It is believed that the number of applications of ARF in the aforementioned and new areas can increase in the future. As one more example of possible applications of ARF in medicine we can mention a promising perspective of transport of nanogels containing drugs to the specific organ. Recent invention of multi-shell hollow nanogels with a responsive shell permeability [38] makes them a convenient object for manipulation by ultrasound. This idea was mentioned in our paper [18] on the basis of modelling particle motion in an acoustic standing wave with mode switching (see Sect. 2.2). The extension of these results for the three-dimensional case can provide a version of acoustic tweezer for the delivery of micro-capsules containing a drug to the required position with the subsequent dissolution of a drug in the process of oscillation in the vicinity of that position (see Fig. 3). For other potential applications of ultrasound in medicine for drug and gene delivery (however, not related directly with the ARF), one can refer to the papers [19, 43].

In spite of a significant progress achieved in the last decades, there remain challenging problems, especially in what concerns collective dynamics of micro objects in the regime of high concentration when they strongly interact in the presence of external acoustic field. The acoustic levitation technique used to lift or drop a separate particle (see Fig. 4) can, apparently, be applied to lift up an ensemble of micro-particles up to creating a fluidised medium when the suspended micro-particles form a fluid-like state. The fluidised granular media have a wide spectrum of applications [49], and their micro-level analogue can supposedly find various applications too (for example, for the material transportation, polishing surfaces, intensification of combustion). A similar technique can be used for an ensemble of microbubbles preventing their floating up in a liquid. This can be used for clearing an upper layer of a fluid from gases or saturate lower layers by oxygen.

A theoretical description of clustering of particle and bubble ensembles needs, in general, the use of a kinetic approach, as it is common, for example, in plasma physics. In particular, for bubbles, the kinetic equation derived in [20] can be exploited. At the same time, the up-to-day computer algorithms allowing to follow trajectories of individual particles in the process of their interaction can be enormously helpful for understanding the behaviour of particle ensembles, including their stochastic motions. Nonetheless, all these results must be based on individual dynamics of particles and bubbles under the ARF action considered above, and we hope that these basic studies will stimulate further development.

References

1. Acoustic Tweezers. https://en.wikipedia.org/wiki/Acoustic_tweezers
2. Angilella, J.-R., Case, D.J., Motter, A.E.: Levitation of heavy particles against gravity in asymptotically downward flows. Chaos **27**(031103), 6p (2017)
3. Aksenov, A.V., Petrov, A.G., Shunderyuk, M.M.: The motion of solid particles in a fluid in a nonlinear ultrasonic standing wave. Doklady Phys. **56**(7), 379–384 (2011)
4. Basset, A.B.: On the motion of a sphere in a viscous liquid. Philos. Trans. R. Soc. Lond. A **179**, 43–63 (1888)
5. Belyaeva, I.Y.: Self-focusing effect in a liquid with gas bubbles. Acoust. Phys. **39**(6), 520–522 (1993)
6. Beyer, R.T.: Lord Rayleigh and nonlinear acoustics. J. Acoust. Soc. Am. **98**, 3032–3034 (1995)
7. Beyer, R.T.: Nonlinear Acoustics. Acoustical Society of America, NY (1997)
8. Bjerknes, V.: Vorlesungen über Hydrodynamische Ferncräfte. Leipzig (1900); Bjerknes, V.F.K.: Fielh of Force. Columbia University Press (1906)
9. Boussinesq, J.: Sur la resistance que oppose un liquide indefini en repos, sans pesanteur, au mouvement varie d'une sphere solide qu'il mouille sur toute sa surface, quand les vitesses restent bien continues et assez faibles pour que leurs carres et produits soient negligeables. Compt. Rend. Acad. Sci. Paris **100**, 935–937 (1885)
10. Brennen, C.E.: Cavitation and Bubble Dynamics. Oxford University Press (1995)

11. Bunkin, F.V., Kravtsov, Yu.A., Lyakhov, G.A.: Acoustic Analogues of Nonlinear-Optics Phenomena, vol. 29, no. 7, pp. 607–619 (1986)
12. Ding, X., Lin, S.-C. S., Li, S., Wang, L., Huang, T.J.: Manipulating single particles using standing surface acoustic waves. In: Proceedings of 16th International Conference on Miniaturized Systems for Chemistry and Life Sciences, Okinawa, Japan, Paper ID No. 1307 (2012)
13. Ding, X., Peng, Z., Lin, S.-C.S., Geri, M., Li, S., Li, P., Chen, Y., Dao, M., Suresh, S., Huang, T.J.: Cell separation using tilted-angle standing surface acoustic waves. PNAS **111**(36), 12992–12997 (2014)
14. Doinikov, A.A.: Mathematical model for collective bubble dynamics in strong ultrasound fields. J. Acoust. Soc. Am. **116**(2), 821–827 (2004)
15. Druzhinin, O.A., Ostrovsky, L.A., Stepanyants, YuA: Dynamics of particles in the steady flows of an inviscid fluid. Chaos **3**, 359–367 (1993)
16. Druzhinin, O.A., Ostrovsky, L.A.: The influence of Basset force on particle dynamics in two-dimensional flows. Phys. D **76**, 34–43 (1994)
17. Gor'kov, L.P.: On the forces acting on a small particle in an acoustic field in an ideal fluid. Sov. Phys. Dokl. **6**, 773–775 (1962)
18. Hassan, H.K., Ostrovsky, L.A., Stepanyants, Y.A.: Particle dynamics in a viscous fluid under the action of acoustic radiation force. Discontinuity Nonlinearity Complexity **6**(3) (2017)
19. Hynynen, K.: Ultrasound for drug and gene delivery to the brain. Adv. Drug. Deliv. Rev. **60**(10), 1209–1217 (2008)
20. Kobelev, YuA, Ostrovsky, L.A.: Nonlinear acoustic phenomena due to bubble drift in a gas-liquid mixture. J. Acoust. Soc. Am. **85**, 621–629 (1989)
21. Kobelev, YuA, Ostrovsky, L.A., Sutin, A.M.: Self-illumination effect for acoustic waves in a liquid with gas bubbles. JETP Lett. **30**, 395–398 (1979)
22. Kuznetsova, L.A., Coakley, W.T.: Applications of ultrasound streaming and radiation force in biosensors. Biosens. Bioelectron. **22**, 1567–1577 (2007)
23. Landau, L.D., Lifshitz, E.M.: Hydrodynamics, 4th edn. Nauka, Moscow (English, p. 1993. Fluid Mechanics, Pergamon Press, Oxford, Transl. 1988)
24. Lamb, H.: Hydrodynamics, 6th edn. Cambridge University Press, Cambridge (1932)
25. Minnaert, M.: On musical air-bubbles and the sound of running water. Philos. Mag. **16**(104), 235–248 (1933)
26. Naugolnykh, K., Ostrovsky, L.: Nonlinear Wave Processes in Acoustics. Cambridge University Press, Cambridge (2006)
27. Nemtsov, B.E.: Effects of radiation interaction of bubbles in a liquid. Sov. Tech. Phys. Lett. **9**, 368–369 (1983)
28. Nyborg, W.L.: Biological effects of ultrasound: development of safety guidelines. Ultrasound Med. Biol. **26**(6), 911–964 (2000)
29. Ochiai, Y., Hoshi, T., Rekimoto, J.: Three-dimensional mid-air acoustic manipulation by ultrasonic phased arrays. PLoS ONE **9**(5), e97590 (2014). https://doi.org/10.1371/journal.pone.0097590
30. Ostrovsky, L.: Concentration of microparticles and bubbles in standing waves. JASA **138**(6), 3607–3612 (2015)
31. Ostrovsky, L., Priev, A., Ponomarev, V., and Barenholz, Y.: Acoustic radiation force for rapid detection of particles in biological liquids. In: Proceedings of Acoustics Meeting, vol. 14 (162 ASA Meeting, San Diego, CA) (2011)
32. Pelekasis, N.A., Tsamopoulos, J.A.: Bjerknes forces between two bubbles. Part 1. Response to a step change in pressure. J. Fluid Mech. **254**, 467–499. Part 2. Response to an oscillatory pressure field. J. Fluid Mech. **254**, 501–527 (1993)
33. Priev, A., Barenholz, Y.: Ultrasonic food quality analyzer based on cylindrical standing waves. In: Proceedings of 20th International Congress on Acoustics, Sydney, Australia, 4 p (2010)
34. Priev, A., Sarvazyan, A.: Cylindrical standing wave resonator for liquid food quality control. J. Acoust. Soc. Am. **125**, 2593 (2009)
35. Sagoff, J.: No magic show: real-world levitation to inspire better pharmaceuticals. Argon National Laboratory, 12 September. http://www.anl.gov/articles/no-magic-show-real-world-levitation-inspire-better-pharmaceuticals (2012)

36. Sarvazyan, A., Ostrovsky, L.: Stirring and mixing of liquids using acoustic radiation force. J. Acoust. Soc. Am. **125**, 3548–3554 (2009)
37. Sarvazyan, P., Rudenko, O.V., Swanson, S.D., Fowlkes, J.B., Emelianov, S.Y.: Shear wave elasticity imaging—a new ultrasonic technology of medical diagnostics. Ultrasound Med. Biol. **24**, 1419–1435 (1998)
38. Schmid, A.J., Dubbert, J., Rudov, A.A., Pedersen, J.S., Lindner, P., Karg, M., Potemkin, I.I., Richtering, W.: Multi-shell hollow nanogels with responsive shell permeability. Sci. Rep. **6**, 22736 (2016)
39. Shimada, T., Kadau, D., Shinbrot, T., Herrmann, H.J.: Swimming in granular media. Phys. Rev. E **80**, 020301R (2009)
40. Silva, G.T., Bruus, H.: Acoustic interaction forces between small particles in an ideal fluid. Phys. Rev. E **90**(063007), 11p (2014)
41. Stepanyants, Y.A., Yeoh, G.H.: Interaction of gaseous bubbles under the action of radiation modified Bjerknes force. In: Proceedings of XXII ICTAM Congress, 25–29 August, 2008, Adelaide, Australia (2008)
42. Stepanyants, Y.A., Yeoh, G.H.: Particle and bubble dynamics in a creeping flow. Eur. J. Mech. - B/Fluids **28**, 619–629 (2009)
43. Timbie, K.F., Mead, B.P., Price, R.J.: Drug and gene delivery across the blood-brain barrier with focused ultrasound. J. Control. Release **219**, 61–75 (2015)
44. Visitskii, Ye.V., Petrov, A.G., Shunderyuk, M.M.: The motion of a particle in a viscous fluid under gravity, vibration and Basset's force. J. Appl. Math. Mech. **73**, 548–557 (2009)
45. Wiklund, M., Hertz, H.M.: Ultrasonic enhancement of beadbased bioaffinity assays. Lab Chip **6**, 1279–1292 (2006)
46. Wilson, T.V.: How acoustic levitation works. How Stuff Works, Science, Science, Physical Science, Acoustics. http://science.howstuffworks.com/acoustic-levitation.htm (2007)
47. Woo, J.: A short history of the development of ultrasound in obstetrics and gynecology. http://www.ob-ultrasound.net/history1.html
48. Wood, R.W., Loomis, A.L.: The physical and biological effects of high frequency sound waves of great intensity. Philos. Mag. **4**, 417–436 (1927)
49. Yang, W.-C.: Handbook of Fluidization and Fluid-Particle Systems. CRC Press, 878 pp (2003)
50. Yosioka, K., Kawasima, Y.: Acoustic radiation pressure on a compressible sphere. Acustica **5**(167–173), 24 (1955)

Nonequilibrium Quantum Dynamics of Many-Body Systems

Lea F. Santos and E. Jonathan Torres-Herrera

1 Introduction

Understanding the properties of interacting many-body quantum systems out of equilibrium is essential to various fields, from atomic, molecular, and condensed matter physics to quantum information and cosmology. Every step forward has the potential to enable new scientific and technological applications. Some examples are listed below.

(i) Nonequilibrium quantum dynamics may reveal new phases of matter that typically do not occur near equilibrium. New phases of matter are tightly connected with the development of new materials, which may revolutionize how we use and produce energy, may lead to new electronic devices, and may give rise to unforeseen innovations.
(ii) Efficient methods to store and transfer many-body quantum coherences are necessary for building analog and digital quantum simulators, developing quantum sensors, and realizing protocols for secure quantum communication.
(iii) One of the most challenging aspects for the development of new electronic devices, such as microchips and hard disks, is the mitigation of local heating. One needs to identify the conditions under which a quantum system can transfer heat rapidly.
(iv) Further progress in spintronic devices, where information can be transferred without any transport of charge, being therefore better protected against dissipation, requires improved techniques for the control and transport of spin excitations.

L.F. Santos (✉)
Department of Physics, Yeshiva University, New York, NY 10016, USA
e-mail: lsantos2@yu.edu

E.J. Torres-Herrera
Instituto de Física, Benemérita Universidad Autónoma de Puebla,
Apt. Postal J-48, Puebla, 72570 Puebla, Mexico
e-mail: etorresh@ifuap.buap.mx

M. Edelman et al. (eds.), *Chaotic, Fractional, and Complex Dynamics: New Insights and Perspectives*, Understanding Complex Systems,
https://doi.org/10.1007/978-3-319-68109-2_12

Since the detection of spin echoes in 1950 [1], the goal of studying nonequilibrium quantum dynamics in controllable scenarios became tangible [2, 3]. A great leap forward came with the Penning and Paul traps, by Hans Dehmelt and Wolfgang Paul, that eventually made possible the study of quantum dynamics of trapped ions [4–6]. Another revolution had as starting point the development and combination of several cooling techniques that culminated in highly controllable experiments with cold atoms, where the evolution of many-body quantum systems is observed for long times [7–9].

The questions that we have been interested in are motivated by those experiments and may also inspire new ones. We have been considering quantum systems with many interacting particles that are well isolated from any sort of environment. By this we mean that the couplings with the surroundings are very weak and can therefore be neglected. Interactions with an environment are unavoidable and do eventually kick in, but we assume that this happens at times much longer than the ones we deal with.

We focus on one-dimensional (1D) many-body quantum systems described by spin-1/2 models. These systems have only two-body interactions and are commonly studied by the experiments mentioned above. The models can also be mapped onto models of spinless fermions and hardcore bosons. In addition, by adjusting the parameters of the Hamiltonian we can cover different regimes (integrable, chaotic, or intermediate between the two), phases (metal vs. insulator, ferromagnetic vs. antiferromagnetic), symmetries, and strength of the interactions, which allows for the investigation of various different interesting scenarios.

We take the system far from equilibrium by perturbing it faster than any of its characteristic times, so that the perturbation can be seen as instantaneous, what is often referred to as "quench". In simple terms, the system is prepared in a nonstationary state. We consider pure states, but the analyses could certainly be extended to mixed states.

Our main goal has been to describe the dynamics of finite lattice many-body quantum systems at different time scales [10–37]. Because these systems are finite, their evolution eventually saturates to an equilibrium point. We were able to obtain a detailed description of the so-called survival probability (probability of finding the system still in its initial state later in time) from the moment the system is taken out of equilibrium all the way to the saturation of its evolution [24–36]. We have not only numerical results, but analytical expressions as well. The survival probability is a simple and insightful quantity. It is part of the equations that compute the temporal evolution of generic physical observables, so having a complete understanding of its evolution provides a better understanding of the dynamics of several observables. This is why we decided to restrict this chapter to this particular quantity and briefly mention others.

Our main findings for the survival probability, presented in this chapter, are enumerated below.

1. The decay of the survival probability depends on the strength of the perturbation. For very strong perturbations, when the energy distribution of the initial state is

unimodal, the decay is Gaussian and therefore faster than the usually expected exponential behavior. The Gaussian decay is related to the Gaussian density of states found in many-body systems with two-body interactions [26–29].
2. Exponential and Gaussian decays are not exclusive to chaotic models and occur also in integrable models perturbed far from equilibrium [26–28].
3. The speed of the decay depends on the energy of the initial state. The decay is faster for initial states with energy close to the middle of the spectrum, where there is a large concentration of eigenstates, than for states with energies near the border of the spectrum [26–28].
4. Decays faster than Gaussian occur when the energy distribution of the initial state is bimodal [28], in which case the quantum speed limit can be reached. Moving away from realistic systems, fast decays can be obtained by increasing the number of particles that interact simultaneously [26–28].
5. After the initial fast (often Gaussian) decay, the dynamics slows down and becomes power-law. The power-law exponent depends on how the spectrum approaches its energy bounds [24, 25] and on the level of delocalization of the eigenstates [31–33].
6. In interacting systems with onsite disorder, the value of the power-law decay exponent detects the transition from chaos to many-body localization. This exponent coincides with the fractal dimension of the system [31–33] and with the slope of the logarithmic growth of the Shannon and entanglement entropies [33].
7. At long times, after the power-law behavior and before saturation, the survival probability shows a dip below its infinite time average [35, 36]. This is known as correlation hole and appears only in systems with level repulsion (that is, not in integrable models). The correlation hole provides a way to detect level repulsion from the dynamics, instead of having to resort to the eigenvalues. This is useful for the experiments mentioned above, which have limited access to the spectra of their systems. Since the correlation hole is a general indicator of the integrable-chaos transition, it serves also as a detector of the metal-insulator transition in interacting systems [33, 35].

Additional highlights of our research, which are not described in this chapter, but may be found in our references, include the following topics.

1. The dynamical behavior of the Shannon entropy and entanglement entropy is equivalent [33, 34]. The first is easier to compute numerically and is potentially accessible experimentally, although it is the second that has been mostly studied theoretically.
2. Effects associated with the correlation hole are observed also in entropies [35] and in experimental observables, such as the spin density imbalance [36].
3. Analytical expressions for the entire evolution of the survival probability, Shannon entropy, and spin density imbalance can be found using full random matrices [34, 36]. Full random matrices are not realistic, but they provide bounds and serve as references to the studies of many-body quantum systems.

4. The behavior of the survival probability may signal the presence of an excited state quantum phase transition. It slows down as one approaches the critical point [20–22].
5. Long-range interactions do not always imply fast dynamics. Depending on the initial state the effects of the long-range couplings may get shielded, resulting in exceedingly slow evolutions [23].
6. Despite isolation, one can still talk about equilibration in isolated finite many-body quantum systems, but in a probabilistic sense. By this we mean that after a transient time, few-body observables simply oscillate around their infinite-time average, being very close to it for most time. To speak of equilibration, these temporal fluctuations need to be small and decrease with system size. In Ref. [37], we show that the size of these fluctuations decrease exponentially with system size in chaotic systems and also in interacting integrable models.
7. When the infinite-time averages of few-body observables are very close to microcanonical averages and the difference between the two decreases with system size, we say that the many-body quantum system has thermalized. We have several studies about how the onset of thermalization depends on the initial state and strength of the interactions [10, 30, 38–42].

The text below is divided in two sections. In Sect. 2 we provide a pedagogical introduction to the 1D spin-1/2 systems that we study and how to distinguish integrable from chaotic models. In Sect. 3, we present our results for the survival probability for short and long times, from perturbation to saturation.

2 Spin-1/2 Models

We investigate a 1D spin-1/2 system. To describe this chain, one uses spin operators $S^{x,y,z} = \sigma^{x,y,z}/2$, where

$$\sigma^x \equiv \begin{pmatrix} 0 & 1 \\ 1 & 0 \end{pmatrix}, \quad \sigma^y \equiv \begin{pmatrix} 0 & -i \\ i & 0 \end{pmatrix}, \quad \sigma^z \equiv \begin{pmatrix} 1 & 0 \\ 0 & -1 \end{pmatrix}$$

are the Pauli matrices and $\hbar$ is set to 1. The quantum state of the spin is represented by a two-component vector (spinor). This state is usually written in terms of the two eigenstates of S^z, which then form the basis. One eigenstate represents the spin pointing up in the z-direction and the other, the spin pointing down. They can be denoted as

$$|\uparrow\rangle = \begin{pmatrix} 1 \\ 0 \end{pmatrix}, \qquad |\downarrow\rangle = \begin{pmatrix} 0 \\ 1 \end{pmatrix}.$$

Since the eigenvalue associated with $|\uparrow\rangle$ is +1/2 and that of $|\downarrow\rangle$ is $-1/2$, we refer to the first as the excitation. The operators S^x and S^y flip the spin up and spin down,

$$S^x | \uparrow\rangle = \frac{1}{2} | \downarrow\rangle \qquad S^x | \downarrow\rangle = \frac{1}{2} | \uparrow\rangle$$

$$S^y | \uparrow\rangle = \frac{i}{2} | \downarrow\rangle \qquad S^y | \downarrow\rangle = -\frac{i}{2} | \uparrow\rangle.$$

Basis

In a chain with several sites, a commonly used basis in which to write the spin-1/2 Hamiltonian matrix corresponds to arrays where on each site the spin either points up or down in the z-direction, as for example $| \downarrow\uparrow\downarrow\uparrow\downarrow\uparrow\downarrow\uparrow \ldots\rangle_z$. This basis is often referred to as natural-basis, computational-basis or site-basis. We use the latter term.

Hamiltonian Terms

One of the terms that we find in spin-1/2 Hamiltonians is

$$H_Z = \sum_k h_k J S_k^z, \tag{1}$$

which appears when each site k is subjected to a different local magnetic field. The fields cause the Zeeman splittings of amplitude $h_k J$ on each site. The parameter J sets the energy scale and we choose $J = 1$. If all sites have $h_k = h$, that indicates a clean system, where a single magnetic field is applied to the entire chain. If only one site has a Zeeman splitting different from the others, we call it the defect site or the impurity of the system. If all sites have different Zeeman splittings, randomly distributed, then the system is disordered.

When more than one spin is present, they may interact. This may happen through the Ising interaction. If the interaction is active between nearest-neighbors (NN) only, that is sites k and $k+1$, it is given by

$$H_{ZZ} = \sum_k J\Delta S_k^z S_{k+1}^z, \tag{2}$$

where $J\Delta$ is the strength of the interaction. This terms causes a pair of adjacent parallel spins to have different energy from a pair of anti-parallel spins, because

$$J\Delta S_k^z S_{k+1}^z | \uparrow_k \uparrow_{k+1}\rangle = +\frac{J\Delta}{4} | \uparrow_k \uparrow_{k+1}\rangle, \tag{3}$$

while

$$J\Delta S_k^z S_{k+1}^z | \uparrow_k \downarrow_{k+1}\rangle = -\frac{J\Delta}{4} | \uparrow_k \downarrow_{k+1}\rangle. \tag{4}$$

The ground state of a Hamiltonian that has only the Ising interaction is ferromagnetic, with all spins aligned in the same direction, when $J\Delta < 0$, and it is antiferromagnetic, with antiparallel neighboring spins, when $J\Delta > 0$. We choose the latter.

Another term that appears in our Hamiltonians is the flip-flop term. It interchanges the position of neighboring up- and down-spins according to

$$J(S^x_k S^x_{k+1} + S^y_k S^y_{k+1})|\uparrow_k\downarrow_{k+1}\rangle = \frac{J}{2}|\downarrow_k\uparrow_{k+1}\rangle.$$

The NN flip-flop term couples site-basis vectors that differ only by the orientation of the spins in two neighboring sites. When the Hamiltonian matrix is written in the site-basis, the flip-flop term constitutes the off-diagonal elements.

Spin-1/2 Hamiltonian

The Hamiltonian that we consider is a combination of the terms described above. It is given by

$$H = dJS^z_{L/2} + \sum_{k=1}^{L} h_k J S^z_k \tag{5}$$
$$+ J\sum_k \left(S^x_k S^x_{k+1} + S^y_k S^y_{k+1} + \Delta S^z_k S^z_{k+1}\right) + \lambda J \sum_k \left(S^x_k S^x_{k+2} + S^y_k S^y_{k+2} + \Delta S^z_k S^z_{k+2}\right).$$

The chain has L sites and we denote by N_{up} the number of up-spins. The amplitude dJ indicates the Zeeman splitting of the defect site. The Zeeman splittings $h_k J$ correspond to onsite disorder caused by random static magnetic fields; h_k are random numbers from a uniform distribution in $[-h, h]$ and h is the strength of the disorder. Δ is the anisotropy parameter; when the Ising interaction and the flip-flop term have the same strength ($\Delta = 1$), the system is isotropic. λ is the ratio between the NN and next-nearest-neighbor (NNN) couplings.

Depending on the boundary conditions, we refer to the chain as open or closed. Open boundary conditions imply that a spin on site 1 can only couple with a spin on site 2 and a spin on site L can only couple with a spin on site $L - 1$. In closed (or periodic) boundary conditions the chain is a ring, where a spin on site 1 can couple with a spin on site 2 and also with a spin on site L. The index in the second and third sums of Eq. (5) runs according to the boundaries.

2.1 Symmetries

Any symmetry of the system is associated with an operator O that commutes with the Hamiltonian. According to Noether's theorem, this operator represents a constant of motion, as seen from $\frac{dO}{dt} = i[H, O]$. For the Hamiltonian in Eq. (5), we identify the following symmetries.

1. H commutes with the total spin in the z-direction, $\mathscr{S}^z = \sum_{k=1}^{L} S^z_k$. The system conserves $\mathscr{S}^z$; it is invariant by a rotation around the z-axis. This means that the eigenstates of H are also eigenstates of $\mathscr{S}^z$, so they have a fixed number of spins pointing up. Each eigenstate $|\psi\rangle$ is a superposition that involves only site-basis vectors with the same number of up-spins. For example, for $L = 4$ and $\mathscr{S}^z = 0$ we have,

$$|\psi\rangle = C_1|1100\rangle + C_2|1010\rangle + C_3|1001\rangle + C_4|0110\rangle + C_5|0101\rangle + C_6|0011\rangle,$$

where C_n's are the probability amplitudes, $n = 1, \ldots \mathscr{D}$, and $\mathscr{D}$ is the dimension of the subspace. The Hamiltonian matrix of a system with L sites written in the site-basis is composed of $L+1$ independent blocks (or subspaces), each with a fixed number of up-spins, $N \in [0, L]$. The dimension of each block is $\mathscr{D} = L!/[(L-N)!N!]$.

2. When $d, h = 0$, Hamiltonian (5) is invariant under reflection, which leads to conservation of parity, that is, H commutes with the parity operator

$$\Pi = \begin{cases} \mathscr{P}_{1,L}\mathscr{P}_{2,L-1}\ldots\mathscr{P}_{\frac{L}{2},\frac{L+2}{2}} & \text{for L = even} \\ \mathscr{P}_{1,L}\mathscr{P}_{2,L-1}\ldots\mathscr{P}_{\frac{L-1}{2},\frac{L+3}{2}} & \text{for L = odd} \end{cases}$$

where $\mathscr{P}_{k,l} = (\sigma_k^x\sigma_l^x + \sigma_k^y\sigma_l^y + \sigma_k^z\sigma_l^z + \mathbb{1}/2$ is the permutation operator and $\mathbb{1}$ is the identity operator. $\mathscr{P}_{k,l}$ swaps the states of the kth and lth sites. For example, for $L = 4$ and a single excitation, $N_{up} = 1$, the probability amplitudes in $|\psi\rangle = a_1|\uparrow\downarrow\downarrow\downarrow\rangle + a_2|\downarrow\uparrow\downarrow\downarrow\rangle + a_3|\downarrow\downarrow\uparrow\downarrow\rangle + a_4|\downarrow\downarrow\downarrow\uparrow\rangle$ are either $a_1 = a_4$ and $a_2 = a_3$ for even parity or $a_1 = -a_4$ and $a_2 = -a_3$ for odd parity.

3. When $d, h = 0$, L is even, and $N_{up} = L/2$, Hamiltonian (5) is invariant under a global π rotation around the x-axis. The operator that represents the rotation is

$$R_\pi^x = \sigma_1^x\sigma_2^x \ldots \sigma_L^x$$

As an example, take $L = 4$ and $N_{up} = 2$. The eigenstate

$$|\psi\rangle = a_1|\uparrow\uparrow\downarrow\downarrow\rangle + a_2|\uparrow\downarrow\uparrow\downarrow\rangle + a_3|\uparrow\downarrow\downarrow\uparrow\rangle + a_4|\downarrow\uparrow\uparrow\downarrow\rangle + a_5|\downarrow\uparrow\downarrow\uparrow\rangle + a_6|\downarrow\downarrow\uparrow\uparrow\rangle$$

has either $a_1 = a_6$, $a_2 = a_5$, and $a_3 = a_4$ or $a_1 = -a_6$, $a_2 = -a_5$, and $a_3 = -a_4$.

4. When $d, h = 0$ and $\Delta = 1$, the total spin $\mathscr{S}_T = \sum_n \mathbf{S}_n$ is conserved.

We can break the symmetries listed above, except for the total spin in the z-direction, as follows. Conservation of total spin can be avoided by choosing $\Delta \neq 1$. Parity and spin reversal can be broken if we deal with an open chain and add an impurity on a site in the border of the chain.

2.2 *Integrable Versus Chaotic Models*

In classical mechanics, if a system with n degrees of freedom has n independent integrals of motion that are Poisson-commuting, then the system is integrable. In this case, the differential equations describing the time evolution can be explicitly integrated using action-angle variables. The solutions display periodic motion on tori in phase space, and ergodicity is nonexistent. In contrast to the classical case, the notion of integrability at the quantum level has been a source of debates [43, 44].

In the case of Hamiltonian (5), we use the term integrability when referring to choices of parameters that allows the Hamiltonian to be solved with the Jordan-Wigner transformation or the Bethe Ansatz [45]. We select the following two cases, respectively.

1. The XX model is a noninteracting integrable model, where $d, h, \Delta, \lambda = 0$.
2. The XXZ model is an integrable interacting model, where $\Delta \neq 0$ and $d, h, \lambda = 0$.

The notion of quantum chaos is another delicate subject. Classical chaos goes back to the studies of Poincaré. It is related to the extreme sensitivity of the dynamics of a system to its initial conditions. The main features of classical chaos can be illustrated with a dynamical billiard. It corresponds to an idealized billiard table that has no friction and where a particle reflects elastically from the boundaries. The motion of the particle is represented in phase space by a trajectory restricted to a surface of constant energy. The shape of the boundaries determines whether the system is chaotic or regular. In the first case, two trajectories with very close initial conditions diverge exponentially in time. The rate of this separation is the Lyapunov exponent. The trajectories may become ergodic, in which case, after a long time, the particle will have visited the entire surface of constant energy and will be equally likely to be found in any point of the accessible phase space.

For quantum systems the notion of phase-space trajectories loses its meaning, since as stated by the Heisenberg uncertainty principle, we can no longer precise the particle's position and momentum at the same time. However, since classical physics is a limit of quantum physics, we could still search for quantum signatures of classical chaos.

The term quantum chaos refers to properties of eigenvalues and eigenstates that are found in the quantum level and indicate whether the system in the classical level is chaotic. It has been conjectured that the spectral fluctuations in the quantum limit of classical system that is chaotic are always the same [46, 47]. This conjecture has been proved in the semiclassical limit. The term has also been extended to include quantum systems without a classical limit, as our spin-1/2 models.

The distribution of the spacings between neighboring energy levels of a quantum system is the most commonly employed tool to distinguish integrable from nonintegrable models, but others exist, such as the level number variance and the spectral rigidity [48], as well as the distribution of the ratio of consecutive level spacings [49]. If the system is chaotic, the energy levels are highly correlated and repel each other; if it is regular (integrable), the energy levels are uncorrelated, randomly distributed, and can cross. The chaotic spin models associated with Hamiltonian (5) include:

1. The defect model with no random disorder ($h = 0$), $0 < \Delta < 1$, $0 < d < 1$, and $\lambda = 0$. The interplay between the defect and the impurity drives the system into the chaotic domain [29, 50].
2. The NNN model with no random disorder ($h = 0$) and $0 < \lambda < 1$ [15, 51–53]. The system remains chaotic whether $\Delta \neq 0$ or $\Delta = 0$ [16].
3. The disordered model with $0 < h < 1$, $\Delta = 1$, $d, \lambda = 0$ [31, 33, 50, 54].

2.2.1 Unfolding Procedure

When studying the level spacing distribution, to be able to compare different systems of different sizes, and also the different parts of the spectrum of the same system, we need to unfold the spectrum. This means that each system's specific mean level density must be removed from the data. It does not make sense to compare local fluctuations from systems with very different average densities. For example, it does not make sense to say that a spectral region with high average density has less repulsion than a spectral region with low average density. We need to separate the local fluctuations from a systematic global energy dependence of the average density. For this, we rescale the energies, so that the mean level spacing is 1. Since the density of states is the number of states in an interval of energy, that is, the reciprocal of the mean level spacing, this renormalization procedure ensures also that the mean local density of states becomes unit.

There are different ways to unfold the spectrum. A simple and good enough recipe is the following [53].

(i) Order the spectrum in increasing values of energy.

(ii) Discard some eigenvalues from the edges of the spectrum, where the fluctuations are large. This is arbitrary, we can discard for example 10% of the eigenvalues.

(ii) Separate the remaining eigenvalues into small sets of eigenvalues.

(iii) For each set, divide the eigenvalues by the mean level spacing of that particular set. The mean level spacing of the new set of renormalized energies is now 1.

Notice that contrary to the level spacing distribution, the distribution of the ratio of consecutive level spacings does not require the unfolding of the spectrum [49].

2.2.2 Level Spacing Distribution of Integrable Models

In integrable models, the eigenvalues are uncorrelated, they are not prohibited from crossing and usually follow Poisson statistics. The distribution $P(s)$ of the neighboring spacings s is given by

$$P_{\mathrm{P}}(s) = e^{-s}. \tag{6}$$

However, deviations from this shape are seen for the XX model due to its the high number of degeneracies. As Δ increases from zero, the excessive degeneracies rapidly fade away and the Poisson distribution is recovered [compare Fig. 1a, b with Fig. 1c, d]. At the root of unit $\Delta = 1/2$, the distribution departs again from Poisson (see Fig. 1e). By changing Δ slightly, for example, by using = 0.48, the Poisson distribution reappears [37].

2.2.3 Level Spacing Distribution of Chaotic Models

The level spacing distribution $P(s)$ of chaotic models is given by the Wigner-Dyson (WD) distribution [48, 55–57]. The specific form of the Wigner-Dyson distribution

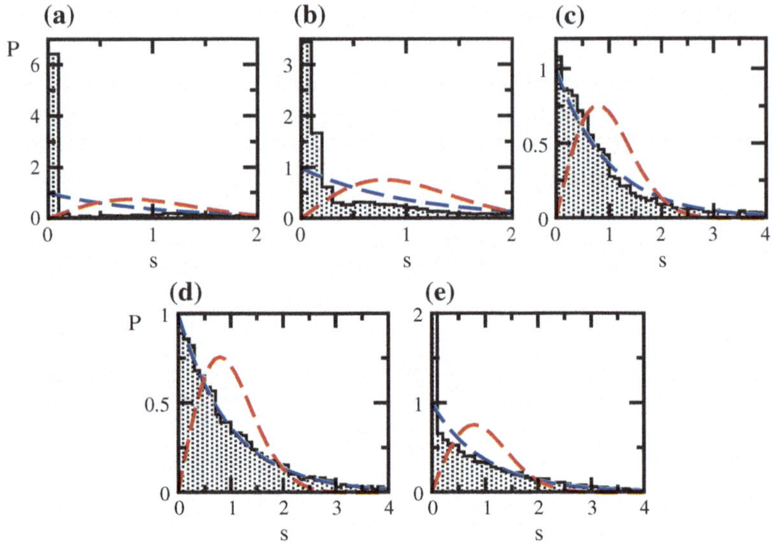

Fig. 1 Level spacing distribution for a single subspace and eigenstates with even parity; $L = 18$, $N_{up} = 6$, $d, h, \lambda = 0$, open boundary conditions. The Poisson and Wigner-Dyson distributions are shown with dashed lines. From **a** to **e**: $\Delta = 0.0, 10^{-3}, 10^{-2}, 0.1, 0.5$

depends on the symmetries of the Hamiltonian. In time-reversal invariant systems with rotational symmetry, the Hamiltonian is represented by real and symmetric matrices, $H_{ij} = H_{ji}$, and the level spacing distribution has the following shape,

$$P_{\rm WD}(s) = \frac{\pi}{2} s \exp\left(-\frac{\pi}{4} s^2\right), \tag{7}$$

which makes evident the level repulsion. This expression was derived exactly for 2×2 matrices and it works extremely well for large matrices [55, 56].

To obtain the level spacing distribution, we need to separate the eigenvalues according to their symmetry sectors. If we mix eigenvalues from different symmetry sectors, we may not achieve a Wigner-Dyson distribution even if the system is chaotic, because eigenvalues from different subspaces are independent, uncorrelated, and have no reason to repel each other [15, 53]. To illustrate this issue, in Fig. 2 we consider two chaotic Hamiltonians with open boundary conditions. They represent clean systems, where $d, h = 0$. The Hamiltonians are chaotic and given by:

$$\text{(a) } H = \sum_{k=1}^{L-1} \left[\left(S_k^x S_{k+1}^x + S_k^y S_{k+1}^y \right) + S_k^z S_{k+1}^z \right] + 0.5 \sum_{k=1}^{L-2} \left[\left(S_k^x S_{k+2}^x + S_k^y S_{k+2}^y \right) + S_k^z S_{k+2}^z \right].$$

$$\text{(b) } H = \sum_{k=1}^{L-1} \left[\left(S_k^x S_{k+1}^x + S_k^y S_{k+1}^y \right) + 0.5 S_k^z S_{k+1}^z \right] + 0.5 \sum_{k=1}^{L-2} \left[\left(S_k^x S_{k+2}^x + S_k^y S_{k+2}^y \right) + 0.5 S_k^z S_{k+2}^z \right].$$

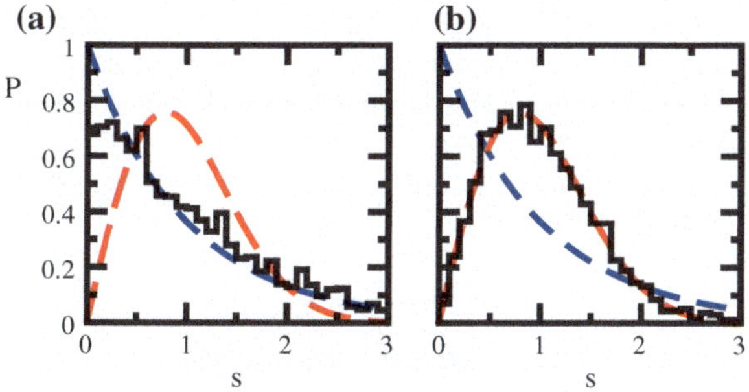

Fig. 2 Level spacing distribution for the chaotic Hamiltonians written in the text. Panel **a**: $L = 14$, $N_{up} = 7$, $\mathscr{S}^z = 0$. Panel **b**: $L = 15$, $N_{up} = 5$ and the eigenvalues are separated by the parity; $P(s)$ is the average of the distributions of the two parity sectors. The Poisson and Wigner-Dyson distributions are shown with dashed lines

In both panels of Fig. 2 we have eigenvalues of a single selected $\mathscr{S}^z$-sector. In panel (b), we avoid the $\mathscr{S}^z = 0$ subspace, where spin reversal symmetry exists, by choosing L odd. We also choose $\Delta \neq 1$ to avoid conservation of total spin. In doing so, the only remaining symmetry is parity, which we do take into account. The expected Wigner-Dyson distribution is found. Contrary to Fig. 2b, a mixes eigenvalues from the three symmetries mentioned above—spin reversal, total spin, and parity—which explains why $P(s)$ becomes so close to a misleading Poisson distribution.

2.2.4 Level Spacing Indicator

To study the crossover from integrability to chaos as a certain parameter is varied, better than plotting the level spacing for each value of the parameter, we can use a quantity that tells us how close we are to a Poisson or to a Wigner-Dyson distribution. An example is the indicator β used to fit $P(s)$ with the Brody distribution [58],

$$P_B(s) = (\beta + 1)bs^\beta \exp\left(-bs^{\beta+1}\right), \quad b = \left[\Gamma\left(\frac{\beta+2}{\beta+1}\right)\right]^{\beta+1}, \tag{8}$$

where Γ is Euler's gamma function. If $\beta = 0$ the distribution is Poisson and $\beta = 1$ indicates a Wigner-Dyson distribution.

Based on heuristic arguments, Izrailev introduced an Ansatz for the level spacing distribution that captures very well the intermediate regime between Poisson and Wigner-Dyson [59, 60],

$$P_I(s) = As^\nu \exp\left[-\frac{\pi^2}{16}\nu s^2 - (C - \frac{\nu}{2})\frac{\pi}{2}s\right], \tag{9}$$

where A and C are constants with values obtained through normalization conditions. The parameter ν in Eq. (9) is related to the degree of localization of the eigenstates. For $\nu = 0$, we can recover the Poisson distribution. For $\nu = 1$, we have the GOE level repulsion, that is $P(s \to 0) \to s$.

Another way to quantify the proximity to the Wigner-Dyson distribution is with the chaos indicator [61]

$$\eta = \frac{\int_0^{s_0} [P(s) - P_{WD}(s)]ds}{\int_0^{s_0} [P_P(s) - P_{WD}(s)]ds}, \tag{10}$$

where s_0 is the first intersection point of $P_P(s)$ and $P_{WD}(s)$. For a Poisson distribution, $\eta \to 1$, and for the Wigner-Dyson, $\eta \to 0$.

Disordered Spin Model

In Fig. 3, we show η as a function of h for the disordered model with $\lambda = 0$, $\Delta = 1$, and closed boundary conditions. η is averaged over several disorder realizations. As the disorder strength h increases from zero (where we have the clean integrable XXZ model), the level spacing distribution first transitions abruptly from Poisson ($\eta \sim 1$) to Wigner-Dyson (small η). For the system sizes considered, it remains Wigner-Dyson for h in the range [0.1, 1], where η plateaus to a small value. As h further increases, the level spacing distribution transitions from Wigner-Dyson back to Poisson. In this second integrable region, the system becomes localized in space [50, 54, 62].

The logarithmic scale of the x-axis in Fig. 3a emphasizes the first transition from the spatially delocalized integrable point to chaos and the linear plot in Fig. 3b stresses the transition to localization in space. The different curves represent different system sizes; they increase from top to bottom. The range of disorder strengths for which η is small increases as L increases. This indicates that in the thermodynamic limit,

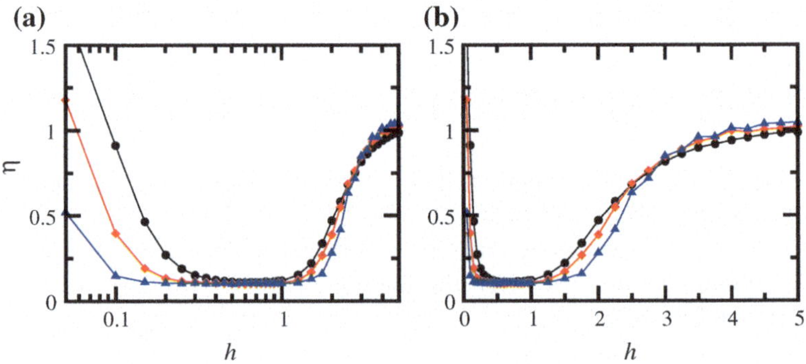

Fig. 3 Chaos indicator η *versus* disorder strength; semilogarithmic scale (**a**) and linear scale (**b**). The system sizes are $L = 12$ (circles), $L = 14$ (diamonds), and $L = 16$ (triangles); $\Delta = 1, d, \lambda = 0$, $N_{up} = L/2$. Average performed over 1082, 291, 77 disorder realizations for $L = 12, 14, 16$, respectively

the two transition regions may disappear, although this is still an open question. An infinitesimally small h may suffice to take the system into the chaotic regime [29, 38]. As for $h > 1$, the transition region may disappear in the thermodynamic limit or persist, in this latter case, maybe as a critical point as one finds in Anderson localization in higher dimensions.

2.3 Density of States and Delocalization Measures

The Wigner-Dyson distribution was first studied in the context of full random matrices. Wigner [63] employed these matrices to describe the spectrum of heavy nuclei. His idea was to ignore the details of the interactions of such complex systems and treat them statistically. Full random matrices are filled with random numbers and their only constraint is to satisfy the symmetries of the system one is trying to describe. In the case of Gaussian orthogonal ensembles (GOE), the full random matrices are invariant with respect to an orthogonal conjugation $O^T HO$, where O is any orthogonal matrix (that is a matrix whose inverse is equal to its transpose). GOE random matrices are real and symmetric, as the Hamiltonian matrices for the spin systems that we study.

The level spacing distribution of GOE random matrices is also given by Eq. (7) and agrees extremely well with the distributions obtained with data from actual nuclei spectra. However, full random matrices are unrealistic, since they assume simultaneous and infinite-range interactions among all the particles of the system that they try to represent. In contrast, our spin models describe realistic systems studied experimentally. They have only two-body and short-range interactions. What are then the properties that clearly distinguish realistic models with two-body interactions from full random matrices?

A distinctive feature between full random matrices and realistic models is the density of states,

$$R(E) = \sum_{\alpha} \delta(E - E_{\alpha}), \tag{11}$$

where E_{α} are the eigenvalues of the Hamiltonian. The density of states of full random matrices follows the standard semicircle distribution [64],

$$R(E) = \frac{2}{\pi \mathscr{E}} \sqrt{1 - \left(\frac{E}{\mathscr{E}}\right)^2}, \tag{12}$$

where $2\mathscr{E}$ is the length of the spectrum, that is $-\mathscr{E} \leq E \leq \mathscr{E}$. The density of states of Hamiltonians with two-body interactions is Gaussian, independent of the regime (integrable or chaotic) of the system. These two cases are illustrated in Fig. 4a, b for full random matrices and the defect model, respectively.

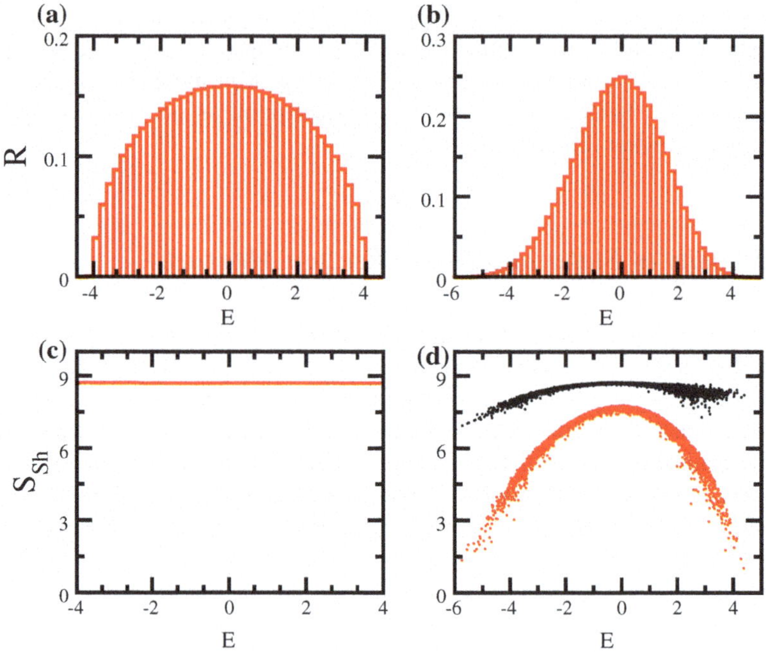

Fig. 4 Left: GOE full random matrix with $\mathcal{D} = 12870$. Right: Defect model with $h, \lambda = 0, \Delta = 0.5$, $d = 0.9, L = 16, N_{up} = 8, \mathcal{D} = 12870$, open boundary conditions. The random numbers of the full random matrix are rescaled so that $\mathcal{E} = 4$. Top: Density of states. Bottom: Shannon entropy for all eigenstates. In (**d**): site-basis (black) and mean-field basis (red)

The Gaussian shape of the density of states of realistic models is reflected into the structure of the eigenstates. The majority of the eigenstates are close to the middle of the spectrum, where strong mixing can then take place and the eigenstates reach their highest level of delocalization. In contrast, the eigenstates closer to the edges of the spectrum are more localized.

There are various ways to quantify how much a state spreads out in a certain basis. One of them is the participation ratio PR. Given an eigenstate $|\psi_\alpha\rangle = \sum_n C_n^\alpha |\phi_n\rangle$ written in a basis $|\phi_n\rangle$,

$$PR^{(\alpha)} = \frac{1}{\sum_n |C_n^\alpha|^4}. \tag{13}$$

A comparable quantity is the Shannon information entropy, defined as

$$S_{Sh}^{(\alpha)} = -\sum_n |C_n^\alpha|^2 \ln |C_n^\alpha|^2. \tag{14}$$

The values of $PR^{(\alpha)}$ and $S_{Sh}^{(\alpha)}$ depend on the chosen basis. In the case of full random matrices, the notion of basis is not well defined. All eigenstates of full random matrices are (pseudo)-random vectors. In the case of GOEs, the coefficients are real random numbers from a Gaussian distribution satisfying the normalization condition. All eigenstates are therefore equivalent and lead to approximately the same values of the participation ratio and of the Shannon entropy [34],

$$PR^{GOE} \sim \mathscr{D}/3, \qquad S_{Sh}^{GOE} \sim \ln(0.48\mathscr{D}). \tag{15}$$

The results above can be obtained by substituting the sum in PR and S_{Sh} by an integral,

$$\sum_n F(C_n) \to \mathscr{D}\int_{-\infty}^{\infty} F(C)P(C)dC.$$

The distribution of the probability amplitudes C_n^α is given by the Gaussian [65],

$$P(C) = \sqrt{\frac{\mathscr{D}}{2\pi}} \exp\left(-\frac{\mathscr{D}}{2}C^2\right),$$

so $\overline{C} = 0$ and $\overline{C^2} = 1/\mathscr{D}$. The latter is obtained by substituting $x = C\sqrt{\mathscr{D}/2}$,

$$\overline{C^2} = \sqrt{\frac{\mathscr{D}}{2\pi}} \int_{-\infty}^{\infty} dCC^2 \exp\left(-\frac{\mathscr{D}}{2}C^2\right) = \sqrt{\frac{\mathscr{D}}{2\pi}} \int_{-\infty}^{\infty} \frac{2}{\mathscr{D}} dx x^2 e^{-x^2} \sqrt{\frac{2}{\mathscr{D}}} = \frac{1}{\mathscr{D}}.$$

Thus, for the inverse of the participation ratio, we have

$$\sum_n |C_n|^4 \to \mathscr{D}\sqrt{\frac{\mathscr{D}}{2\pi}} \int_{-\infty}^{\infty} dCC^4 \exp\left(-\frac{\mathscr{D}}{2}C^2\right) = \mathscr{D}\sqrt{\frac{\mathscr{D}}{2\pi}} \int_{-\infty}^{\infty} dx \frac{4}{\mathscr{D}^2} x^4 e^{-x^2} \sqrt{\frac{2}{\mathscr{D}}}$$
$$= \frac{4}{\mathscr{D}\sqrt{\pi}} \frac{3\sqrt{\pi}}{4} = \frac{3}{\mathscr{D}}$$

and for the Shannon information entropy,

$$S_{Sh}^{GOE} \sim -\mathscr{D}\sqrt{\frac{\mathscr{D}}{2\pi}} \int_{-\infty}^{\infty} \exp\left(-\frac{\mathscr{D}C^2}{2}\right) C^2 \ln C^2 dC$$
$$= -2 + \ln 2 + \gamma_e + \ln \mathscr{D} \sim \ln(0.48\mathscr{D}),$$

where γ_e is Euler's constant.

In Figs. 4c, d, we show the Shannon entropy for full random matrices and the defect model, respectively. For the first, apart from small fluctuations, $S_{Sh} = S_{Sh}^{GOE}$. For the realistic model, we show S_{Sh} for the eigenstates written in two different basis representation. The choice of basis depends on the problem we are interested in. For studies of localization in real space, the site-basis is a natural choice. Another alternative, often used to distinguish the regular from the chaotic region, is the mean-field basis, which corresponds to the integrable (regular) part of the Hamiltonian. In the case of the defect model, a reasonable choice for the mean-field basis corresponds to the eigenstates of the XXZ model [18, 19]. Both cases are shown in Fig. 4d.

The energy dependence of the structure of the eigenstates of a realistic system has consequence for its dynamics [18, 19, 24–28, 30, 32, 34, 66] and viability of thermalization [10, 12, 38, 39, 42]. The dynamics is slower for an initial state with energy close to the edge of the spectrum than for an initial state close to the middle of the spectrum. Thermalization is expected for chaotic systems, but it may not occur for initial states with energies very close to the border of the spectrum.

3 Dynamics: Survival Probability

Now that we have a general idea about the spectrum and the structure of the eigenstates of systems with two-body interactions, as those described by spin-1/2 models [Eq. (5)], we proceed with the analysis of their dynamics. We assume that the system is prepared in an initial state $|\Psi(0)\rangle$ that is an eigenstate of a certain initial Hamiltonian H_0. The dynamics starts with the sudden change (quench) of a parameter of the Hamiltonian that brings it to a new final Hamiltonian H,

$$H_0 \xrightarrow{quench} H = H_0 + gV, \tag{16}$$

where g is the strength of the perturbation.

There are various quantities that we can use to analyze the evolution of the system. We look here at the simplest one: the probability of finding the system at time t still in state $|\Psi(0)\rangle$, which is known as the survival probability and is given by

$$W_{n_0}(t) = |\langle\Psi(0)|\Psi(t)\rangle|^2 = |\langle\Psi(0)|e^{-iHt}|\Psi(0)\rangle|^2. \tag{17}$$

$W_{n_0}(t)$ is also known as nondecay probability, return probability, or fidelity, but it is incorrect to call it Loschmidt echo, since we have only evolution forward, there is no time reversal ("echo") involved.

By writing the initial state in the eigenstates $|\psi_\alpha\rangle$ of H, Eq. (17) becomes

$$W_{n_0}(t) = \left|\sum_\alpha |C_{n_0}^\alpha|^2 e^{-iE_\alpha t}\right|^2 = \left|\int dE\, e^{-iEt} \rho_0(E)\right|^2, \tag{18}$$

where $C_{n_0}^\alpha = \langle\psi_\alpha|\Psi(0)\rangle$ are the overlaps and

$$\rho_0(E) \equiv \sum_\alpha |C_{n_0}^\alpha|^2 \delta(E - E_\alpha) \tag{19}$$

is the energy distribution of the initial state, referred to as local density of states (LDOS). The survival probability is the absolute square of the Fourier transform of the LDOS. If we have detailed information about $\rho_0(E)$ we can predict the behavior of $W_{n_0}(t)$. The mean and variance of the LDOS are respectively the energy of the initial state,

$$E_0 = \langle \Psi(0)|H|\Psi(0)\rangle = \sum_\alpha |C_{n_0}^\alpha|^2 E_\alpha, \tag{20}$$

and

$$\sigma_0^2 = \sum_\alpha |C_{n_0}^\alpha|^2 (E_\alpha - E_0)^2. \tag{21}$$

3.1 Short Times

The decay of $W_{n_0}(t)$ shows different behaviors at different time scales. At extremely short times, $t \ll \sigma_0^{-1}$, the decay is quadratic. This is a universal behavior that does not depend on H_0 or H, but simply on σ_0. It is obtained by Taylor expanding the phase factor in Eq. (18),

$$\begin{aligned} W_{n_0}(t) &\approx \left| e^{-iE_0 t} \left[\sum_\alpha |C_{n_0}^\alpha|^2 - i \sum_\alpha |C_{n_0}^\alpha|^2 (E_\alpha - E_0) t - \frac{1}{2} \sum_\alpha |C_{n_0}^\alpha|^2 (E_\alpha - E_0)^2 t^2 \right] \right|^2 \\ &\approx 1 - \sigma_0^2 t^2. \end{aligned} \tag{22}$$

But we are actually interested in what happens after the quadratic decay.

3.1.1 Exponential and Gaussian Decay

After the universal quadratic behavior, the decay depends on the strength of the perturbation, which determines the shape of the LDOS. The LDOS is close to a delta function for $g \to 0$ and its Fourier transform leads to a very slow decay of $W_{n_0}(t)$. The two left top panels of Fig. 5 show LDOS for $g \to 0$ and the two right top panels present the corresponding $W_{n_0}(t)$. The first and third columns of the figure are obtained for the integrable XXZ model and the initial state is an eigenstate of the XX model with E_0 far from the edges of the spectrum. The second and fourth columns of the figure show the results for the chaotic NNN model and the initial state is an eigenstate of the XXZ model with E_0 away from the borders of the spectrum.

The LDOS broadens as the strength of the perturbation increases. When the perturbation gV is larger than the mean level spacing (Fermi golden rule regime), the LDOS becomes Lorentzian [27, 29],

$$\rho_0(E) = \frac{1}{2\pi} \frac{\Gamma_0}{(E_0 - E)^2 + \Gamma_0^2/4}, \tag{23}$$

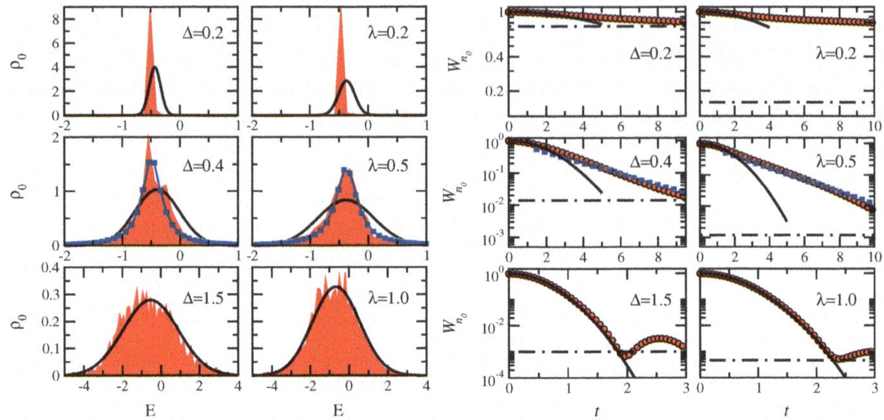

Fig. 5 Local density of states (two left columns) and survival probability (two right columns). First and third columns: XXZ model. Second and fourth columns: NNN model with $\Delta = 0.5$. The values of Δ for the XXZ model and of λ for the NNN model are indicated in the figure. The initial state has energy far from the edge of the spectrum; third column: $|\Psi(0)\rangle$ is an eigenstate of the XX model; fourth column: $|\Psi(0)\rangle$ is an eigenstate of the XXZ model. Blue squares: Lorentzian fit and exponential decay. Black solid line: Gaussian LDOS and Gaussian decay with σ_0 from Eq. (21). Red shaded area and circles: numerical results. $L = 18$, $N_{up} = 6$, $h, d = 0$, open chain. Horizontal dot-dashed lines indicate the saturation point [Eq. (32)]

where Γ_0 is the width of the distribution. The Fourier transform of the Lorentzian gives the exponential decay

$$W_{n_0}(t) = \exp(-\Gamma_0 t). \tag{24}$$

Lorentzian LDOS and exponential decays are shown in the middle panels of Fig. 5.

As the perturbation further increases, the LDOS widens even more and eventually becomes Gaussian [26–28, 67, 68],

$$\rho_0(E) = \frac{1}{\sqrt{2\pi\sigma_0^2}} \exp\left[-\frac{(E-E_0)^2}{2\sigma_0^2}\right], \tag{25}$$

This shape reflects the Gaussian density of states. In this case, the decay of the survival probability is Gaussian,

$$W_{n_0}(t) = \exp(-\sigma_0^2 t^2). \tag{26}$$

It is important to stress that these very fast decays of the survival probability, exponential and even Gaussian, are not exclusive to chaotic systems. As we show in Fig. 5 for the XXZ model (first and third columns), fast evolutions can also happen for integrable models. The speed of the dynamics depends on the strength of the perturbation, not on the regime, integrable or chaotic, of the Hamiltonian [18, 19, 24–28, 30, 32, 34, 66].

Similarities in the time evolution of integrable and chaotic models perturbed far from equilibrium can be captured also with other dynamical quantities, such as the Shannon information entropy and the von Neumann entanglement entropy. In fact, as shown in Refs. [33, 34], there is a clear parallel between the behaviors of both entropies.

We take this opportunity to mention that the equation for the evolution of observables contain the survival probability explicitly. For an observable O, we have

$$\begin{aligned} O(t) &= W_{n_0}(t)O(0) \\ &+ \sum_{n\neq n_0} \langle \Psi(0)|e^{iHt}|\Psi(0)\rangle O_{n_0,n}\langle \phi_n|e^{-iHt}|\Psi(0)\rangle \\ &+ \sum_{n\neq n_0} \langle \Psi(0)|e^{iHt}|\phi_n\rangle O_{n,n_0}\langle \Psi(0)|e^{-iHt}|\Psi(0)\rangle \\ &+ \sum_{n,m\neq n_0} \langle \Psi(0)|e^{iHt}|\phi_n\rangle O_{n,m}\langle \phi_m|e^{-iHt}|\Psi(0)\rangle, \end{aligned} \tag{27}$$

where $O_{n,m} = \langle n|O|m\rangle$ and $|\phi_n\rangle$ are the eigenstates of the initial Hamiltonian that defines the initial state. The analysis of the evolution of observables is more demanding than the study of the survival probability, since they depend on the overlaps between $|\Psi(t)\rangle$ and the other basis vectors of the initial Hamiltonian and on the details of the observables.

3.1.2 Faster Than Gaussian and Quantum Speed Limit

There are scenarios where the decay of $W_{n_0}(t)$ can be even faster than Gaussian. This happens, for example, when the LDOS is bimodal (or multimodal), in which case the speed of the evolution becomes controlled by the distance between the peaks [28]. This can be achieved by preparing the system in an eigenstate of the XXZ model and evolving it with the defect model for $d \gg 1$. When the amplitude of the magnetic field on the defect site is very large, the density of states and consequently also the LDOS splits in two separated Gaussian peaks, as shown in Fig. 6a. If both peaks have the same width σ_G, the Fourier transform of $\rho_0(E)$ gives

$$W_{n_0}(t) \simeq \cos^2(\sigma_0 t)\exp(-\sigma_G^2 t^2), \tag{28}$$

where σ_0 is now approximately $d/2$. One sees that for $t < \pi/(2\sigma_0)$, the survival probability approaches the bound associated with the energy-time uncertainty relation, $W_{n_0}(t) \simeq \cos^2(\sigma_0 t)$ [69–71]. For $t > \pi/(2\sigma_0)$ there are revivals. The envelope of the decay of these oscillations is Gaussian and controlled by σ_G. The expression in Eq. (28) matches very well the decay of the survival probability shown in Fig. 6b.

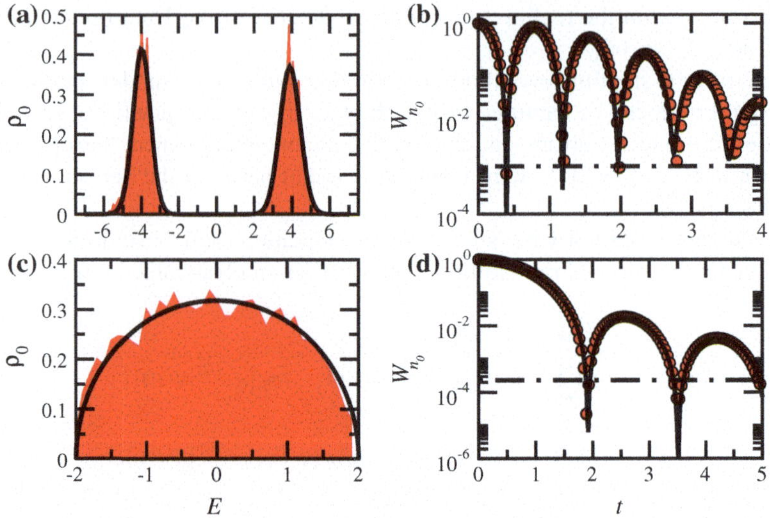

Fig. 6 Local density of states (left) and survival probability (right). In **a** and **b**: defect model, $d = 8.0$, $\Delta = 0.48$, $h, \lambda = 0$, $L = 16$, $N_{up} = 8$. Initial state in **b** is an eigenstate of the XXZ model with E_0 in the middle of the spectrum. In **c** and **d**: Full random matrix from GOE, $\mathscr{D} = 12870$, rescaled energies so that $\mathscr{E} \sim 2$. Initial state in **d** is an eigenstate of another GOE full random matrix. Red shaded areas and circles: numerical results. Black solid lines on the right: analytical expressions (28) and (29). Horizontal dot-dashed lines indicate the saturation point [Eq. (32)]

Another example of a decay faster than Gaussian occurs for systems with random and simultaneous interactions among many particles. The extreme case is that of full random matrices. In Fig. 6c, we show the LDOS for an initial state corresponding to an eigenstate of a GOE full random matrix that is evolved with another GOE full random matrix. The LDOS has a semicircle shape [26–28, 64], as the density of states for full random matrices [Eq. (12)],

$$\rho_0(E) = \frac{1}{\pi\sigma_0}\sqrt{1-\left(\frac{E}{2\sigma_0}\right)^2}, \tag{29}$$

where $\sigma_0^2 = \int_{-\mathscr{E}}^{\mathscr{E}} \rho_0(E)E^2 dE = \mathscr{E}/2$. The Fourier transform of the semicircle gives the following analytical expression for the survival probability [26, 27, 34]

$$W_{n_0}(t) = \frac{[\mathscr{J}_1(2\sigma_0 t)]^2}{\sigma_0^2 t^2}, \tag{30}$$

where $\mathscr{J}_1$ is the Bessel function of the first kind. This expression agrees with the numerical results in Fig. 6d. The decay up to $t \sim \mathscr{E}$ is faster than Gaussian. Later, it shows oscillations that decay as a power-law $\propto t^{-3}$. Indeed, for $t \gg \sigma_0^{-1}$, Eq. (30) leads to

$$W_{n_0}(t \gg \sigma_0^{-1}) \to \frac{1 - \sin(4\sigma_0 t)}{2\pi \sigma_0^3 t^3}. \tag{31}$$

The onset of the power-law decay for longer times, as depicted in Fig. 6d, prompts the question of what happens to the survival probability for the spin models at times longer than those shown in Fig. 5. If we wait long enough, since the studied systems are finite, the dynamics eventually saturates to the infinite-time average,

$$\overline{W}_{n_0} = \lim_{t\to\infty} \frac{1}{t} \int_0^t d\tau\, F(\tau) = \sum_\alpha |C_{n_0}^\alpha|^4 = IPR_{n_0}, \tag{32}$$

where IPR_{n_0} is the inverse of the participation ratio of the initial state written in the energy eigenbasis. Our question is whether there is some other well defined behavior between the initial exponential or Gaussian decays and the saturation to $\overline{W}_{n_0}$. This is the subject of the next subsection.

3.2 *From Short to Long Times: Strong Perturbation*

Since our systems are finite and relatively small, the analysis of long-time dynamics is subjected to finite size effects. To circumvent this problem, we focus now on the disordered Hamiltonian (5) with $d, \lambda = 0$, random uniform numbers $h_k \in [-h, h]$, $\Delta = 1$, and closed boundary conditions, that is,

$$H = \sum_{k=1}^{L} h_k S_k^z + \sum_{k=1}^{L} \left(S_k^x S_{k+1}^x + S_k^y S_{k+1}^y + S_k^z S_{k+1}^z \right). \tag{33}$$

Hamiltonian (33) is paradigmatic in the studies of many-body localization (MBL) [50, 54, 62, 72, 73]. MBL refers to localization in face of the interplay between interaction and disorder. It is an extension to the Anderson localization, where interaction is absent. Without the Ising interaction, the eigenstates of the disordered noninteracting 1D system are exponentially localized in configuration space for any value of h. The question that has been discussed more intensely since the beginning of the millennium is whether localization may still take place when interaction is added. Although not precise, the value $h_c \approx 3.5$ for the disorder strength in (33) is supposed to determine the critical point for the transition from the ergodic (chaotic) to the MBL phase.

By varying the disorder strength, we can study the survival probability at long times close to the clean integrable point ($h = 0$), in the chaotic regime, in the intermediate region between ergodicity and localization, and in the MBL phase, which is another integrable point [see Fig. 3]. In the chaotic regime, the eigenstates away from the border of the spectrum are highly delocalized and similar to random vectors [as in Fig. 4d]. We refer to these states as chaotic or ergodic states, although it is important to keep in mind that ergodicity in the sense of full random matrices, where the eigenstates are random vectors, does not exist in realistic systems. As the disorder strength increases and we move from the chaotic to the MBL phase, passing through the intermediate region, the eigenstates become less spread out in space. As we discussed in [33], they remain extended in this intermediate region, but are no longer ergodic. This reduction in the level of delocalization of the eigenstates naturally slows down the dynamics.

We take as initial states, single site-basis vectors. This is equivalent to a quench, where the initial Hamiltonian is only the Ising part of Hamiltonian (33) and the final Hamiltonian that dictates the evolution is the complete H (33). In view of Eq. (16), this case corresponds to a strong perturbation.

We perform averages over initial states and disorder realizations. This reduces finite-size effects and unveils features of the dynamics that could otherwise be hidden by sample to sample fluctuations. The average is done over $0.1\mathscr{D}$ initial states with energies close to the middle of the spectrum and over enough disorder realizations to have a total of $\sim 10^5$ statistical data. The average is represented with the symbol $< . >$. We choose initial states with energy close to the middle of the spectrum ($E_0 \approx 0$), because there localization is more difficult, due to the large concentration of energy levels. If localization occurs at $E_0 \approx 0$, then it is certain to have taken place at other regions of the spectrum.

Figure 7 depicts the time evolution of the averaged survival probability, $\langle W_{n_0}(t)\rangle$, from very short to very long times. The disorder strength h ranges from $h = 0.2$ (chaotic regime) to $h = 4.0$ where the system is likely already in the MBL phase.

According to Eq. (22), the dynamics at very short times ($t \ll \sigma_0^{-1}$) depends only on σ_0. If we write the Hamiltonian matrix in the site-basis (denoted by $|\phi_n\rangle$), we can show that

$$\sigma_0 = \sum_{\alpha} |C_{n_0}^{\alpha}|^2 (E_\alpha - E_0)^2 = \sqrt{\sum_{n \neq n_0} |\langle \phi_n | H | \phi_{n_0} \rangle|^2}\,, \qquad (34)$$

where $|\phi_{n_0}\rangle = |\Psi(0)\rangle$. In the site-basis, the disorder appears only in the diagonal elements of the Hamiltonian matrix. Thus, the dynamics at very short times is completely independent of the presence of disorder.

The subsequent evolution is purely Gaussian, as expected from the Gaussian envelope of the LDOS [see Fig. 8]. The evolution in this time scale agrees very well with the analytical expression $\langle W_{n_0}(t)\rangle = \exp(-\sigma_0^2 t^2)$ discussed in Eq. (26). This is illustrated with circles in Fig. 7. When h becomes large, the time interval of the Gaussian decay shortens, and possibly only the quadratic part of the decay persists.

3.2.1 Power-Law Decays

After the fast Gaussian evolution, oscillations emerge. These are not fluctuations that could be reduced with a large number realizations, as those after equilibrium. These oscillations may in fact belong to the power-law decays that become evident in Fig. 7 for $t > 2$.

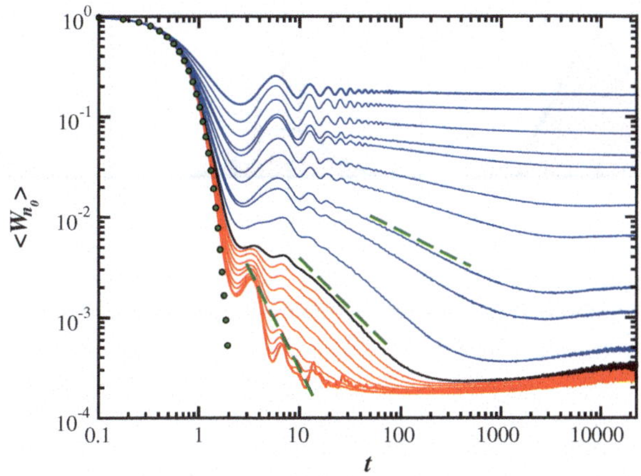

Fig. 7 Survival probability. From bottom to top, $h = 0.2, 0.3, \ldots 0.9$, $h = 0.95, 1, 1.25, \ldots 3$, $h = 3.5, 4$. Circles: analytical Gaussian decay $\langle W_{n0}(t)\rangle = \exp(-\sigma_0^2 t^2)$. Dashed lines are, from bottom to top, $\gamma = 2$, $\gamma = 1$, and $\gamma = 0.5$ for $h = 0.2$, $h = 1.0$ and $h = 1.75$, respectively. Averages over 10^5 data of disorder realizations and initial states with $E_0 \sim 0$; $L = 16$, $N_{up} = 8$

The power-law exponent γ in $\langle W_{n_0}(t)\rangle \propto t^{-\gamma}$ depends on the disorder strength h. The two different colors in Fig. 7 (red and blue curves) indicate two different causes of the power-law decay, as we discuss next.

Chaotic eigenstates

The red curves in the bottom of Fig. 7 are associated with the results for the system in the chaotic domain. According to Fig. 3, this occurs for $0.1 < h < 1$. In this region, the LDOS is well filled as seen in Fig. 8a for $h = 0.5$. The analysis of the participation ratio of the initial state confirms ergodicity, $\langle \mathrm{PR}_{n_0}\rangle \propto \mathscr{D}$. In this region, we expect γ to be close to 2, as is indeed obtained with the curve for $h = 0.2$ that is shown in the bottom of Fig. 7 together with the dashed line that represents $\langle W_{n_0}(t)\rangle \propto t^{-2}$. The exponent $\gamma = 2$ is caused by the so-called Khalfin effect. It refers to the emergence of the power-law decay of the survival probability due to the unavoidable presence of bounds in the spectrum [74–77]. The phenomenon has been extensively studied for continuous systems. We have argued that similar analyses can be extended to the discrete spectra of finite lattice many-body quantum systems when the LDOS is ergodically filled [24, 25].

Notice, however, that as h increases above 0.2 up to 1, γ decreases from 2 and approaches 1, as seen in Fig. 7. These intermediate values, $1 \le \gamma < 2$, are probably caused by a competition between the effects of energy bounds and minor correlations between the eigenstates.

Multifractal eigenstates

The black curve in Fig. 7 marks the borderline between the chaotic region (red) and the intermediate region (blue). In the latter, the eigenstates become multifractal.

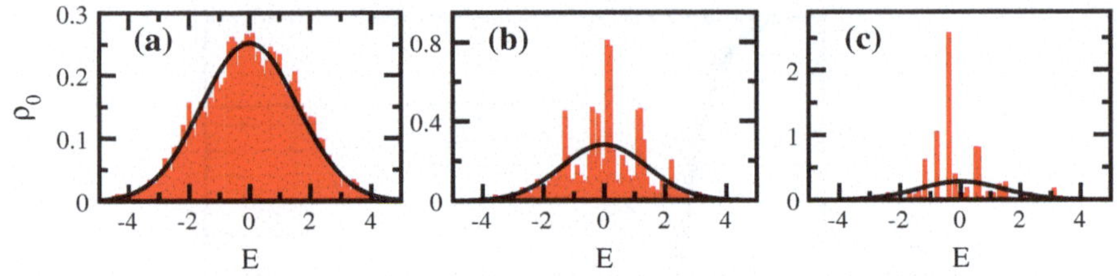

Fig. 8 Local density of states for a single disorder realization; $h = 0.5$ (**a**), $h = 1.5$ (**b**) and $h = 2.7$ (**c**); $L = 16$, $N_{up} = 8$. The envelopes (solid lines) of the distributions are Gaussians with center E_0 [Eq. (20)] and width σ_0 [Eq. (34)], which is consistent with the situation of strong perturbation

Multifractality implies that the sums of the moments M of the components $|C_n^\alpha|^2$ of the eigenstates $|\psi_\alpha\rangle = \sum_n C_n^\alpha |\phi_n\rangle$ written in the site-basis $|\phi_n\rangle$ show multifractal scaling with the dimension of the Hilbert space $\mathscr{D}$, that is

$$\langle M_q \rangle = \sum_n |C_n^\alpha|^{2q} \sim \mathscr{D}^{-(q-1)D_q}, \tag{35}$$

where D_q is the fractal dimension. Multifractality occurs when D_q depends nonlinearly on q, instead of being a constant, as in the metallic ($D_q = 1$) or in the insulating ($D_q = 0$) phase. Most of our studies have concentrated on the second moment M_2 for the eigenstates written in the site-basis and for the initial states (which are site-basis vectors) written in the energy eigenbasis. Our focus has therefore been on D_2. The second moment is nothing but the participation ratio, $PR^{(\alpha)}$ for the eigenstates and PR_{n_0} for the initial states. We calculated D_1 in [33] and other q's have been recently studied as well [78].

Our scaling analyses for $PR^{(\alpha)}$ and PR_{n_0} suggest that both lead to the same value of D_2. This value is ~ 1 in the chaotic region and < 1 in the intermediate region. The intermediate region is therefore characterized by eigenstates that are not yet localized, but are not chaotic either. In this region, D_2 decreases as h increases. The fractality of the states also leads to the sparsity of the LDOS, as seen in Fig. 8c, d.

We got excited when we realized that in the intermediate region, the value of D_2 coincides with exponent of the power-law decay γ, that is $\langle W_{n_0}(t) \rangle \propto t^{-D_2}$. This agreement is better understood if one writes the survival probability in terms of the correlation function $\mathscr{C}(E) = \sum_{\alpha_1,\alpha_2} |C_{n_0}^{(\alpha_1)}|^2 |C_{n_0}^{(\alpha_2)}|^2 \delta(E - E_{\alpha_1} + E_{\alpha_2})$ as follows

$$W_{n_0}(t) = \int_{-\infty}^{\infty} dE e^{-iEt} \mathscr{C}(E). \tag{36}$$

The long-time behavior of $W_{n_0}(t)$ is determined by small E. A power-law decay with exponent D_2 emerges for long t when [79–85]

$$\mathscr{C}(E \to 0) \propto E^{D_2 - 1}. \tag{37}$$

This is analogous to what has been found in noninteracting disordered systems at least as early as in the studies by Chalker [79, 80].

During the time interval where the power-law decays with $\gamma = D_2 < 1$ are seen for the survival probability, we observe also a logarithmic growth of the Shannon entropy and entanglement entropy controlled by the same fractal dimension as $S \sim A + D_2 \ln t$ (A is a constant).

3.2.2 Correlation Hole

After the power-law decay, there is still one more interesting feature in the decay of the survival probability before it finally saturates to $\overline{W}_{n_0}$. The survival probability may fall below the saturation value and then raise to $\overline{W}_{n_0}$. This dip is known as correlation hole [86–89]. It is an explicit dynamical manifestation of level repulsion; it only appears in nonintegrable finite systems [33, 35]. Thus, by studying the evolution of the survival probability at long times, we gain information about level statistics. The correlation hole is visible in Fig. 7, being clearly deeper in the chaotic region. As the disorder strength increases and the system approaches the MBL phase, the hole fades away and eventually disappears.

We can use the depth of the correlation hole to quantify how close or far the system is to the chaotic region. To measure the depth, we compute

$$\kappa = \frac{\overline{W}_{n_0} - \langle W_{n_0}^{min} \rangle}{\overline{W}_{n_0}}. \tag{38}$$

In full random matrices from GOE, $W_{n_0}^{min} \sim 2/\mathscr{D}$ [88] and $\overline{W}_{n_0} \sim 3/\mathscr{D}$, so the maximum value that κ can have is $1/3$.

In Fig. 9, we show κ as a function of the disorder strength. It approaches the maximum value $1/3$ in the chaotic region. It decreases for small h, since the system gets closer to the integrable clean point, and for large h, as the system approaches localization. Similarly to D_2, κ is another alternative to detect the transition from chaos to spatial localization. But notice that κ can in fact detect and integrable-chaos transition.

The correlation hole is seen also in observables, such as the spin density imbalance. Using full random matrices we were able to find exact analytical expressions for the survival probability and for the density imbalance from $t = 0$ to saturation [36]. These expressions helped us to propose functions that matched very well the entire evolution of the survival probability of spin systems deep in the chaotic region, including the correlation hole, and that captured very well different behaviors of the imbalance at different time scales [36].

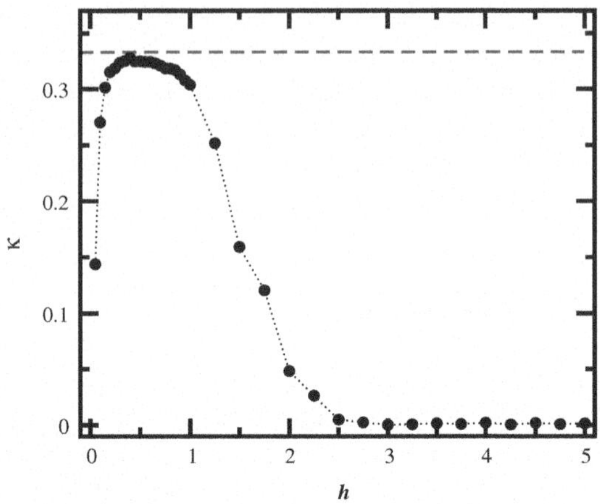

Fig. 9 Depth κ of the correlation hole *versus* disorder strength h. Dashed line corresponds to the result for full random matrices, $\kappa^{FRM} = 1/3$. $L = 16$, $N_{up} = 8$

4 Conclusions

We close this chapter with a very brief discussion about future plans. In addition to the immediate goal of extending the level of details that we have obtained for the survival probability to other physical observables, we intend to explore how our studies are affected by couplings with an environment. As we have shown, the dynamics of isolated many-body quantum systems depends on several factors, such as the energy of the initial state, the strength of the perturbation that takes the system out of equilibrium, the regime of the Hamiltonian (whether integrable or chaotic), the presence of disorder, the strength of the interactions, and the number of particles that interact simultaneously. Despite these many factors and the different behaviors at different time scales, we have been able to extract general features. What should happen to our picture when external interactions are also included? What will follow from the interplay between internal and external interactions? Which will be the dominant elements controlling the dynamics and can they have different roles at different time scales?

Acknowledgements LFS was supported by the NSF grant No. DMR-1603418. EJTH acknowledges funding from PRODEP-SEP and Proyectos VIEP-BUAP, Mexico.

References

1. Hahn, E.L.: Spin echoes. Phys. Rev. **80**, 580 (1950)
2. Ramanathan, C., Cappellaro, P., Viola, L., Cory, D.G.: Experimental characterization of coherent magnetization transport in a one-dimensional spin system. New J. Phys. **13**, 103015 (2011)
3. Wei, K.X., Ramanathan, C., Cappellaro, P.: Exploring localization in nuclear spin chains. arXiv:1612.05249

4. Leibfried, D., Blatt, R., Monroe, C., Wineland, D.: Quantum dynamics of single trapped ions. Rev. Mod. Phys. **75**, 281 (2003)
5. Jurcevic, P., Lanyon, B.P., Hauke, P., Hempel, C., Zoller, P., Blatt, R., Roos, C.F.: Quasiparticle engineering and entanglement propagation in a quantum many-body system. Nature **511**, 202 (2014)
6. Richerme, P., Gong, Z.X., Lee, A., Senko, C., Smith, J., Foss-Feig, M., Michalakis, S., Gorshkov, A.V., Monroe, C.: Non-local propagation of correlations in quantum systems with long-range interactions. Nature **511**, 198 (2014)
7. Bloch, I., Dalibard, J., Zwerger, W.: Many-body physics with ultracold gases. Rev. Mod. Phys. **80**, 885 (2008)
8. Schreiber, M., Hodgman, S.S., Bordia, P., Lüschen, H.P., Fischer, M.H., Vosk, R., Altman, E., Schneider, U., Bloch, I.: Observation of many-body localization of interacting fermions in a quasirandom optical lattice. Science **349**, 842 (2015)
9. Kaufman, A.M., Eric Tai, A.L.M., Rispoli, M., Schittko, R., Preiss, P.M., Greiner, M.: Quantum thermalization through entanglement in an isolated many-body system. Science **353**, 794 (2016)
10. Borgonovi, F., Izrailev, F.M., Santos, L.F., Zelevinsky, V.G.: Quantum chaos and thermalization in isolated systems of interacting particles. Phys. Rep. **626**, 1 (2016)
11. He, K., Santos, L.F., Wright, T.M., Rigol, M.: Single-particle and many-body analyses of a quasiperiodic integrable system after a quench. Phys. Rev. A **87**, 063637 (2013)
12. Rigol, M., Santos, L.F.: Quantum chaos and thermalization in gapped systems. Phys. Rev. A **82**, 011604(R) (2010)
13. Santos, L.F.: Transport control in low-dimensional spin-1/2 Heisenberg systems. Phys. Rev. E **78**, 031125 (2008)
14. Rego, L.G.C., Santos, L.F., Batista, V.S.: Coherent control of quantum dynamics with sequences of unitary phase-kick pulses. Annu. Rev. Phys. Chem. **60**, 293 (2009)
15. Santos, L.F.: Transport and control in one-dimensional systems. J. Math. Phys. **50**, 095211 (2009)
16. Santos, L.F., Mitra, A.: Domain wall dynamics in integrable and chaotic spin-1/2 chains. Phys. Rev. E **84**, 016206 (2011)
17. Santos, L.F., Dykman, M.I.: Quantum interference-induced stability of repulsively bound pairs of excitations. New J. Phys. **14**, 095019 (2012)
18. Santos, L.F., Borgonovi, F., Izrailev, F.M.: Chaos and statistical relaxation in quantum systems of interacting particles. Phys. Rev. Lett. **108**, 094102 (2012)
19. Santos, L.F., Borgonovi, F., Izrailev, F.M.: Onset of chaos and relaxation in isolated systems of interacting spins-1/2: energy shell approach. Phys. Rev. E **85**, 036209 (2012)
20. Santos, L.F., Pérez-Bernal, F.: Structure of eigenstates and quench dynamics at an excited-state quantum phase transition. Phys. Rev. A **92**, 050101 (2015)
21. Santos, L.F., Távora, M., Pérez-Bernal, F.: Excited-state quantum phase transitions in many-body systems with infinite-range interaction: localization, dynamics, and bifurcation. Phys. Rev. A **94**, 012113 (2016)
22. F. Pérez-Bernal, L.F. Santos: Structure of eigenstates and quench dynamics at an excited-state quantum phase transition. arXiv:1604.06851 (accepted at Fortschr. Phys.)
23. Santos, L.F., Borgonovi, F., Celardo, G.L.: Cooperative shielding in many-body systems with long-range interaction. Phys. Rev. Lett. **116**, 250402 (2016)
24. Távora, M., Torres-Herrera, E.J., Santos, L.F.: Inevitable power-law behavior of isolated many-body quantum systems and how it anticipates thermalization. Phys. Rev. A **94**, 041603 (2016)
25. Távora, M., Torres-Herrera, E.J., Santos, L.F.: Power-law decay exponents: a dynamical criterion for predicting thermalization. Phys. Rev. A **95**, 013604 (2017)
26. Torres-Herrera, E.J., Santos, L.F.: Quench dynamics of isolated many-body quantum systems. Phys. Rev. A **89**, 043620 (2014)
27. Torres-Herrera, E.J., Vyas, M., Santos, L.F.: General features of the relaxation dynamics of interacting quantum systems. New J. Phys. **16**, 063010 (2014)
28. Torres-Herrera, E.J., Santos, L.F.: Nonexponential fidelity decay in isolated interacting quantum systems. Phys. Rev. A **90**, 033623 (2014)

29. Torres-Herrera, E.J., Santos, L.F.: Local quenches with global effects in interacting quantum systems. Phys. Rev. E **89**, 062110 (2014)
30. Torres-Herrera, E.J., Kollmar, D., Santos, L.F.: Relaxation and thermalization of isolated many-body quantum systems. Phys. Scr. T **165**, 014018 (2015)
31. Torres-Herrera, E.J., Santos, L.F.: Dynamics at the many-body localization transition. Phys. Rev. B **92**, 014208 (2015)
32. Torres-Herrera, E.J., Távora, M., Santos, L.F.: Survival probability of the néel state in clean and disordered systems: an overview. Braz. J. Phys. **46**, 239 (2016)
33. Torres-Herrera, E.J., Santos, L.F.: Extended nonergodic states in disordered many-body quantum systems. Ann. Phys. (Berlin). 1600284 (2017)
34. Torres-Herrera, E.J., Karp, J., Távora, M., Santos, L.F.: Realistic many-body quantum systems vs. full random matrices: static and dynamical properties. Entropy **18**, 359 (2016)
35. Torres-Herrera, E.J., Santos, L.F.: Dynamical manifestations of quantum chaos: correlation hole and bulge. arXiv:1702.04363 (to appear in the Phil. Trans. R. Soc. A)
36. Torres-Herrera, E.J., García-García, A., Santos, L.F.: Generic dynamical features of quenched interacting quantum systems: survival probability, density imbalance and out-of-time-ordered correlator. arXiv:1704.06272
37. Zangara, P.R., Dente, A.D., Torres-Herrera, E.J., Pastawski, H.M., Iucci, A., Santos, L.F.: Time fluctuations in isolated quantum systems of interacting particles. Phys. Rev. E **88**, 032913 (2013)
38. Santos, L.F., Rigol, M.: Onset of quantum chaos in one-dimensional bosonic and fermionic systems and its relation to thermalization. Phys. Rev. E **81**, 036206 (2010)
39. Santos, L.F., Rigol, M.: Localization and the effects of symmetries in the thermalization properties of one-dimensional quantum systems. Phys. Rev. E **82**, 031130 (2010)
40. Santos, L.F., Polkovnikov, A., Rigol, M.: Entropy of isolated quantum systems after a quench. Phys. Rev. Lett. **107**, 040601 (2011)
41. Santos, L.F., Polkovnikov, A., Rigol, M.: Weak and strong typicality in quantum systems. Phys. Rev. E **86**, 010102 (2012)
42. Torres-Herrera, E.J., Santos, L.F.: Effects of the interplay between initial state and Hamiltonian on the thermalization of isolated quantum many-body systems. Phys. Rev. E **88**, 042121 (2013)
43. Sutherland, B.: Beautiful Models. World Scientific, New Jersey (2005)
44. Caux, J.S., Mossel, J.: Remarks on the notion of quantum integrability. J. Stat Mech: Theory Exp. **2011**, P02023 (2011)
45. Bethe, H.A.: On the theory of metal i. eigenvalues and eigenfunctions of a linear chain of atoms. Z. Phys. **71**, 205 (1931)
46. Casati, G., Valz-Gris, F., Guarnieri, I.: On the connection between quantization of nonintegrable systems and statistical theory of spectra, Lett. Nuovo Cimento (1971–1985) **28**, 279 (1980)
47. Bohigas, O., Giannoni, M.J., Schmit, C.: Characterization of chaotic quantum spectra and universality of level fluctuation laws. Phys. Rev. Lett. **52**, 1 (1984)
48. Guhr, T., Mueller-Gröeling, A., Weidenmüller, H.A.: Random matrix theories in quantum physics: common concepts. Phys. Rep. **299**, 189 (1998)
49. Atas, Y.Y., Bogomolny, E., Giraud, O., Roux, G.: Distribution of the ratio of consecutive level spacings in random matrix ensembles. Phys. Rev. Lett. **110**, 084101 (2013)
50. Santos, L.F.: Integrability of a disordered Heisenberg spin-1/2 chain. J. Phys. A **37**, 4723 (2004)
51. Hsu, T.C., d'Auriac, J.C.A.: Level repulsion in integrable and almost-integrable quantum spin models. Phys. Rev. B **47**, 14291 (1993)
52. Kudo, K., Deguchi, T.: Level statistics of xxz spin chains with discrete symmetries: analysis through finite-size effects. J. Phys. Soc. Jpn. **74**, 1992 (2005)
53. Gubin, A., Santos, L.F.: Quantum chaos: an introduction via chains of interacting spins 1/2. Am. J. Phys. **80**, 246 (2012)
54. Santos, L.F., Rigolin, G., Escobar, C.O.: Entanglement versus chaos in disordered spin systems. Phys. Rev. A **69**, 042304 (2004)
55. Mehta, M.L.: Random Matrices. Academic Press, Boston (1991)
56. Haake, F.: Quantum Signatures of Chaos. Springer, Berlin (1991)

57. Reichl, L.E.: The transition to chaos: conservative classical systems and quantum manifestations. Springer, New York (2004)
58. Brody, T.A., Flores, J., French, J.B., Mello, P.A., Pandey, A., Wong, S.S.M.: Random-matrix physics—spectrum and strength fluctuations. Rev. Mod. Phys **53**, 385 (1981)
59. Izrailev, F.: Quantum localization and statistics of quasienergy spectrum in a classically chaotic system. Phys. Lett. A **134**, 13 (1988)
60. Izrailev, F.M.: Simple models of quantum chaos: spectrum and eigenfunctions. Phys. Rep. **196**, 299 (1990)
61. Jacquod, P., Shepelyansky, D.L.: Emergence of quantum chaos in finite interacting fermi systems. Phys. Rev. Lett. **79**, 1837 (1997)
62. Dukesz, F., Zilbergerts, M., Santos, L.F.: Interplay between interaction and (un)correlated disorder in one-dimensional many-particle systems: delocalization and global entanglement. New J. Phys. **11**, 043026 (2009)
63. Wigner, E.P.: On a class of analytic functions from the quantum theory of collisions. Ann. Math. **53**, 36 (1951)
64. Wigner, E.P.: Characteristic vectors of bordered matrices with infinite dimensions. Ann. Math. **62**, 548 (1955)
65. Zelevinsky, V., Brown, B.A., Frazier, N., Horoi, M.: The nuclear shell model as a testing ground for many-body quantum chaos. Phys. Rep. **276**, 85 (1996)
66. Torres-Herrera, E.J., Santos, L.F.: In: AIP Proceedings, ed. by Danielewicz, P., Zelevinsky, V. (APS, East Lansing, Michigan, 2014)
67. Flambaum, V.V., Izrailev, F.M.: Unconventional decay law for excited states in closed many-body systems. Phys. Rev. E **64**, 026124 (2001)
68. Izrailev, F.M., Castañeda-Mendoza, A.: Return probability: exponential versus gaussian decay. Phys. Lett. A **350**, 355 (2006)
69. Bhattacharyya, K.: Quantum decay and the Mandelstam-Tamm-energy inequality. J. Phys. A **16**, 2993 (1983)
70. Ufink, J.: The rate of evolution of a quantum state. Am. J. Phys. **61**, 935 (1993)
71. Giovannetti, V., Lloyd, S., Maccone, L.: Quantum limits to dynamical evolution. Phys. Rev. A **67**, 052109 (2003)
72. Santos, L.F., Dykman, M.I., Shapiro, M., Izrailev, F.M.: Strong many-particle localization and quantum computing with perpetually coupled qubits. Phys. Rev. A **71**, 012317 (2005)
73. Nandkishore, R., Huse, D.: Many-body localization and thermalization in quantum statistical mechanics. Annu. Rev. Condens. Matter Phys. **6**, 15 (2015)
74. Khalfin, L.A.: Contribution to the decay theory of a quasi-stationary state. Sov. Phys. JETP **6**, 1053 (1958)
75. Muga, J.G., Ruschhaupt, A., del Campo, A.: Time in Quantum Mechanics, vol. 2. Springer, London (2009)
76. Urbanowski, K.: General properties of the evolution of unstable states at long times. Eur. Phys. J. D **54**, 25 (2009)
77. del Campo, A.: Exact quantum decay of an interacting many-particle system: the calogero-sutherland model. New J. Phy. **18**, 015014 (2016)
78. Serbyn, M., Papić, Z., Abanin, D.A.: Thouless energy and multifractality across the many-body localization transition (2016). arXiv:1610.02389
79. Chalker, J.T., Daniell, G.J.: Scaling, diffusion, and the integer quantized Hall effect. Phys. Rev. Lett. **61**, 593 (1988)
80. Chalker, J.: Scaling and eigenfunction correlations near a mobility edge. Physica A **167**(1), 253 (1990)
81. Ketzmerick, R., Petschel, G., Geisel, T.: Slow decay of temporal correlations in quantum systems with cantor spectra. Phys. Rev. Lett. **69**, 695 (1992)
82. Huckestein, B., Schweitzer, L.: Relation between the correlation dimensions of multifractal wave functions and spectral measures in integer quantum Hall systems. Phys. Rev. Lett. **72**, 713 (1994)

83. Huckestein, B., Klesse, R.: Wave-packet dynamics at the mobility edge in two- and three-dimensional systems. Phys. Rev. B **59**, 9714 (1999)
84. Cuevas, E., Kravtsov, V.E.: Two-eigenfunction correlation in a multifractal metal and insulator. Phys. Rev. B **76**, 235119 (2007)
85. Kravtsov, V.E., Ossipov, A., Yevtushenko, O.M.: Return probability and scaling exponents in the critical random matrix ensemble. J. Phys. A **44**, 305003 (2011)
86. Leviandier, L., Lombardi, M., Jost, R., Pique, J.P.: Fourier transform: a tool to measure statistical level properties in very complex spectra. Phys. Rev. Lett. **56**, 2449 (1986)
87. Guhr, T., Weidenmüller, H.: Correlations in anticrossing spectra and scattering theory. Analytical aspects. Chem. Phys. **146**, 21 (1990)
88. Alhassid, Y., Levine, R.D.: Spectral autocorrelation function in the statistical theory of energy levels. Phys. Rev. A **46**, 4650 (1992)
89. Gorin, T., Seligman, T.H.: Signatures of the correlation hole in total and partial cross sections. Phys. Rev. E **65**, 026214 (2002)

Multi-jittering Instability in Oscillatory Systems with Pulse Coupling

Vladimir Klinshov, Leonhard Lücken, Serhiy Yanchuk and Vladimir Nekorkin

Abstract In oscillatory systems with pulse coupling regular spiking regimes may destabilize via a peculiar scenario called "multi-jitter instability". At the bifurcation point numerous so-called "jittering" regimes with distinct inter-spike intervals emerge simultaneously. Such regimes were first discovered in a single oscillator with delayed pulse feedback and later were found in networks of coupled oscillators. The present chapter reviews recent results on multi-jitter instability and discussed its features.

1 Introduction

Time delays naturally arise in physical systems of various nature [1–4]. The most common origin of time delays is the inertness of systems and finite speed of signal propagation in media. As a consequence, the current state of the system may influence not only its present, but future dynamics as well. A prototypical example of a system with time delays is a neural network [2]. When a neuron exceeds the excitation threshold it emits a spike which does not immediately effect the peer neurons. It takes some time for the spike to propagate along the axon, and extra time to

V. Klinshov (✉) · V. Nekorkin
Institute of Applied Physics of the Russian Academy of Sciences,
46 Ul'yanov Street, 603950 Nizhny Novgorod, Russia
e-mail: vladimir.klinshov@ipfran.ru

V. Nekorkin
e-mail: vnekorkin@neuron.appl.sci-nnov.ru

L. Lücken
German Aerospace Center (DLR), Institute of Transportation Systems,
Rutherford str. 2, 12489 Berlin, Germany
e-mail: leonhard.luecken@dlr.de

S. Yanchuk
Technical University of Berlin, Institute of Mathematics,
Straße des 17. Juni 136, 10623 Berlin, Germany
e-mail: yanchuk@math.tu-berlin.de

M. Edelman et al. (eds.), *Chaotic, Fractional, and Complex Dynamics: New Insights and Perspectives*, Understanding Complex Systems,
https://doi.org/10.1007/978-3-319-68109-2_13

activate the synapse. The combination of these two processes leads to the emergence of nonzero time delay between the spike emission and the resultant perturbation of the target neuron. Typically, time delays in neural networks do not exceed several milliseconds, whereas the timescales of the neuronal dynamics range from tens to hundreds of milliseconds. However, in other systems delays may be comparable and ever larger than the characteristic time scales. Such situations are typical in laser physics where the impact of delays is significant (e.g. see [4, 5] and references within). Particularly, the time-delay feedback or coupling is known to be an effective tool for dynamics control, and very long delay lines are sometimes used [4, 6]. Moreover, in some cases the delays may be non-constant but depend on time or system state. For example, in metal cutting the cutting force depends dynamical displacements of the structure at the present and the previous cut, which leads to the emergence of time-varying delays [7].

Numerous theoretical research demonstrate that introduction of delays causes significant change in dynamics. Delay-induced synchronization and desynchronization were reported, emergence of multistable synchronous solutions was often observed, and the properties of existing synchronous regimes were shown to be altered after the delay introduction (see the reviews [8]). The influence of delays may result in new dynamical phenomena, such as multistability [9–12], oscillations death [13], transition to chaos [14, 15], and other complicated regimes [3, 4, 16–18]. Theoretical results inspired experimental study of synchronization of oscillators with delayed coupling. Various configurations of interacting lasers were studied with coupling delays arising from the finite speed of light propagation along fibers [19–22]. Electronic [14, 23, 24] and chemicals [25] oscillators have also been considered with delayed coupling realized with the help of digital delay lines.

From the mathematical point of view, systems with time delays correspond to delay-differential equations. The simplest form of such an equation is the following:

$$\frac{dx(t)}{dt} = f(x(t), x(t-\tau)), \tag{1}$$

Here, $x \in \mathbb{R}^n$ is the system state vector and τ is the delay. The r.h.s of (1) depends not only on the current state of the system, but also on its state in the past. The consequence is that the phase space of (1) is the space of functions $[-\tau, 0] \mapsto \mathbb{R}^n$ that has infinite dimension [26, 27]. The infinite dimension of the phase space makes DDEs much more complicated objects than ODEs for both analytic and numerical study. However, in certain special cases systems with time delays may be reduced to finite dimension which simplifies their analysis and allows to apply well developed techniques developed for the finite-dimensional systems. The present chapter is devoted to one of such cases, namely the case of pulse-mediated interactions.

In the present chapter we consider the dynamics of periodic oscillators with delayed pulse interactions. Periodic behavior is typical for a variety of physical, chemical, biological, and other systems. Independently of its nature, the state of a periodic oscillator can be naturally determined by its phase [28], that is, the single variable indicating the position of the system within its cycle. The concept of the phase proved to be exceptionally useful for the study of driven and coupled oscillators [28–30]. In the absence of external stimulation, the phase φ of the oscillator grows uniformly with $d\varphi/dt = \omega$, where $\omega = T^{-1}$ is the natural frequency of the oscillator, T its natural period. The phase changes on the circle [0,1] with $\varphi = 0$ and $\varphi = 1$ identified.

Interaction via pulse-like signals is important in neuron populations [31–35], biological [28, 36, 37], optical and opto-electronic systems [38–41]. To define pulse-coupled systems, let us first consider a dynamical system under external forcing.

In order to describe the response of phase oscillators to pulse stimulation the so-called phase response curve (PRC) is widely used [28, 35, 42–45]. The PRC defines the oscillator's response to a single short stimulus (pulse). It determines the resulting phase shift caused by a pulse given that the latter arrived at the phase φ. The PRC can be calculated numerically or measured experimentally for oscillatory systems of different origin. These properties make it a useful tool for the study of forced or coupled oscillators [35, 37, 43, 46–49], and it is especially appropriate in neuroscience where the interactions are mediated by pulses.

In this chapter we study phase-reduced models and describe each oscillator of the network by its phase φ_j. The dynamics of a network of phase oscillators with pulse delayed coupling is governed by

$$\frac{d\varphi_i}{dt} = \omega_i + a_{ji} Z_{ji}(\varphi_i) \sum_{t_j^s} \delta(t - t_j^s - \tau_{ij}), \tag{2}$$

where a_{ji} is the adjacency matrix of the network, $Z_{ji}(\varphi)$ is the PRC for the i-th oscillator under the pulse arriving from the j-th one, and t_j^s are the moments when the j-th oscillator produces spikes. The moments t_j^s of spike emission are determined from the intrinsic dynamics of the j-th oscillator. Namely, we say that the oscillator emits a spike when it phase reaches one and resets to zero after that.

Note that the delayed terms in (2) arise only as the arguments of the delta-functions, i.e. the delayed terms differ from zero only at discrete moments of the pulses arrival. The dynamics of the network therefore only depends on the values of the phases at those very time moments, while the history of the phase evolution between those moments is of no importance. This observation is the key point, which allows a dimensional reduction of the network dynamics. In [50] it was shown that the network (2) of oscillators with delayed pulsatile coupling can be reduced to a finite-dimensional discrete map under quite general conditions. Below in Sects. 2 and 5 we will derive such maps for certain particular cases and use them for the analysis.

The dynamics of (2) depends on many aspects–the shape and the magnitude of the PRCs, the network topology and the value of the delays. In the following, we will present the results on an interesting type of dynamics described recently for networks with pulse delayed coupling: the so-called "jittering regimes". These regimes occur when regular spiking activity, where all oscillators generate spikes with equal interspike intervals (ISIs), destabilizes, and spikes are emitted with distinct inter-spike intervals. It turns out that for the emergence of such the regimes all the aspects of the network are important: the PRC shape, the delays, and the network structure. However, although the role of the first two, the PRC shape and the delay, have been studied in detail, the role of the network structure is much less clear. The jittering regimes have been studied in detail for a single oscillator with delayed feedback and a feed-forward ring of oscillators [51–53]. As for more complex setups, similar regimes have been discovered in some of them, but the detailed study is still lacking.

The rest of the chapter is organized as follows. Sections 2–3 are devoted to the simplest system in which the jittering instability is observed, namely a single oscillator with delayed feedback. The results reviewed in these sections were first reported in [51, 52]. In Sect. 2 we derive the discrete map that describes the dynamics of the oscillator. In Sect. 3 we use this map to study the jittering regimes describe their features. Sections 5–6 are devoted to feed-forward rings and summarize the results from [53]. In Sect. 5 we study regular rotating waves in feed-forward rings of oscillators, and in Sect. 6 we show how they destabilize and give birth to jittering waves. In the last Sect. 7 we discuss the general features of the multi-jitter instability and give preliminary results which demonstrate its presence in networks with more complex structure.

2 Dynamics of One Oscillator with Pulse Delayed Feedback

The simplest possible case for (2) is when only one oscillator with a self-feedback is included. In this case the system is described by

$$\frac{d\varphi}{dt} = 1 + Z(\varphi) \sum_{t_s} \delta(t - t_s - \tau), \tag{3}$$

where t_s are the instants when the pulses are emitted.

First, let us calculate the inter-spike interval (ISI) of the oscillator subject to the pulse input. Suppose the oscillator produced a spike at $t = t_s$ and let us calculate the moment of the next spike emission t_{s+1}. The derivation is illustrated in Fig. 1. After the spike at $t = t_s$, the the phase grows uniformly from $\varphi(t_s) = 0$ until it reaches

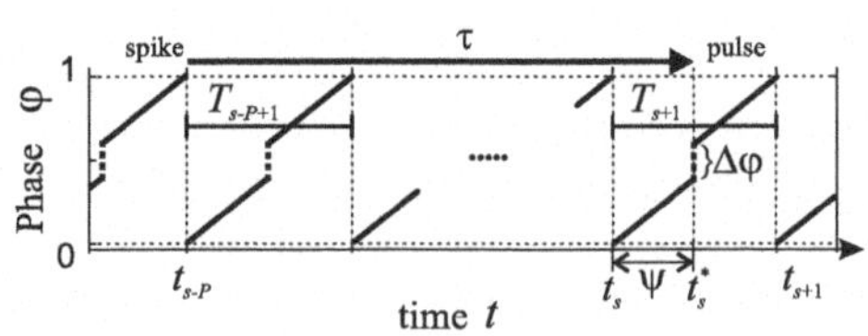

Fig. 1 Derivation of the map (5): the new ISI T_{s+1} depends on the pulse emitted at $t = t_{s-P}$. ψ is the phase at which the pulse arrives to the oscillator, and $\Delta\varphi$ is the phase shift it causes

unity, except for the moments when pulses arrive, which were generated previously. Assuming that only one pulse arrives per ISI we can determine the arrival time $t_s^* \in (t_s, t_{s+1})$ of a pulse emitted at some prior spike time t_{s-P} as

$$t_s^* = t_{s-P} + \tau = t_s + \tau - \sum_{k=s-P+1}^{s} T_k.$$

Here, $T_k = t_k - t_{k-1}$, and P is the number of spikes emitted by the oscillator between the time t_{s-P} when the spike is emitted and t_s^* when it arrives at the oscillator. The phase value $\psi > 0$ at the moment of the pulse arrival equals

$$\psi = t_s^* - t_s = \tau - \sum_{k=s-P+1}^{s} T_k. \tag{4}$$

Hence the oscillator's phase immediately after the pulse arrival is $\varphi(t_s^* + 0) = \psi + \Delta\varphi$, where the phase shift $\Delta\varphi = Z(\psi)$. Further the phase grows uniformly until it reaches unity, and the ISI $T_{s+1} = t_{s+1} - t_s$ is determined from the condition $\varphi(t_{s+1}) = 1$:

$$\begin{aligned} 1 = \varphi(t_{s+1}) &= \psi + Z(\psi) + t_{s+1} - t_s^* \\ &= T_{s+1} + Z\left(\tau - \sum_{k=s-P+1}^{s} T_k\right). \end{aligned}$$

Hence, we obtain

$$T_{s+1} = 1 - Z\left(\tau - \sum_{k=s-P+1}^{s} T_k\right). \tag{5}$$

Equation (5) determines the next inter-spike interval in dependence of the P previous ones. In fact, it is a discrete map describing the dynamics of the oscillator. The dimensionality of the map equals P, since it is defined on the space of vectors $\mathbf{T}_s = (T_{s-P+1}, \ldots, T_s) \in \mathbb{R}^P$. The number P equals the number of spikes stored in the memory of the delay line, and is proportional to the value of the delay.

Let us first study the fixed points of the map (5). They correspond to dynamical regimes when all the inter-spikes intervals are equal, i.e. the oscillator emits spikes periodically with $T_s = T$. This implies, that there is exactly one spike within each inter-spike interval and the map (5) is valid. We call such the regime regular spiking, and its period T is given as

$$T = 1 - Z(\tau - PT), \tag{6}$$

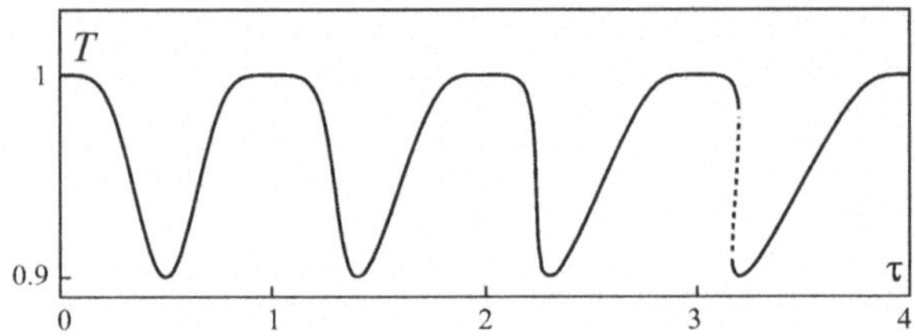

Fig. 2 One-dimensional bifurcation diagram for the period T of regular spiking solutions versus delay τ according to (7). Red solid lines correspond to stable, blue dashed lines to unstable regular spiking regimes. The squares indicate fold bifurcations

Fig. 3 The phase response curve (12) for $q = 5$ (dashed line) and $q = 28$ (solid line)

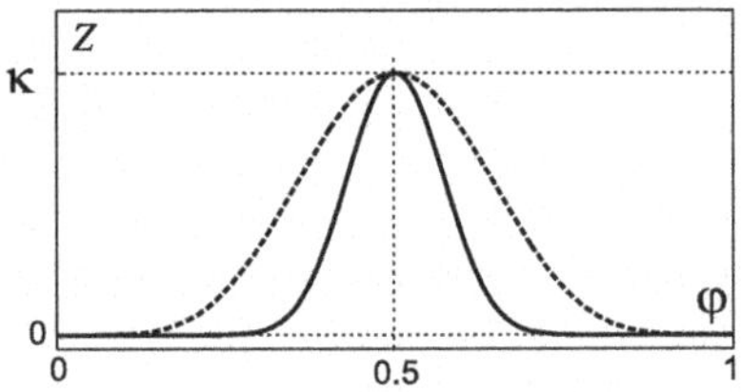

where $P = [\tau/T]$ is the number of periods within the delay. Here $[\tau/T]$ denotes the integer part of the expression τ/T, i.e. the number of ISIs contained in τ. Thus, $\tau = PT + \psi$ with $\psi := \tau \mod T$, and we can write Eq. (6) in the parametric form:

$$\begin{aligned} T &= 1 - Z(\psi), \\ \tau &= P(1 - Z(\psi)) + \psi. \end{aligned} \tag{7}$$

This form allows an explicit representation of all regular spiking regimes as a function of τ by substituting $P = 0, 1, 2, \dots$ in (7) and varying ψ within the interval [0,1]. In Fig. 2 the period of these regimes is plotted versus the delay for the PRC $Z(\varphi) = \kappa \sin^q(\pi\varphi)$ and $\kappa = 0.1$, $q = 5$ [see Fig. 3]. As one may see, at least one regular spiking solution exists for each delay. For large delays folding takes place leading to an emergence of several solution branches.

The stability of the regular spiking can be also analyzed with the help of the map (5). For this sake small perturbations δ_s are introduced to the initial conditions so that $T_s = T + \delta_s$. The regime is stable if the perturbations are damped with time. Small perturbations do not violate the property that only one spike occurs within each inter-spike interval, so the map (5) can be used to study their evolution. Substitute $T_s = T + \delta_s$ into (5) and its linearization leads to the map for the perturbations

$$\delta_{s+1} = Z'(\psi) \sum_{k=s-P+1}^{s} \delta_k, \tag{8}$$

where $\psi = \tau - PT$. The stability of the linear map (8) may be determined by the roots of the characteristic polynomial

$$\chi_{P,\alpha}(\lambda) = \lambda^P - \alpha \sum_{k=0}^{P-1} \lambda^k \tag{9}$$

where $\alpha := Z'(\psi)$. These roots are the characteristic multipliers of the regular spiking regime and describe how small perturbations are either damped (if all roots fulfill $|\lambda| < 1$) or amplified ($|\lambda| > 1$ for at least one root). Thus, a change of stability induced by a change of the parameter α must be accompanied by at least one multiplier λ having an absolute value equal to one. Note that the parameters of the original system, which determine the value of α are the PRC and the delay time τ.

To study the roots of (9) it is convenient to multiply $\chi_{P,\alpha}(\lambda)$ by $(\lambda - 1)$ and study the extended characteristic equation

$$\tilde{\chi}_{P,\alpha}(\lambda) = (\lambda - 1)\,\chi_{P,\alpha}(\lambda) = \lambda^{P+1} - (1+\alpha)\lambda^P + \alpha = 0. \tag{10}$$

The set $\tilde{\Lambda}$ of roots of $\tilde{\chi}_{P,\alpha}(\lambda)$ contains all roots $\Lambda = \{\lambda_1, \ldots, \lambda_P\}$ of $\chi_{P,\alpha}(\lambda)$ and the root $\lambda_{P+1} = 1$, i.e. $\Lambda = \tilde{\Lambda}\backslash\{1\}$ if not $1 \in \Lambda$. The case $1 \in \Lambda$ can only occur for $\alpha = 1/P$ as we show below.

Critical roots of $\tilde{\chi}_{P,\alpha}(\lambda)$ are characterized by $|\lambda| = 1$ and can therefore be written as $\lambda = e^{i\varphi} \in \mathbb{C}$ with $\varphi \in \mathbb{R}$. Substituting $\lambda = e^{i\varphi}$ into (10) we obtain

$$e^{i(P+1)\varphi} + \alpha = (1+\alpha)e^{iP\varphi}. \tag{11}$$

Taking the absolute value on both sides of (11) yields

$$\left|e^{i(P+1)\varphi} + \alpha\right| = |1+\alpha|,$$

which, for $\alpha \neq 0$, implies $e^{i(P+1)\varphi} = 1$. This means $\varphi = 2\pi k/(P+1)$ for some $k \in \mathbb{Z}$. Substituting this into (11) gives

$$1 + \alpha = (1+\alpha)e^{i2\pi kP/(P+1)}.$$

For $\alpha \neq -1$, this requires $k \in (P+1)\mathbb{Z}$. Thus, for $\alpha \notin \{-1, 0\}$, $\lambda = 1$ is the only solution of (10) with $|\lambda| = 1$. It corresponds to a critical multiplier of (9) only for $\alpha = 1/P$, where it is a double root of (10). Indeed, substituting $\lambda = 1$ into (9) one obtains $1 - P\alpha = 0$. For $\alpha = -1$, (10) reduces to

$$\lambda^{P+1} = 1.$$

Hence, P critical multipliers $\lambda_k = e^{i2\pi k/(P+1)}$, $k = 1, \ldots, P$, appear simultaneously at $\alpha = -1$. For $\alpha = 0$, (9) reduces to $\lambda^P = 0$, which obviously exhibits no critical multipliers.

Finally, we can formulate the properties of the spectrum of (9) as follows. The spectrum is stable for $-1 < \alpha < 1/P$. A critical multiplier $\lambda = 1$ emerges at $\alpha = 1/P$ and remains unstable for $\alpha > 1/P$. This scenario corresponds to the fold bifurcation, as illustrated in Fig. 2. A much more peculiar scenario takes place at $\alpha = -1$ when P critical multipliers $\lambda_k = e^{i2\pi k/(P+1)}$, $k = 1, \ldots, P$, emerge simultaneously. For $\alpha < -1$, these P multipliers remain unstable. Note that the condition for this bifurcation depends only on the slope α of the PRC function and does not depend on the number P. Increasing the delay one may make P arbitrary large, so that the bifurcation seems to become very degenerate. However, it occurs generically in our setup if only the PRC contains intervals with slope < -1. In the following we will study corresponding bifurcation scenario in more detail.

3 Jittering Regimes in a Single Oscillator

For numerical illustrations we consider the PRC

$$Z(\varphi) = \kappa \times (\sin(\pi\varphi))^q , \tag{12}$$

which is shown in Fig. 3. Here $\kappa = 0.1$ is the feedback strength, and $q > 1$ is a parameter that controls the steepness of the PRC, which appears to be an important quantity for the studied dynamical phenomenon. However, our analysis is valid for an arbitrary shape of the PRC.

When q the steepness of $Z(\varphi)$ is moderate. In particular it fulfills $Z'(\psi) > -1$ everywhere. Thus, the only possible regime of (3) is regular spiking. Increasing q until it exceeds a critical value q^*, leads to the appearance of two points $\psi_{A,B} \in (0, 1)$ for which $Z'(\psi_{A,B}) = -1$. This means that for appropriate values of the delay time τ, such that

$$\psi = \tau \bmod T \in \{\psi_A, \psi_B\},$$

there are $P = [\tau/T]$ multipliers located on the unit circle simultaneously. From (7) the corresponding delay times are determined as

$$\tau^P_{A,B} = P(1 - Z(\psi_{A,B})) + \psi_{A,B}. \tag{13}$$

At the points $\tau = \tau^P_{A,B}$ we can observe an interesting bifurcation structure for increasing values of P. Figure 4a shows a bifurcation diagram obtained from numerical simulations of the system (3) for a range of values $\tau \in [0,4]$ and with $Z(\varphi)$ given by (12) with $q = 28 > q^*$. For $P = 1$, only one multiplier $\lambda = -1$ of (9) crosses the unit circle at τ^1_A, resp. τ^1_B. This gives rise to a pair of supercritical period doubling bifurcations enclosing the region $[\tau^1_A, \tau^1_B]$ where the regular spiking is unstable and a stable spiking regime exists, which exhibits two different ISI lengths denoted by Θ_1 and Θ_2, see Fig. 4b and d. From the perspective of the discrete mapping (5) this corresponds to a stable period-2 solution In the following we will denote such a solution as $(\overline{\Theta_1, \Theta_2})$, where the periodicity of the ISI sequence is indicated by an overline. Note that map (5) is one-dimensional for $P = 1$, and $\Theta_{1,2}$ must satisfy

$$\Theta_{1,2} = 1 - Z\left(\tau - \Theta_{2,1}\right) = 1 - Z\left(\tau - 1 - Z(\tau - \Theta_{1,2})\right). \tag{14}$$

For $P \geq 2$, P multipliers become critical simultaneously at $\tau = \tau^P_{A,B}$ and the regular spiking changes its stability at $\tau \in \{\tau^P_A, \tau^P_B\}$. Numerical studies show that various irregular spiking regimes appear in this interval. The observed solutions are constituted by distinct inter-spike intervals and have high periods. For moderate supercritical steepness $Z'(\psi) \lesssim -1$ their period typically equals $P + 1$, and the ISIs are close to the period of regular spiking The property that the ISIs are not constant anymore but "jitter" with time around an average value led us to adopt the term "jittering" regimes.

Figure 4e shows an example of a "jittering" solution $\overline{(T_1, T_1, T_2, T_2)}$ involving two different ISIs. Interestingly, despite of the high period, most of the emergent jittering solutions consist of only two distinct inter-spike intervals. The typical $(P + 1)$-periodicity is a consequence of the irregular ordering of these two ISIs. As a result, each solution corresponds to only two, and not $P + 1$, points in the bifurcation diagram. In the following we call such solutions "bipartite". For larger P, a variety of different bipartite solutions with $(P + 1)$-periodic ISI sequences can be observed in a neighborhood of the interval (τ^P_A, τ^P_B) where the regular spiking is unstable. The stability regions of solutions alternate and may overlap leading to regions of multistability.

The origin of the bipartite solutions is combinatorial and intrinsically related to the period-2 solution that emerges at $P = 1$. Namely, it is possible to construct all

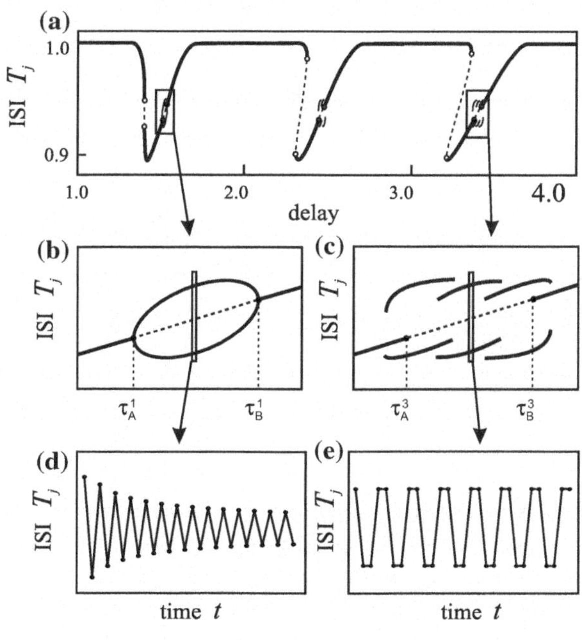

Fig. 4 **a** One-dimensional, numerical bifurcation diagram of (3), resp. (5), for the ISIs T_j versus the delay τ with $q = 28$. Red dots correspond to attractors, blue dashed lines corresponds to unstable regular spiking as obtained from (5). Fold bifurcations are indicated by blue hollow circles, while solid blue circles indicate the multi-jitter bifurcations arising at $Z'(\psi) = -1$, see main text. **b** and **c** are zooms of (**a**). **d** A stable period-2 solution for $\tau = 1.5$. **f** A stable period-4 solution $\overline{(T_1, T_1, T_2, T_2)}$ for $\tau = 3.38$

bipartite solutions from the period-2 solution$\overline{(\Theta_1, \Theta_2)}$. Indeed, consider an arbitrary $(P+1)$-periodic sequence of ISIs $\overline{(T_1, T_2, \ldots, T_{P+1})}$, where each T_j equals one of the solutions $\Theta_{1,2}$ of (14) for some delay $\tau = \tau_0 \in \left[\tau_A^1, \tau_B^1\right]$. Let $n_1 \geq 1$ and $n_2 \geq 1$ be the number of ISIs equal to Θ_1 and to Θ_2 respectively. Then it is readily checked that the constructed sequence is a solution of (5) at the feedback delay time

$$\tau_{n_1,n_2} = \tau_0 + (n_1 - 1)\,\Theta_1 + (n_2 - 1)\,\Theta_2. \tag{15}$$

Indeed, let us calculate the next ISI consulting (5) and (15). Note that,

$$\begin{aligned} T_{s+1} &= 1 - Z\left(\tau - (n_1\Theta_1 + n_2\Theta_2) + T_{s-P}\right) \\ &= 1 - Z\left(\tau_0 - \Theta_1 - \Theta_2 + T_{s-P}\right). \end{aligned}$$

Due to (14) this yields $T_{s+1} = T_{s-P}$. Therefore, the sequence is in fact a $(P+1)$-periodic point of (5).

Although bipartite solutions are the most common jittering solutions, others can also be observed. Interestingly, some of them may also be obtained analytically. To do this let us consider an equation

$$1 - T = Z(T - \theta), \tag{16}$$

where $\theta > 0$ is a constant and $T \in [\theta, \theta + 1]$. As will be shown below, this equation can be used as a basis for construction of solutions of the map (5). Indeed, whenever (16) has a solution Θ it implies immediately the existence of RS with period Θ for the delay values $\tau = \Theta - \theta + P\Theta$, $P \in \mathbb{N}$. This observation is not surprising since it corresponds to the well known mechanism of periodic solution reappearance in delay differential equations [9]. However, more intricate is the case when for some value of θ Eq. (16) has two or more solutions Θ_k. In this case one can construct an arbitrary $(P+1)$-periodic sequence $\overline{(T_1, T_2, \ldots, T_{P+1})}$, $T_j \in \{\Theta_k\}$ for arbitrary large P, and each such sequence will be a solution of the map (5) for

$$\tau = \left(\sum_{k=1}^{P+1} T_k\right) - \theta. \tag{17}$$

This statement is readily confirmed by a direct check. Indeed, for each item T_{j+1} of the sequence $\overline{(T_1, T_2, \ldots, T_{P+1})}$ we have

$$T_{s+1} = 1 - Z\left(\tau - \sum_{k=s-P+1}^{s} T_k\right), = T_{s-P} = 1 - Z(T_{s-P} - \theta)$$

which coincides with (5).

Figure 5 illustrates possible solutions to Eq. (16) in the for a PRC Z as defined in (12). If the slope of the right hand side of (12), that is of Z, is not less than -1 for all values of T, only one intersection and only one root Θ_1 of (16) exists (Fig. 5a).

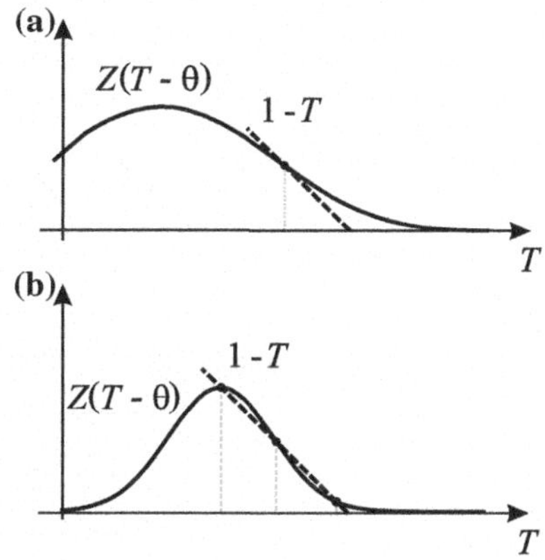

Fig. 5 Solutions of Eq. (16) for **a** the slope > -1 and **b** < -1 The red line corresponds to the right hand side, the blue line to the left hand side of (16)

But if the PRC is steep enough, namely has the slope < -1, three different solutions Θ_1, Θ_2 and Θ_3 of (16) exist within a certain interval of θ (Fig. 5b). In this case, the values of θ corresponding to the emergence of the new roots can be found from the condition that the left and the right parts of (16) contact. These values equal

$$\theta_{A,B} = 1 - \psi_{A,B} - Z(\psi_{A,B}),$$

where $\psi_{A,B}$ are the points where the slope of the PRC equals $Z'\left(\psi_{A,B}\right) = -1$. Equation (16) has three different solutions inside the interval $\theta \in [\theta_A; \theta_B]$ and only one solution outside of this interval. It can not have more than three different solutions for the PRC (24).

This explains the emergence of bipartite solutions, and also tripartite solutions which consist of three dfferent ISIs. These solutions exist for the supercritical steepness, i.e. when Eq. (16) can have different roots Θ_k. Each periodic sequence $(\overline{T_1, \ldots, T_{P+1}})$ of these roots, which has a period that divides $P+1$, corresponds to a periodic solution of the map (5) with τ given by Eq. (17). To construct all solutions for a given P one should compose all possible periodic sequences of period $\leq P+1$ for all values of $\theta \in [\theta_A; \theta_B]$.

A series of the bipartite and tripartite solution branches obtained from (16) is shown in Fig. 6 for $q = 28$ and various P. The obtained solutions coincide with the attractors from the bifurcation diagrams in Fig. 4 and complement the diagrams by parts which are difficult to obtain by direct simulation like unstable and tripartite solutions. Each observed bipartite solution corresponds to a pair of points (τ, T_1) and (τ, T_2) on the diagram; each tripartite solution corresponds to three such points. Note that solution branches which contain the same quantities of each contained ISI coincide. For instance, in the case $P = 3$ the branches corresponding to bipartite solutions of the form $(\overline{T_1, T_2}) \equiv (\overline{T_1, T_2, T_1, T_2})$ and $(\overline{T_1, T_1, T_2, T_2})$ lie on top of each other in Fig. 6c.

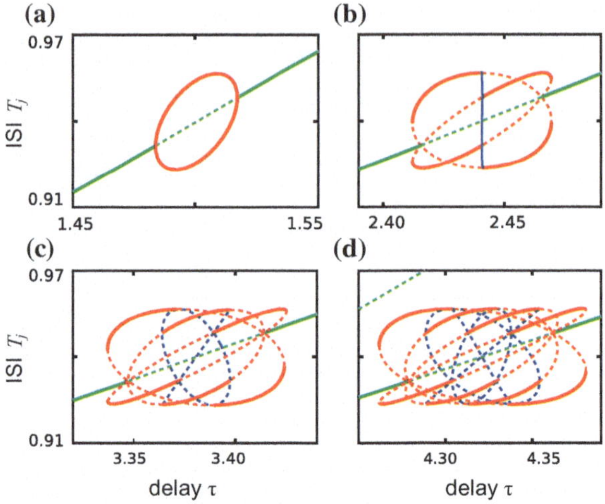

Fig. 6 **a–d** Branches of regular spiking (green), bipartite (red) and tripartite (blue) solutions for $P = 1, \dots, 4$, in system (1) with PRC (12) and $q = 28$. Stable parts of the branches are shown by solid, and unstable by dashed lines

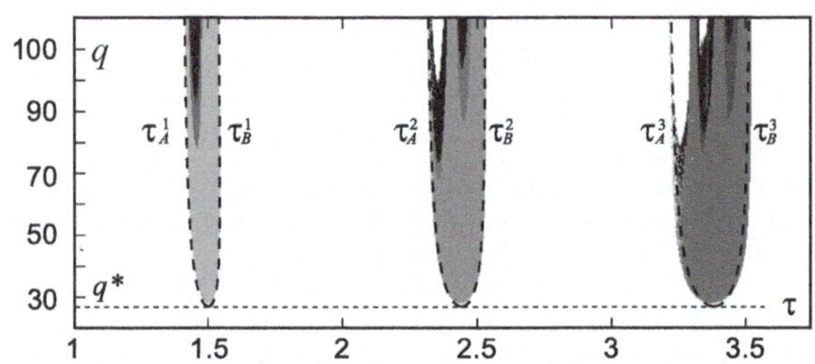

Fig. 7 Numerically obtained two-dimensional bifurcation diagram for system (1) with PRC (12). Free parameters are the delay time τ and the steepness parameter q. The maximal observed attractor period is coded in color. White corresponds to period-1 (regular spiking), shades of gray correspond to finite numbers larger than one (bipartite and multipartite spiking), black corresponds to quasi-periodic or chaotic spiking, or a period larger than 100. Red solid curves consist of multi-jitter bifurcation points (corresponding to curves $\tau^P_{A,B}(q)$), and dashed red line indicates the critical steepness with slope$= -1$

Equation (16) allows to obtain bipartite and tripartite solutions analytically but it does not guarantee that the obtained solutions are the only ones, which can exist in the system. Indeed, when the criticality, i.e. the maximal value of $-Z'(\varphi) > 1$, grows, the bi- and tripartite $(P + 1)$-periodic solutions may bifurcate again and give birth to more complex behavior. This is illustrated in Fig. 7 where a two-dimensional, numerical bifurcation diagram in the parameters τ and q is shown. The diagram was obtained by simulating (1) with $N = 20$ different random initial values for each of grid point, and the maximal observed period after a transient time is plotted. If the period exceeded 100, the observed regime was considered aperiodic (black points).

The large white area in the figure corresponds to regular spiking, i.e. period one. For supercritical steepness, "islands" of irregularity appear, each corresponding to one value of $P = 1, 2, 3 \dots$. The internal structure of these islands is quite complicated and includes areas with solutions of different periods, often connected via period doubling bifurcations, as well as areas with quasiperiodic and chaotic solutions. Close to the border of each island there are bipartite period-$(P+1)$ solutions. Deeper in the interior, solutions of higher periods emerge, as well as quasiperiodic and chaotic solutions. However, we also observe windows of regularity inside the irregularity islands.

4 High Multistability of Jittering Regimes

Let us estimate a number of different bipartite solutions for given P. For each possibility to write $P+1 = n_1 + n_2$ with positive integers n_1 and n_2, we obtain $\binom{P+1}{n_1}$ different sequences of length $P+1$ consisting of n_1 entries equal to T_1 and n_2 entries equal to T_2. Not all of these sequences correspond to different solutions of (5), since some of them might be transformable to others by a periodical shift. Both sequences correspond to the same periodic solution if and only if this is possible. Therefore we can estimate that the total number of different solutions containing exactly n_1 ISIs T_1 in their sequential representation is bounded from below by $\binom{P+1}{n_1}/(P+1)$. In general this number does not equal the number of different solutions because the quotient $(P+1)$ disregards possible shift duplicates. Summing up over n_1 gives an estimate for the total number N_P of solutions for a given value of P as

$$N_P \geq \frac{1}{(P+1)} \sum_{k=1}^{P} \binom{P+1}{k} = \frac{2^{P+1}-2}{(P+1)}.$$

Notice that all these bipartite solutions exist for the same value of P but possibly for different ranges of the delay τ. However, easy to show that all of them form a continuous branch emerging and vanishing in the bifurcation points τ_A^P and τ_B^P. To this end consider the period-2 solutions when τ_0 approaches the border of the interval $\left[\tau_A^1; \tau_B^1\right]$, say $\tau_0 \to \tau_A^1 = 1 - \kappa Z(\psi_A) + \psi_A$. In this case the roots $\Theta_1$and Θ_2 tend to the same value $T_0 = 1 - \kappa Z(\psi_A)$ which is the period of the RS at the bifurcation point. Then if we construct a bipartite solution for some P as it is described above, the value of the delay for which it exists tends to the value $\tau \to \tau_0 + (n_1 + n_2 - 2)T_0 = P(1 - \kappa Z(\psi_A)) + \psi_A = \tau_{A,B}^P$. Easy to notice that this limit is exactly the delay in the points of bifurcation. This finding is confirmed by the diagrams in Fig. 6 in which all branches of bipartite solutions start from the bifurcation points on the RS branch.

Thus, we have proved that in jitter bifurcation points a large number of bipartite solutions branch off the regular spiking. The number of different solutions emerging in the dimension explosion points grows exponentially with the delay. This feature results in quite sophisticated structure of the neighborhood of the bifurcation points.

Many of the emergent jittering regimes are stable which implies high multistability. Particularly, although we lack a proof, it seems that all the bipartite solutions with same values of $n_1$and n_2 are stable simultaneously, see also [52]. High multistability of the emergent jittering regimes leads us to call the corresponding scenario a "multi-jitter" bifurcation.

High multistability is a well-known property of systems with delays. As already mentioned, an important mechanism in this context is the so-called reappearance of periodic solution in delay differential equations [9]. However, though similar in certain aspects, the phenomenon reported here is different and the induced degree of multistability is exponentiated. To clearly draw the distinction we first briefly describe the phenomenon of the common periodical reappearance. Consider a general equation

$$\frac{dx}{dt}(t) = f\left(x\left(t\right), x\left(t-\tau\right)\right), \tag{18}$$

and suppose it has a T-periodic solution $x(t) = x(t+T)$ for one value $\tau = \tau_0$ of the delay. Then it is clear from the periodicity of the solution $x(t)$ that it also solves the equation (18) for the delay value $\tau_P = \tau + PT$, where $P \in \mathbb{N}$. It means that the same periodic solution reappears at infinitely many delay values. It is noteworthy that the intervals of existence of reappearent solutions grow linearly with P. This results in overlapping of these intervals for large delays and may cause multistability with a number of coexisting solutions proportional to the delay.

In the system (1) multistability develops much faster, namely, the number of coexisting solutions grows exponentially with the delay. This shows that the involved mechanism is quite different from the reappearance of periodical solutions. However, a certain type of reappearance still takes place. Although the new solutions constructed with the help of period-2 solution (14) are not the same as the initial solution, they still consist of the same ISIs. In this sense we observe a "reappearance of ISIs" in the bipartite solutions.

5 Ring of Oscillators with Pulse Delayed Coupling

We have studied the multi-jitter bifurcation and jittering regimes in a single oscillator with delayed feedback. However, it turns out that multi-jitter instability can be observed in network systems also. In the current section we study it in a ring of N oscillators with pulse delayed coupling. The system is governed by

$$\frac{d\varphi_j}{dt} = \omega_j + Z(\varphi_j)\sum_{t_s^{j-1}} \delta(t - t_s^{j-1} - \tau_j). \tag{19}$$

where $j = \overline{1, N}$ is the oscillator id, and each oscillator is described by its phase $\varphi_j \in [0, 1]$. Each j-th oscillator receives input from its previous neighbour, the $(j-1)$-st

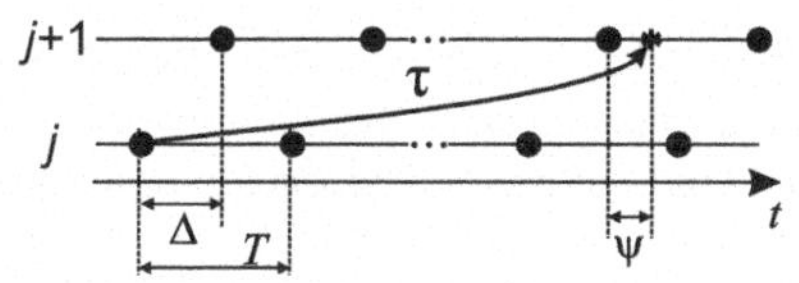

Fig. 8 Dynamics of the ring demonstrating a rotating wave with a period T and lag Δ. Points denote the time instants when the oscillators produce spikes. A spike emitted at a certain moment of time by an oscillator arrives to the next oscillator after the delay τ

oscillator (the first one receives input from the last one, so we identify 0 and N). This means that each spike produced by the $(j-1)$-st oscillator at t_s^{j-1} results in a pulse arriving to the j-th one after the delay τ_j. We will further consider identical oscillators with $\omega_j = 1$ and set all the delays equal $\tau_j = \tau$ (which can be easily generalized, see [49]).

The basic dynamical regimes observed in the ring are rotating waves characterised by the same period of all the oscillators and equal time lag between the neighbors. Suppose that (19) demonstrates such a regime with the period T and the lag Δ, as depicted in Fig. 8b. Then, each oscillator receives one pulse per period at phase $\psi = (\tau - \Delta) \mod T$, which allows to determine the period as $T = 1 - Z(\psi)$. The total time lag over the whole ring must be a multiple of the period, i.e. $\Delta = RT/N$, where $R = 0, \ldots, N-1$. Taking this all into account, the equations for rotating waves can be written in parametric form as

$$T = 1 - Z(\psi), \ \Delta = RT/N, \ \tau = PT + \Delta + \psi. \tag{20}$$

Here, $\psi \in [0; 1]$ is the phase at which oscillators receive input. R is the wavenumber which characterizes the type of the rotation wave and changes from zero to $N-1$. For example, $R = 0$ corresponds to complete synchronization of the ring, $R = 1$ to the splay state, and so on. P is the integer number controlling the value of the delay. Its maximal value is unbounded, and its minimal value is determined by non-negativity of the resultant delay τ.

Further we study the local stability of a rotation wave. For this sake we introduce a small perturbation and study its dynamics. We consider a perturbed solution in the form $t_s^j = sT + j\Delta + \delta_s^j$, where $\delta_s^j \ll T$ are deviations from the periodical regime. Then, the inter-spike interval $T_s^j = t_{s+1}^j - t_s^j$ can be determined as $T_s^j = 1 - Z(\psi_s^j)$, where ψ_s^j is the phase at which the pulse arrives. It is influenced by the timing of a spike emitted by the $(j-1)$-st oscillator P periods earlier: $\psi_s^j = t_{s-P}^{j-1} + \tau - t_s^j$. Thus, we come to an equation

$$\delta_{s+1}^j = \delta_s^j + 1 - T - Z\left(\psi + \delta_{s-P}^{j-1} - \delta_s^j\right). \tag{21}$$

This equation is in fact a map that governs the deviations dynamics. Note that the deviations on the next period $(s+1)$ depend on the deviations on the current period s and also on the deviations P periods ago. Thus, the map has dimension

$N(P+1)$. For small deviations, (21) can be linearized and its solutions found in the form $\delta_s^j \sim \exp(i\omega s + ikj)$, where ω is the frequency and k the wavenumber (not to be confused with R). This leads to the characteristic equation

$$e^{i(P+1)\omega} - (1+\alpha)e^{iP\omega} + \alpha e^{-ik} = 0, \tag{22}$$

where $\alpha = Z'(\psi)$. Note that the periodical boundary conditions on the ring allow only a limited set of wavenumbers $k = 2\pi n/N$, where $n = 0, \dots, N-1$.

In the phase space of (19), the rotation wave is a limit cycle with period T and multipliers $\lambda = \exp(i\omega)$. Thus, perturbations growth or damping depends on the imaginary part of ω. Let us study the critical case corresponding to $\mathrm{Im}\,\omega = 0$. For this sake it is convenient to fix k and consider the points $x = e^{i(P+1)\omega} + \alpha e^{-ik}$ on a complex plane. These points lie on a circle C_1 with the radius one and the center at the distance $|\alpha|$ from the origin. Easy to show from (22) that

$$|x| = |e^{i(P+1)\omega} + \alpha e^{-ik}| = |(1+\alpha)e^{iP\omega}| = |1+\alpha|.$$

For $\alpha \neq -1$, the points with $|x| = |1+\alpha|$ form a circle C_2 with the radius $|1+\alpha|$ and the center in the origin. For $\alpha \neq 0$, the two circles C_1 and C_2 contact at the only point $x = (1+\alpha)e^{-ik}$ corresponding to $e^{i(P+1)\omega} = e^{-ik}$. Thus, only such ω satisfy (22). Plugging this into (22) one obtains

$$(1+\alpha)e^{i(P+1)\omega} = (1+\alpha)e^{iP\omega}.$$

For $\alpha \neq -1$ this implies $\exp(i\omega) = 1$. Plugging this into (22) results in $\alpha = \alpha\exp(-ik)$, which for $\alpha \neq 0$ implies $k = 0$. Thus, for $\alpha \notin \{0, -1\}$ the only critical multiplier $\lambda = 1$ exists corresponding to the wavenumber $k = 0$. This multiplier $\lambda = 1$ is always present in the spectrum and corresponds to the neutral stability along the limit cycle. To check if $\lambda = 1$ may be a multiple root for $k = 0$ we divide (22) by $(\lambda - 1)$ and obtain

$$\lambda^P - \alpha \sum_{j=0}^{P-1} \lambda^j = 0. \tag{23}$$

Easy to see that $\lambda = 1$ is a root of (23) only for $\alpha = 1/P$. Thus, for $k = 0$ the root $\lambda = 1$ is simple for $\alpha \neq 1/P$ and has multiplicity 2 for $\alpha = 1/P$.

Other critical Lyapunov exponents may emerge only in the two cases, $\alpha = 0$ or $\alpha = -1$. For $\alpha = 0$, (22) implies $\exp(i\omega) = 1$, or $\lambda = 1$. For $\alpha = -1$, (22) implies $\exp(i(P+1)\omega) = \exp(-ik)$ or $\omega = -ik + i2\pi m/(P+1)$, where $m = 0, \dots, P$.

Let us summarize the results for the characteristic equation (22). Its spectrum has the form $\Lambda \cup \{\lambda_0\}$, where $\lambda_0 = 1$ is the trivial multiplier corresponding to the neutral stability along the limit cycle. Stability of the limit cycle is defined by the set Λ which includes critical multipliers only in the following cases:

$\alpha = 1/P$: one critical multiplier $\lambda = 1$ for $k = 0$.

$\alpha = 0$: $(N - 1)$ critical multipliers $\lambda = 1$ for $k \neq 0$, i.e. $k = 2\pi n/N$, where $n = 1, \ldots, N - 1$.

$\alpha = -1$: $N(P + 1) - 1$ critical multiplers $\lambda = \exp(-ik + i2\pi m/(P + 1))$, where $m = 0, \ldots, P$ for $k \neq 0$ and $m = 1, \ldots, P$ for $k = 0$.

Thus, only at the points $\alpha \in -1, 0, 1/P$ the limit cycle may change its stability.

To determine at which intervals of α the limit cycle is stable consider the case $\alpha = 0$. For such α, (22) has the form $\lambda^P(\lambda - 1) = 0$, and the set Λ consists of PN zero multipliers and $(N - 1)$ critical multipliers $\lambda_k = 1$ corresponding to $k \neq 0$. Let us study how these critical multipliers change at small variation of α. For $|\alpha| \ll 1$, the multipliers remain close to unity, i.e. $\lambda_k = 1 + \beta_k$, where $|\beta_k| \ll 1$. Plugging the multiplier in this form into (22) and keeping only the first order terms one obtains $\beta_k = \alpha\,(1 - \exp(-ik)) + o(\alpha)$. This implies $|\lambda_k|^2 = 1 + 2\alpha(1 - \cos k) + o(\alpha)$. Thus, for $k \neq 0$, the multipliers $|\lambda_k| < 1$ for $\alpha < 0$. This means that the limit cycle is stable in the parameter interval $-1 < \alpha < 0$ and loses its stability on the boundaries of this interval.

At $\alpha = 0$, the emergence of $N - 1$ critical multipliers $\lambda_k = 1$ marks occurrence of pitch-fork bifurcations in the manifolds corresponding to each $k \neq 0$. As a result, asymmetrical spiking regimes are born which are not in the focus of the current study.

6 Jittering Waves and Their Relation to Jittering Regimes of a Single oscillator

A much more remarkable scenario is observed at $\alpha = -1$ where all the multipliers become critical at once. Thus, it is the same multi-jitter bifurcation as in the case of single oscillator. In the bifurcation point, the wave loses its stability, and the jittering waves with distinct inter-spike intervals emerge. Below we carry out the detailed numerical study of the bifurcation and the emergent jittering waves.

For this sake we consider the PRC in the form

$$Z(\varphi) = \frac{\kappa}{2}\left(1 - \cos(2\pi\varphi^2)\right) \tag{24}$$

Here, κ is the coupling strength which controls the magnitude and the slope of the PRC. We used $\kappa = 0.185$ for which the PRC is depicted in Fig. 9. For this values two points φ exist with $Z'(\varphi) = -1$, and an interval with $Z'(\varphi) < 1$ is between them. This interval is marked red on the plot. According to our theory, the multi-jitter bifurcations should be observed on the borders of this interval, and jittering regimes inside.

We varied the delay and simulated system (19) directly starting from 20 different random initial conditions for each value of τ. The obtained numerical results for $N = 6$ are depicted in Fig. 10a by dots. The color of each dot corresponds to the wavenumber of the established regime (see the legend). Gray dots correspond to asymmetric regimes. By thin dashed lines are plotted the branches obtained theoret-

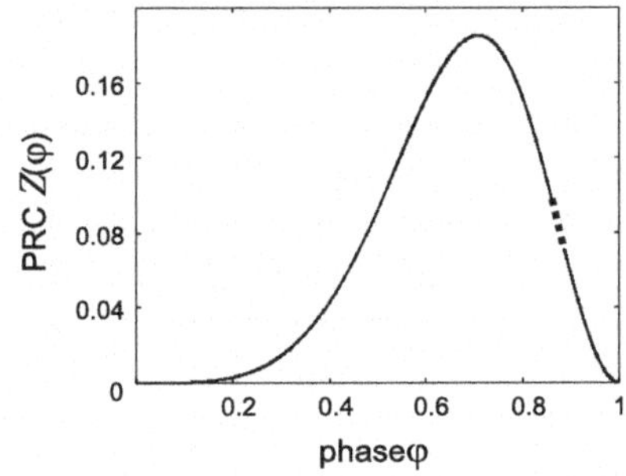

Fig. 9 The phase response curve (24) for $\kappa = 0.185$

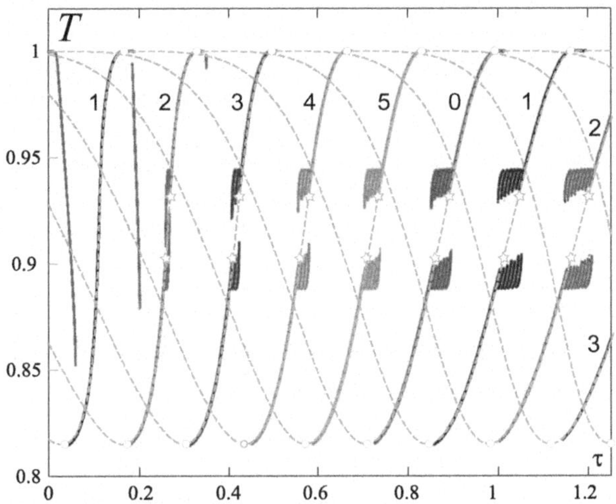

Fig. 10 Numerical bifurcation diagram for the ring of $N = 6$ oscillators, the observed inter-spike intervals versus the delay. Dots of different colors correspond to different wavenumbers, see the legend. Thin dashed lines correspond to theoretical branches, circles to the pitch-fork bifurcations, stars to the multi-jitter bifurcations

ically according to (20). The points of multi-jitter bifurcations are marked by stars. One can see that the numerically obtained dots lie precisely on the stable parts of the theoretical branches. Other observed regimes are the asymmetric ones which are not considered herein, and the jittering ones which are described in more details below.

Jittering regimes are characterized by distinct inter-spike intervals. On the bifurcation diagram, such the regimes correspond to several points of the same color for the same value of τ. Such regimes are observed in the intervals where the rotating wave is destabilized through the multi-jitter bifurcation. Moreover, the jittering regimes may be observed in even wider intervals which suggests that the bifurcation may be subcritical. Each wave is close to the rotation wave from which it is born. However, the intervals between the consecutive spikes of each oscillator are not constant anymore, but constitute a periodical sequence of short and long intervals. Noteworthy is that all the short and all the long intervals are equal, and their sequences are identical for all the oscillators. Thus, we see bipartite regimes, as in the case of a single oscillator.

The prominent similarity between the properties of the multi-jittering instability in a single oscillator and a ring of oscillators calls for explanation. To provide it, let us first consider a general oscillator with delayed feedback

$$\frac{dx}{dt} = f\left(x(t), x(t-\tau)\right). \tag{25}$$

Suggest it has a periodical solution $h(t)$ with the period T_0 at $\tau = \tau_0$ so that $h(t) = h(t + T_0)$. Now consider a ring of N oscillators with delayed unidirectional coupling

$$\frac{dx_n}{dt} = f(x_n(t), x_{n-1}(t-\tau)). \tag{26}$$

Let us look for solutions of (26) in the form of a rotating wave $x_n(t) = h(t + n\theta)$. Plugging this solution into (26) one obtains

$$\frac{dh(t+n\theta)}{dt} = f\left(h(t+n\theta), h\left(t+n\theta-\theta-\tau\right)\right). \tag{27}$$

Since $h(t)$ is a T_0-periodic solution of at $\tau = \tau_0$, it satisfies (27) for $\tau = \tau_0 - \theta$. The boundary conditions on the ring imply $x_n(t) \equiv x_{n+N}(t)$ or $h(t) \equiv h(t + N\theta)$, which is fulfilled for $\theta = MT_0/N$, $M \in Z$. Thus, the existence of the solution $x = h(t)$ of (25) with period T_0 at $\tau = \tau_0$ implies the existence of the solution $x_n(t) = h(t + n\theta)$ of (26) at $\tau = \tau_0 - \theta$ for $\theta = RT_0/N$, $R \in Z$ the wavenumber. This feature may be characterized as "reappearance" of periodical solutions of a single oscillator as rotating waves in a ring.

The notion of "reappearance" of solutions of a single oscillator as rotating waves explains the prominent similarity between the jittering solutions in a single oscillator and jittering waves in a ring of oscillators. Indeed, assume that we have a jittering solution in a single oscillator, then the corresponding jittering wave will be observed in a ring of oscillators. Bipartite jittering solutions correspond to bipartite jittering waves, which explains the similarity of the multi-jitter scenario in a single oscillator and in a ring of oscillators.

7 Discussion and Conclusions

We have reported the multi-jitter instability in a single oscillator with pulse delayed feedback [Eq. (3)] and a ring of pulse oscillators with delayed coupling [Eq. (19)]. In such systems, the basic dynamical regimes are regular spiking and regular rotating waves, correspondingly. We have shown that both regimes may destabilize in a multi-jitter bifurcation. In such a scenario, all the multipliers of the basic regimes become critical simultaneously, and jittering regimes, or jittering waves, with distinct inter-spike intervals emerge. Near the bifurcation these jittering regimes are typically bipartite, i.e. consist of two different values of inter-spike intervals, which are arranged in distinct sequences for the different regimes.

Bipartite regimes can be encoded by binary sequences, and in both cases the period of the observed sequences is proportional to the delay. An important feature is that for an arbitrary binary sequence of a given period, a parameter interval does exist where the corresponding jittering regime is present and stable. Further, regimes with the same number of long and short ISIs are stable in the same parameter interval. This leads to an immense multistability, which increases exponentially as the delay grows.

In spite of the similarity between the single oscillator system and the ring-coupled system, the multi-jitter bifurcation in rings has an important distinguishing feature: For one oscillator, the value of the delay must be large compared to the oscillator's natural period. Specifically, the multi-jitter bifurcation giving birth to jittering regimes of period Π takes place at the delay

$$\tau_{\Pi} = (\Pi - 1)\left(1 - Z\left(\psi^*\right)\right) + \psi^*,$$

where ψ^* is the phase with $Z'(\psi^*) = -1$. Since the PRC $Z(\varphi)$ is typically small in amplitude, the period of jittering regimes is roughly the delay divided over the natural period. Thus, to obtain jittering regimes with long periods one needs delays several times larger than the natural period.

For rings the situation is different. As follows from (20), multi-jitter bifurcations take place at delays $\tau = (P + R/N)(1 - Z(\psi^*)) + \psi^*$, where $P, R \in \mathbb{Z}$ which implies that jittering regimes with period Π emerge at the delay

$$\tau_{\Pi} = (\Pi/N - 1)\left(1 - Z\left(\psi^*\right)\right) + \psi^*.$$

Thus, the period of the emergent jittering solutions is roughly proportional to the delay multiplied by the number of oscillators, or the total delay along the ring. As a consequence, even short coupling delays may result in higher-periodical jittering regimes if the number of oscillators is large enough.

The condition for the bifurcation in the both systems is that the PRC $Z(\varphi)$ is steep enough, namely, its has intervals with slope $Z'(\varphi) < -1$. In this case the map $F(\varphi) = \varphi + \kappa Z(\varphi)$ describing the phase shift under the action of a pulse becomes non-monotonous. This means that the non-monotonicity of the phase transition map is inevitable for the multi-jitter bifurcation. In realistic oscillator models, the non-monotonicity of the phase transition map $F(\varphi)$ corresponds to a bended shape of the isochrons in the phase space. In this situation, an external pulse may "switch" the order of the points on the limit cycle, as illustrated in Fig. 11 for the FitzHugh-Nagumo model.

The reported multi-jitter bifurcation seems to be very degenerate. In higher dimensional, and real systems the bifurcation usually unfolds into a sequence of single bifurcations, although one can still observe high multistability of jittering solutions [52]. However, the bifuraction emerges generically in the systems (3) and (19) as soon as its PRC is steep enough. How can one resolve this contradiction between the degeneracy

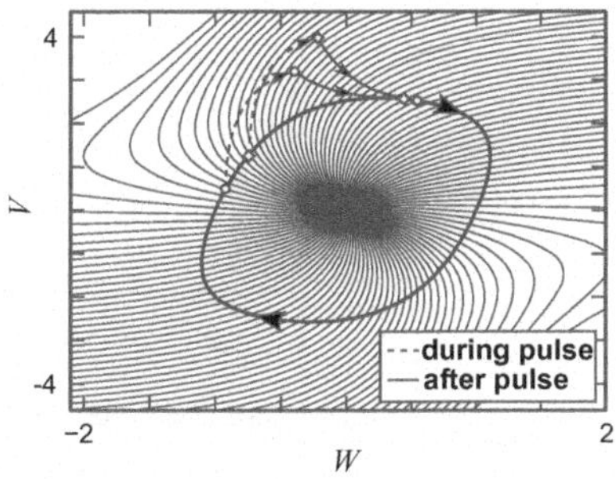

Fig. 11 The isochrons of the FHN model show the structure necessary for a supercritical PRC steepness $Z'(\varphi) < -1$. The phase order of two different points on the limit cycle, x and y, is reversed by the corresponding pulse. This means they are carried to new points x_{new} and y_{new} which lie on isochrons with a reversed phase order

of the bifurcation and its generic emergence? Let us elaborate on this question by considering the map (5). This map transforms a set $\mathbf{x} = (T_{s-P+1}, \ldots, T_s)$ of P ISIs into a new set $\mathbf{x}_{\text{new}} = (T_{s-P+2}, \ldots, T_{s+1})$. Thus, the phase space of the map has dimension P. However, along the most of the components of the vector $\mathbf{x}$ the map is just a coordinate shift, so that $x_{\text{new},1} = x_2$, $x_{\text{new},2} = x_3, \ldots, x_{\text{new},P-1} = x_P$ and so on. Only the last coordinate of the vector is transformed non-trivially: $x_{\text{new}P} = 1 - \kappa Z(\psi)$, where $\psi = \tau - \sum_{k=1}^{P} x_k$. This results in a very specific structure of the map Jacobian

$$D = \begin{pmatrix} 0 & 1 & 0 & 0 \\ \vdots & \ddots & \ddots & 0 \\ 0 & \ldots & 0 & 1 \\ \alpha & \ldots & \ldots & \alpha \end{pmatrix}, \tag{28}$$

where $\alpha = \kappa Z'\left(\tau - \sum_{k=1}^{P} x_k\right)$, which in turn leads to the characteristic equation (9). The important point is that under any admissible perturbations of the system (3) the structure of the Jacobian preserves. The transformation of the coordinates $x_1 \ldots x_{P-1}$can not have any other form but the one given above because otherwise it would mean that the intervals of time in the past become longer or shorter while we move into the future. Similarly, the time ψ between the pulse emission and the pulse arrival can not be calculated in a different way, otherwise that would mean that duration of the time interval does not equal the sum of durations of sub-intervals. These two are fundamental properties of time which can not be violated by any meaningful perturbation. Possible perturbations can only affect the delay τ, the feedback strength κ or the PRC shape. All this may result in change of the coefficient α while the structure of the Jacobian remains as in (28). Thus, in a certain sense generic emergence of multi-jitter bifurcations rely on fundamental properties of time.

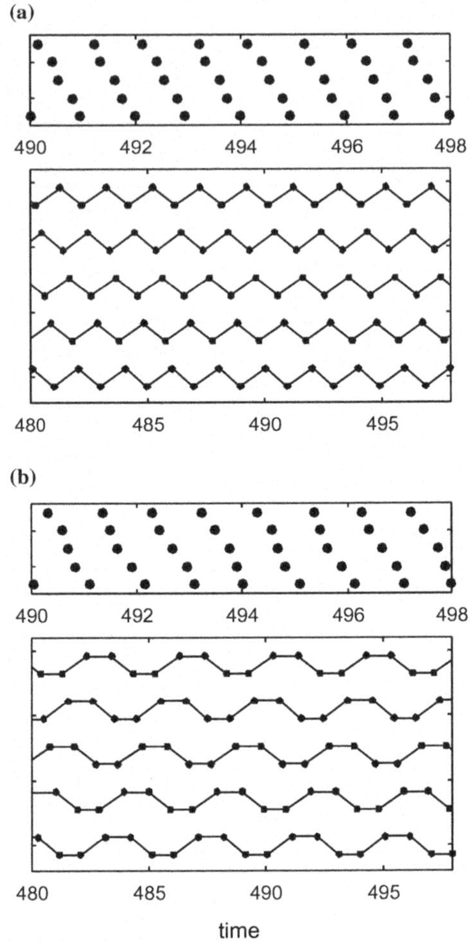

Fig. 12 Two different jittering rotating waves observed in a ring (26) with $N = 5$ and $\tau = 0$. The top panels depict the moments of spike generation, the bottom inter-spike intervals. The PRC $Z(\varphi)$ is selected as the one of the electronic FitzHugh-Nagumo oscillator [51, 53]

An interesting and open question is the necessary conditions for the emergence of jittering. When the multi-jitter bifurcation was first discovered, it was believed that the presence of a delay is an inevitable component. However, later we were able to find jittering waves in rings of oscillators without delays, which is illustrated in Fig. 12. Here, the symmetry related to the equivalence of time intervals in the delay line is replaced by the symmetry related to the identity of the units in the ring. Thus, the multi-jitter bifurcation and the emergence of numerous jittering solutions at the bifurcation point still may take place.

Another question is the whether the ring topology is the only one which supports multi-jitter instability or it may be observed in networks of more complex structure. Recently we have started the search for multi-jittering instability in networks with various typologies different from the ring. Some preliminary results of the study are shown in Fig. 13a which provides an example of the jittering regime observed in a network of four identical oscillators which are coupled all-to-all, including self-coupling. The coupling delays are all the same and equal $\tau = 1.55$, whereas the PRC is chosen in the form $Z(\varphi) = -\mu \sin 2\pi\varphi$. For $\mu \approx 0.03$ the two-cluster periodical regime destabilizes and gives birth to the jittering regime with distinct inter-spike intervals. The oscillators are still split into two clusters, but the time intervals between the clusters are not constant anymore but change with time. Several features of the emergent dynamics draws the attention. First, the emergent jittering regime is not bipartite, but consists of various inter-spike intervals. Second, the regime is not periodic but quasi-periodic. The both features are observed starting from the bifurcation point, in contrast with the case of a single oscillator where they develop only for large criticality. However, if the criticality grows, the observed regimes get more complicated, including the cluster's destruction, as depicted in Fig. 13b. It is also noteworthy that the jittering regimes are observed for the PRC whose slope does not equal -1 in any point.

To conclude, the multi-jitter instability is a novel dynamical phenomenon recently discovered in systems with pulse coupling. It is characterized by the destabilization of regular spiking regimes and the emergence of the so-called jittering regimes characterized by irregular inter-spike intervals. The most fascinating feature is the combinatorial accumulation of co-existent jittering regimes. In systems with delay this leads to an exponential growth of multistability with the delay time. First discovered in a single oscillator with delayed feedback, the multi-jitter instability was later observed in more complex networks. These findings allow to assume a universal character for this type of dynamics. A central task for the future research in this topic is to determine the class of systems where a multi-jitter instability can be observed and to obtain the conditions for its emergence.

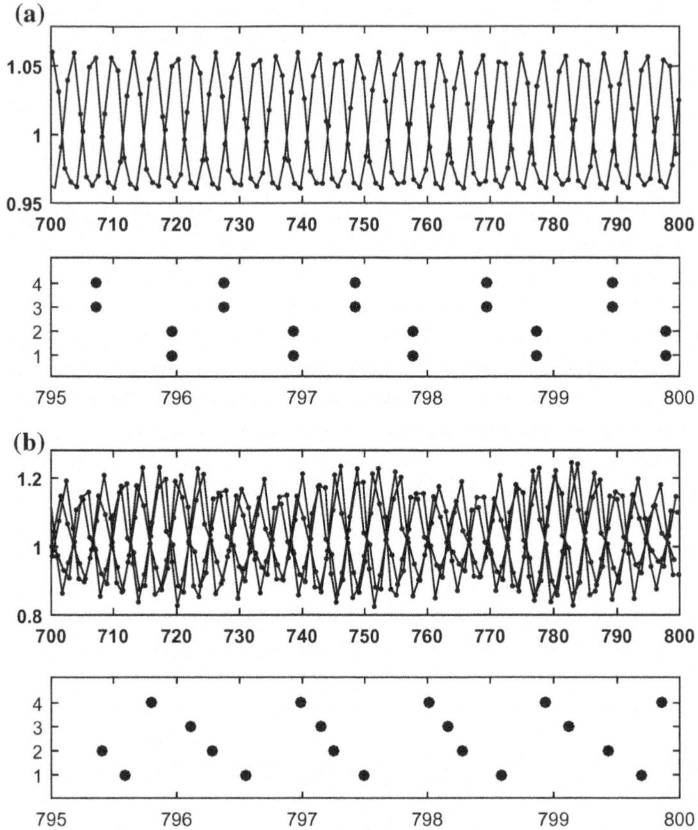

Fig. 13 Jittering regimes in a network of four all-to-all coupled oscillators, the explanation and the parameters values in the text

References

1. Manor, Y., Koch, C., Segev, I.: Biophys. J. **60**
2. Wu, J.: Introduction to Neural Dynamics and Signal Transmission Delay. Walter de Gruyter, Berlin, Boston (2001)
3. Erneux, T.: Applied Delay Differential Equations. Springer, New York, Berlin (2009)
4. Soriano, M.C., García-Ojalvo, J., Mirasso, C.R.: I. Fischer, Rev. Mod. Phys. **85**
5. Junges, L. Gallas, J.A.: JOSA B **33**
6. Yanchuk, S., Giacomelli. G.J.: Phys. A **50**
7. Otto, A., Rauh, D., Kolouch, M., Radons, G.: Int. J. Machine Tools Manufacture **82**
8. Klinshov, V.V., Nekorki, V.I.: Phys. Usp. **56**
9. Yanchuk, S., Perlikowski, P.: Phys. Rev. E **79**
10. Schuster, H.G., Wagner, P.: Progr. Theor. Phys. **81**
11. Ernst, U., Pawelzik, K., Geisel, T.: Phys. Rev. E **57**
12. Klinshov, V.V., Nekorkin, V.I.: Chaos Solitons Fractals **44**
13. Reddy, D.V.R., Sen, A., Johnston, G.L.: Phys. Rev. Lett. **80**

14. Heiligenthal, S., et al.: Phys. Rev. Lett. **107**
15. Junges, L., Gallas, J.A.: Phys. Lett. A **376**
16. Izhikevich, E.M.: Neural Comput. **18**
17. Popovych, O.V., Yanchuk, S., Tass, P.A.: Phys. Rev. Lett. **107**
18. Kantner, M., Schöll, E., Yanchuk, S.: Sci. Rep. **5**
19. Heil, T., et al.: Phys. Rev. Lett. **86**
20. Wunsche, H.J., et al.: Phys. Rev. Lett. **94**
21. Vicente, R., et al.: Phys. Rev. E **73**
22. Fischer, I., et al.: Phys. Rev. Lett. **97**
23. Jungling, S., et al.: Phys. Rev. E **84**
24. Weicker, S., et al.: Phys. Rev. E **89**
25. Horvath, V., et al.: Angew. Chem. Int. Ed. **51**
26. Smith, H.: An Introduction to Delay Differential Equations with Applications to the Life Sciences. Springer, Science Business Media (2010)
27. Hale, J.K.: Theory of Functional Differential Equations. Springer (1977)
28. Winfree, A.T.: The Geometry of Biological Time. Springer, New York (2001)
29. Kuramoto, Y.: Chemical Oscillations Waves and Turbulence. Springer, Berlin Heidelberg (1984)
30. Pikovsky, A., Rosenblum, M., Kurths, J.: Synchronization: A Universal Concept in Nonlinear Sciences. Cambridge University Press (2003)
31. Izhikevich, E.M.: Dynamical Systems in Neuroscience: The Geometry of Excitability and Bursting. MIT Press, Cambridge, MA (2005)
32. Abott, L., van Vreeswijk, C.: Phys. Rev. E **48**
33. Maass, W., Schmitt, M.: Inform. Comput. **153**
34. Zillmer, R., Livi, R., Politi, A., Torcini, A.: Phys. Rev. E **76**
35. Canavier, C.C., Achuthan, S.: Math. Biosci. **226**
36. Buck, J.: Q. Rev. Biol. **63**
37. Anumonwo, J.M., Delmar, M., Vinet, A., Michaels, D.C., Jalife, J.: Circ. Res. **68**
38. Colet, P., Roy, R.: Opt. Lett. **19**
39. Nizette, M. Rachinskii, D., Vladimirov, A. Wolfrum, M.: Phys. D **218**
40. Boyd, R.W., Gauthier, D.J.: Sci. **326**
41. Rosin, D.P., Rontani, D., Gauthier, D.J., Schöll, E.: Phys. Rev. Lett. **110**
42. Kuramoto Y.: Chemical Oscillations Waves and Turbulence, Springer, Berlin (1984)
43. Ermentrout, B.: Neural Comput. **8**
44. Tass, P.: Phase Resetting in Medicine and Biology. Stochastic Modelling and Data Analysis. Springer (1999)
45. Lücken, L., Yanchuk, S.: Phys. D **24**
46. Glass, L., Guevara, M.R., Belair, J., Shrier, A.: Phys. Rev. A **29**
47. Galán, R.F., Ermentrout, G.B.: N. N. Urban. Phys. Rev. Lett. **94**
48. Ermentrout, G.B., II Beverlin, B., Troyer, T., Netoff, T.I.: J. Comput. Neurosci. **31**
49. Lücken, L., Yanchuk, S., Popovych, O.V., Tass, P.A.: Front. Comput. Neurosci. **7**
50. Klinshov, V.V., Nekorkin, V.I.: Commun. Nonlinear Sci. Numer. Simul. **18**
51. Klinshov, V., Lücken, L., Shchapin, D., Nekorkin, V., Yanchukm, S.: Phys. Rev. Lett. **114**
52. Klinshov, V., Lücken, L., Shchapin, D., Nekorkin, V., Yanchukm, S.: Phys. Rev. Lett. E **92**
53. Klinshov, V., Shchapin, D., Yanchuk, S., Nekorkin, V.: Phys. Rev. E **94**

Power-Grids as Complex Networks: Emerging Investigations into Robustness and Stability

J.M.V. Grzybowski, Elbert E.N. Macau and T. Yoneyama

Abstract Power grids are ubiquitous engineering systems composed of tens or even hundreds of interconnected subsystems. Such systems resemble a complex network in the sense that both the link structure and the node dynamics are influential to its overall behavior. Several decades of intensive research on power grids were not enough to uncover the intricacies of stability issues triggered by the structure-dynamics interplay. In this context, several attempts have been made to approach these issues. In this chapter, we review a number of recent results in the topic of robustness and stability in power grids, developed within the framework of the Theory of Complex Networks, especially those concerned with the description of node dynamics by means of the second-order Kuramoto model.

1 Introduction

Power networks are very large-scale distributed systems in which electrical machines and devices interact with each other by means of electric and magnetic interconnections. The AC (alternating current) nature of modern power networks powered mostly by synchronous generators connected in parallel requires synchronous evolution in time. Therefore, strict frequency synchronization must be maintained despite of sudden changes in a system load or unanticipated faults in any of its components [1, 25,

J.M.V. Grzybowski (✉)
UFFS - Federal University of Fronteira Sul, Rodovia RS 135 km 72,
Erechim, RS, Brazil
e-mail: jose.grzybowski@uffs.edu.br

Elbert E.N. Macau
INPE - National Institute for Space Research, Av. dos Astronautas,
1758, Jd. Granja, São José dos Campos, SP, Brazil
e-mail: elbert@inpe.br

T. Yoneyama
ITA - Aeronautics Institute of Technology, Pça. Mal. Eduardo Gomes,
50-Vila das Acácias, São José dos Campos, SP, Brazil
e-mail: takashi@ita.br

M. Edelman et al. (eds.), *Chaotic, Fractional, and Complex Dynamics: New Insights and Perspectives*, Understanding Complex Systems,
https://doi.org/10.1007/978-3-319-68109-2_14

28]. Those systems are designed aiming to be stable and robust on fails. In special, it is desired that the system's synchronized state is preserved even under severe disturbances. Usually, this goal is sought by increasing connectivity and interdependence among a system's elements. However, this procedure also implies intrincacy of the interactions at various scales, which by its turns may entail stability problems under perturbations. Given their interconnected structure and dynamics, power networks can be regarded as complex networks of oscillators in which the power lines are the edges while the electrical machines and devices are the nodes. The interacting dynamics among the constitutive elements results in the emergence of collective dynamics with peculiar and specific properties [4, 36, 47, 52].

In the context of power systems engineering, classical techniques have been long applied in the design of power networks [25]. Still, there are several open issues related to stability and robusteness of these systems. The climax of these issues happens when a blackout occur, which may leave millions of people in large regions, even countries, without eletricity [18, 19]. Recently, the concept of smart grids, in which different types of power sources with distintic capacities and availability in time are increasingly incorporated into power grids, brings an unprecedented complexity to the stability and robustness analysis. Therefore, in recent years, this aspect of power grids has received increasing attention from the Physics and Complex Networks communities [4, 5, 12, 22, 33, 36, 50, 51, 55]. They aim to exploit concepts and techniques developed in their research areas to gain insights into the mechanisms underlying the functioning and malfunctioning of power networks. Several recent publications have brought remarakble contributions to this theme from diverse and complementary viewpoints. We hope to cover some of these contributions in this chapter.

The chapter is organized as follows. In the remaining of Sect. 1, we briefly comment on stability and robustness in power networks and then cover basic topics such as the different types of stability that are of interest in the context of power grids; the swing equation, which is the basic equation for the modelling of a power generator and the concepts of topological model, reduced model, structure-preserving model and generator-motor model, which are the most common approaches taken into consideration to the modelling of power networks. In Sect. 2, we cover some alternative approaches to the assessment, understanding and enhancement of stability and robustness in power networks, as seen from the viewpoint of Complex Networks Theory. In Sect. 3, we focus on some current limitations of such alternative approaches and the challenges involved in rendering them applicable in practice. We also prospect probable features of future power networks and the corresponding demands they shall pose.

Stability in power networks

Power grids are frequently described as country or continent-wide electric circuits designed to deliver electric power in an economical and reliable manner. The proper functioning of a power system requires that power utilities deliver uninterrupted electrical power on demand, i.e., generation should meet the consumption load at all times. This mode of operation is called *load following* and it has to deal with dynamical fluctuations on load demands. These small and unpredictable variations in load cause small perturbations in the steady state of the power network. As the the actual generation is mostly performed by synchronous machines, small perturbations resulting from this stochastic-like load variations must be properly dealt with by the system, such that all generators remain synchronous over time.

Besides following load demand, power utilities have to deal with unanticipated power faults, which contribute to originate instabilities and disturbances in the synchronous state of the grid. Power faults can be caused most commonly by short circuits, which disturb the systems' steady state and can cause sudden changes in the network topology and states. In such occasions, a transient period follows in which generators can experience significant rotor angle excursions away from their original steady state. Under a severe disturbance, a power network can eventually reach a new steady state after a transient period, and such new steady state can be significantly different from the pre-disturbance one. An acceptable steady state is one in which all the network variables lie within acceptable limits in the post-disturbance period. If the power grids fails to stabilize in such way, several undesired conditions can emerge, such as frequency excursion, voltage collapse, generator tripping and load shedding, among others. In this sense, stability means the ability of the power network to regain synchrony and settle to an acceptable steady-state after being subject to a large disturbance. For a detailed account on the subject, we refer to [25, 26].

Robustness in power networks

A central matter regarding electrical networks and network structures in general is the extent to which their topology can support the continuity of the processes they perform and remain operational under the failure or malfunctioning of a subset of its nodes or links. Real power networks are designed such that they fulfill the $N-1$ criterion, which means that the power network should be able to operate upon the unanticipated loss of any of its components, including major generating units. This means that the sudden loss of any network node must be accommodated by the remaining nodes. Such structural ability to handle shortcomings and remain operational is termed robustness. Robustness is generally related to the availability of redundant paths connecting nodes, which facilitates the redistribution of power flow upon a line or node failure [11, 37].

1.1 Background and Motivation

Complete frequency synchronization is a stringent requirement in power networks since they are powered mostly by synchronous generators. This means that the grid components behave as coupled oscillators and must operate in step relatively to each other. This is done by interconnecting generators to an steady electrical frequency that should be maintained over time despite of changes in load demands. By interconnecting synchronous generators, one can increase the rotational inertia of the system, thus in practice making it less vulnerable to destabilizing disturbances [25, 28].

The rotating nature of synchronous generators requires that a synchronization process take place every time a generator is connected to the power-grid in order to avoid the occurrence of large electro-mechanical transients. The synchronization process consists of matching the frequency, voltage and the phase angle of the generator's electrical terminals to those of the oscillating grid [56]. As synchronization is not handled properly, several issues may arise: the generator can be subject to mechanical stress due to rapid acceleration or deceleration, the power system can be subject to voltage deviations from normal, which can be rather large if the generator is synchronizing to a relatively weak power system [56]. These facts concur to the destabilization of the power network and, as such, are undesirable. Thereby, the generator ideally joins the grid as synchronization requirements are met concerning its frequency, voltage and phase angle, relatively to those of the network. Similarly, as a generator goes out of synchrony, it shall be disconnected from the grid to avoid the propagation of instabilities generated by its desynchronized dynamics. This task is accomplished through manual or automatic control mechanisms that put into operation circuit breakers that isolate the generator until it is brought back to normal operation [56].

Understanding the interplay between topology and dynamics of power grids is one way in which power-grid design can benefit form results from Complex Networks. As the process of interest is modelled and analyzed as a dynamical complex network of oscillators, one can unveil relevant aspects of the dynamics of instability and also their effect upon the power-grid functioning and efficiency. Furthermore, it allows the investigation of innate features of the network that contribute to the stability and robustness of the power-grid against failures and perturbations. Towards that end, different frameworks have been proposed in the literature as a collective effort to unveil and characterize the relations between synchronization stability and the network physical topology (given by its physical connections or wires) or electrical topology (given by its electrical connections or admittances). In addition, several studies focused on the topology itself and their robustness.

Considering a network of coupled generators, the synchronous operation is read as

$$\omega_1(t) = \omega_2(t) = \cdots = \omega_N(t), \tag{1}$$

meaning that angular velocities of all the oscillators are equal. This synchronous manifold is frequently invoked in theoretical studies of frequency synchronization in power networks [4, 13, 21, 36].

Before we cover some recent contributions to the understanding of stability and robustness in power networks from the viewpoint of complex networks, we present basic concepts of stability as found in classic power system stability theory [26]. Thereby, the next section briefly covers the concept of stability and its variants in the context of power networks.

1.2 Basic Concepts: Stability

1.2.1 Rotor Angle Stability

Small-signal stability

Small-signal stability is the capability of the power system to handle small disturbances and sustain synchronous behavior upon them. The term *small disturbance* refers to minor changes in the generation/load balance upon which the system settles to a new stable state very near from the original one. It can be caused by the loss of a load, for instance, which implies the generation/load balance to be upset. Under small disturbances the small-signal stability of the system can be usually assessed by means of equations of the power system linearized about the steady-state [25].

Transient stability

Transient stability is the capability of the power system to handle acute short-lived disturbances and return to synchronous behavior upon them. The term *acute disturbance* refers to considerable changes in the generation/load balance, such as the loss of a large load, the loss of a generator or a fault on a key transmission line. In this case, the range of excursion observed in the power system state variables is large enough to forbid the linearization of power system equations [25]. Thus, the nonlinear equations must be handled in order to assess the stability of the power system. An illustration of transient stability for the simplified power grid in Fig. 1 is shown in Fig. 2.

1.2.2 Voltage Stability

Voltage stability is the capability of the power system to sustain steady voltage at all buses within prescribed tolerable margins [25]. The adequate functioning of the power grid relies on this requirement being met both during normal operation and after a disturbance. The study of voltage stability includes the assessment of the behavior of a power system as subjected to faults, loss of generation and sudden increase in load, among others.

1.3 Basic Machine Model: The Swing Equation

The net torque or accelerating torque T_a acting on the rotor is given by the difference between the mechanical torque T_m provided by the prime mover and the electromagnetic torque T_e produced by power transfer. This yields

$$T_a = T_m - T_e \tag{2}$$

Note that as $T_m > T_e$, the rotor accelerates, while it deccelerates as $T_m < T_e$. Under a balanced generation/demand condition, $T_m = T_e$. The effect of the net torque T_a upon the rotor is expressed as a the rate of change of its angular velocity

$$J\frac{d\omega}{dt} = T_a, \tag{3}$$

where J is the total moment of inertia of generator and turbine [25]. While some authors consider the model (3) in their studies, it is often acknowledged that an explicit expression for the damping torque is desirable. The damping torque T_d is modeled as a torque proportional to the rotor angular velocity, that is

$$T_d = D\omega, \tag{4}$$

where D is the damping torque constant. In this case, the accelerating torque is redefined as $T_a = T_m - T_e - T_d$ and, as a result, the swing equation explicitly accounting for the damping torque can be written as

$$J\frac{d\omega}{dt} = T_m - T_e - D\omega \tag{5}$$

The electrical power transfer T_e between two buses connected by a lossless line modeled as an inductive reactance X can be assumed to be equivalent to the real

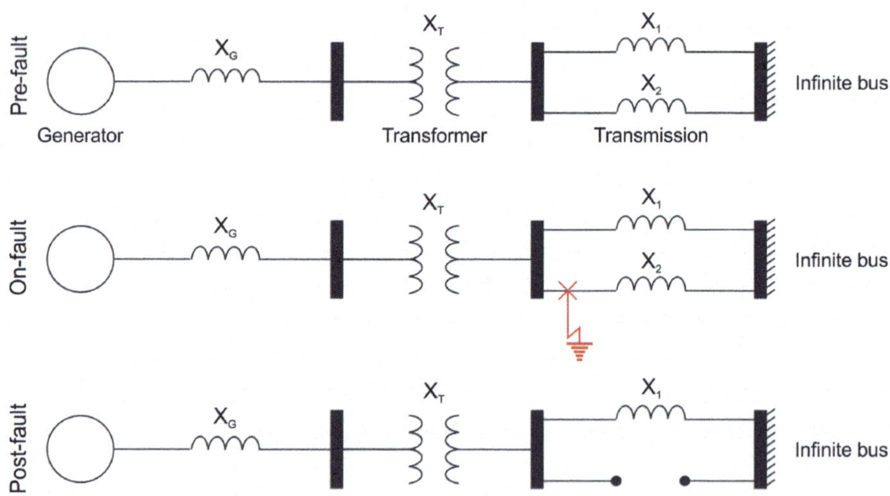

Fig. 1 Scheme of a generator-to-infinite-bus power system. The infinite bus approximates a large utility whose voltage, angle and frequency remain constant despite changes in power flow from the generator. During the pre-fault stage (top), the grid operates in steady-state. In the on-fault stage (middle), a short-circuit to ground occur in a power line and disturbs the steady-state of the system. In the postfault stage (bottom), the fault is cleared by opening the faulted power line

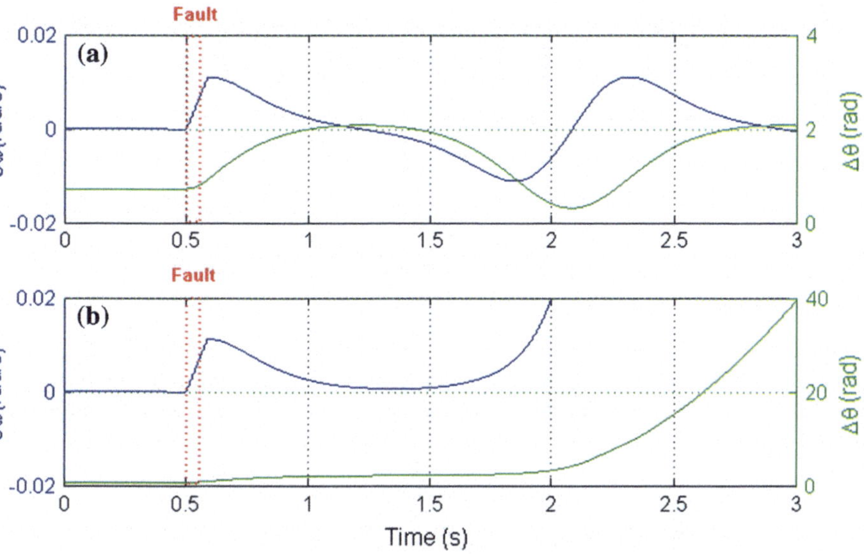

Fig. 2 Illustration of transient stability in power networks: the power network is operating under steady-state (pre-fault) when a line fault F occurs at $t_F = 0.5s$; the faulted line is opened after a clearing time t_C and the transient dynamics follows for $t > t_C$. Scenario **a** illustrates the case when $t_C \leq t_F + \Delta t_{crit}$, as the rotor angle deviation remains bounded and the angular velocity deviation settles at zero; **b** illustrates $t_C > t_F + \Delta t_{crit}$. The critical clearing time Δt_{crit} is the largest time period a given fault can last before the system is rendered transient unstable

power transferred between the buses. The apparent power in the receiving end is given as

$$S_R = P_R + jQ_R = E_R I^* = E_R \left(\frac{E_S - E_R}{jX}\right)^* = E_R \left(\frac{E_S cos\delta + jE_S \sin\delta - E_R}{jX}\right)^*, \quad (6)$$

which gives the expression for real power transfer

$$P_R = \frac{E_R E_S}{X} \sin\delta = P_e \quad (7)$$

As the equations are rewritten in per unit values, the torque and power quantities can be used interchangeably. Thus, the swing equation with damping can be expressed as

$$J\frac{d^2\delta}{dt^2} + D\frac{d\delta}{dt} + P_{max}\sin\delta = P_m, \quad (8)$$

where

$$P_{max} = \frac{E_R E_S}{X} \quad (9)$$

In a network, the maximum power transfer between two buses i, j can be expressed as $P_{i,j}$ and the swing equation for the ith node can be written as

$$J_i \frac{d^2\delta_i}{dt^2} + D_i \frac{d\delta_i}{dt} + \sum_j P_{i,j} \sin\left(\delta_i - \delta_j\right) = P_{m_i}, \tag{10}$$

where

$$P_{i,j} = \frac{E_i E_j}{X_{i,j}} \tag{11}$$

and $P_{m_i}$is the mechanical input power in the ith node. Dividing the equation of the ith node by its moment of inertia J_i, the swing equation can be rewritten in the more usual form as

$$\frac{d^2\delta_i}{dt^2} = -\alpha_i \frac{d\delta_i}{dt} + \frac{1}{J_i}\left(P_{m_i} - \sum_j P_{i,j} \sin\left(\delta_i - \delta_j\right)\right), \tag{12}$$

where $\alpha_i = D_i J_i$. Alternatively to Eq. 12, the swing equation is often described as a system of first-order equations in the form

$$\begin{aligned} \dot{\delta}_i &= \omega_i \\ \dot{\omega}_i &= -\alpha_i \omega_i + \tfrac{1}{J_i}\left(P_{m_i} + \sum P_{i,j} \sin\left(\delta_j - \delta_i\right)\right) \end{aligned}, \tag{13}$$

where $P_{i,j}$ is the entry in the ith row and jth column in the matrix of maximum power transfers that results from the solution of power flow equations. Without loss of generality, ω_i can be regarded as the deviation of the angular velocity of the ith oscillator relatively to the reference angular velocity of the power system. The reference electrical frequency f_0, commonly set to 50 or 60 Hz in real power systems, serves as a reference rotating frame whose angular velocity is given by $2\pi f_0$. This means that the angular velocity of each oscillator is expressed relatively to a rotating frame at reference angular velocity $2\pi f_0$. As such, the actual angular velocity of the ith oscillator is given by $\omega_i + 2\pi f_0$.

1.4 Modeling Power-Grids as Complex Networks

The basis to the understanding of a power network as a complex network is the fact that power networks are arranged according to a given topology that is highly influential on its overall behavior. The topology of the power network can be represented by a graph that expresses the interconnections among nodes. In this context, either the physical network or the electrical network, i.e., the impedances or admittances among nodes, can be considered, depending on the goal of the study. In this subsection, the most common approaches to the modeling of power grids are presented on the basis of Refs. [5, 12, 22, 27, 39].

Topological model

Topological models of power systems are known to neglect node dynamics and seek relevant knowledge of the power grid by analyzing the structure of its corresponding graph. In this context, the statistics of the network and its behavior under random error or intentional attack are considered [6, 12, 22, 35, 44, 50, 51]. The study of topological features of power systems can help enhance the understanding of how cascading failures build up from a structural point of view (see for instance Refs. [7, 44, 55, 58, 59]).

Effective network model

In the effective network model, the system loads are assumed to be constant impedances, such that the original network of generators and loads can be reduced to a network of generators only. The idea is that the remaining network will only have nodes where currents are injected. As such, in the reduced network the generators are interconnected by effective impedances that take into account the equivalent impedance of transmission lines and loads. This formulation assumes that the power consumption at any load is a function of voltage magnitude at the load node. Thus, for a network with n_G generators and n_L loads, such that $n = n_G + n_L$, the original network impedance matrix of size $n \times n$ is reduced to the so-called effective network of size $n_G \times n_G$ by matrix partitioning or Kron reduction (we refer to [15]). After the reduction process, the loads become part of the matrix of effective impedances among generators. The generators can be modeled by the swing equation in Eq. 12 and the values of $P_{i,j}$, δ_i, δ_j are collected from the solution of the power-flow equations for the reduced network. A clear disadvantage of network reduction is that the original network topology is lost. As pointed in Ref. [2], this can conceal the influence of the network topology upon stability. Further, load nodes frequently have dynamics of their own that add important elements to the evaluation of stability in a power system [2, 25].

Structure-preserving model

The *structure-preserving model,* as presented in Ref. [2, 39], proposes a dynamical model of load buses in which it is assumed that node voltages are constant and frequency variations about the initial operating point $P^0_{i_i}$ are small, where P_i is the real power drawn by the *ith* load bus. Under these conditions, in a network with generators and loads, the load dynamics is modeled as

$$P_i = P_i^0 + D_i\dot{\delta}_i, \tag{14}$$

where $D_i > 0$ is the frequency coefficient of load that represents the behavior of power-frequency control systems. Note that, as $D_i \to 0^+$, one obtains a constant load model. Rewriting Eq. 14, the equation for loads can be expressed as

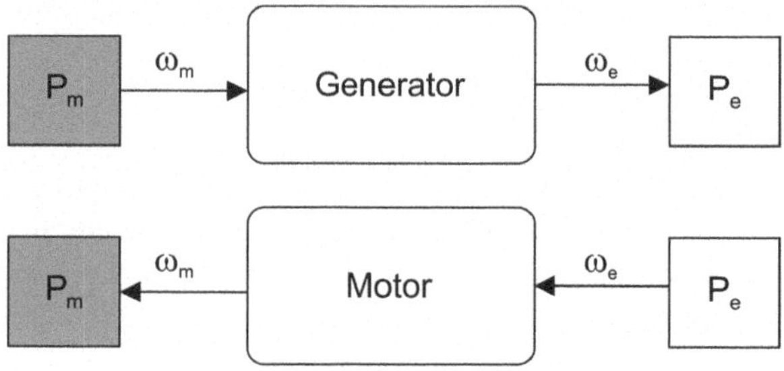

Fig. 3 Simplified in-out scheme of an electrical machine as it works as a generator or a motor

$$D_i\dot{\delta}_i = -P_i^0 + \sum_j P_{i,j} \sin\left(\delta_j - \delta_i\right) \tag{15}$$

The generator nodes can be modeled by the swing equation and the values of $P_{i,j}$, δ_i, δ_j are collected from the solution of the power-flow equations for the original network.

Synchronous motors model

The synchronous motor model, as considered in Refs. [4, 17, 20, 21], assumes that all the network loads behave as synchronous motors. As such, they can be modeled by the swing equation, exactly as generators, except for the fact that instead of delivering power to the network as a generator does ($P_i > 0$), the synchronous motor will withdraw power from it ($P_i < 0$). In essence, the synchronous motor model is more general than the load model considered in the structure-preserving and the effective network models. Recall from the formulation of Eq. 10 that the swing equation describing the synchronous motor model reduces to the load model in Eq. 15 as $J_i = 0$. This approach of modeling power networks is based on the the fact that a synchronous machine can work either as a generator or as a motor. This fact is further illustrated in Fig. 3.

2 Emerging investigations into Robustness and Stability in Power Systems

A number of studies regarding the stability of power networks considered synchronization and stability in the second-order Kuramoto model [4, 8, 9, 13, 16, 17, 20, 43, 48, 52], which can be formally mapped to the classical swing equation [17, 20]. Not only the characterization of the phase transition is of interest, but also admissible classes of initial configurations and natural frequency distributions which lead to synchronization [9], sufficient conditions for initial setups leading to asymptotic complete phase-frequency synchronization [9], critical coupling for the onset of synchronization [21], parameter optimization for stability [36], the limits of existence of coherent and incoherent solutions and observations about the transition from incoherence to coherence as a function of system size and inertia [43] and the role of hysteresis in the phase transition [43]. In the following, a summary of recent contributions to the topic is presented.

2.1 *Basin Stability, Detours, Dead Trees and Transient Stability*

Basin stability and transient stability

The concept of basin stability was applied in a number of works as an interesting approach to unveil the basin of attraction of an equilibrium point [33, 34, 47, 53]. The basin stability establishes (i) the size of the basin of attraction of the equilibrium point within the portion of the state-space around it to which the system can be pushed due to a disturbance and (ii) the probability that the system will return to steady state after a disturbance. In the face of a perturbation that moves the systems' trajectories from its steady-state, the basin stability allows to determine whether the system will settle or not as a function of its post-disturbance state. Since this methodology does not require the linearization of the systems' equations, it allows the evaluation of large perturbations, as it happens to be the case of transients in power networks.

The basin stability can be defined in the context of synchronization in power-grid models. Define $\mathscr{B}$ as the set of points that belong to the basin of the synchronous state and, following [33], define the indicator function

$$\chi_{\mathscr{B}}(\theta, \omega) = \begin{cases} 1, & \textit{if } (\theta, \omega) \in \mathscr{B} \\ 0, & \textit{otherwise}. \end{cases}, \tag{16}$$

which sorts the points that belong to the basin stability $\mathscr{B}$. Further, consider a probability density function $\rho(\theta, \omega)$ that assigns a probability of a post-disturbance initial state (θ, ω). Further, $\int \rho(\theta, \omega)\, d\theta d\omega = 1$ indicates that all the system states that may result from a perturbation are considered. The basin stability $\mathscr{S}$ can be defined as

$$\mathscr{S}(\mathscr{B}) = \int \chi_{\mathscr{B}}(\theta, \omega)\, \rho(\theta, \omega)\, \mathrm{d}\theta \mathrm{d}\omega \tag{17}$$

and it consists of a numerical value $\mathscr{S} \in [0, 1]$ that gives the likelihood that the system will return to steady-state after a perturbation. Figure 4 illustrates the concept in the study of a generator-to-infinite-bus power grid model (in Fig. 5) using the classical swing equation (Eq. 12), as presented in Ref. [33]. In these simulations, we present the basin of attraction of the equilibrium point $(\theta - \theta_S, \omega) = (0, 0)$, where $\theta_S = \theta_\infty + \arctan(\frac{P_m}{P_{max}})$ is the angle of the generator relatively to the infinite bus, assumed as $\theta_\infty = 0$. After suffering a perturbation, the generator can either return to steady-state behavior in the case when the perturbation drives trajectories somewhere into the basin of attraction of $(\theta - \theta_S, \omega)$ or not, in the case it drives the trajectories

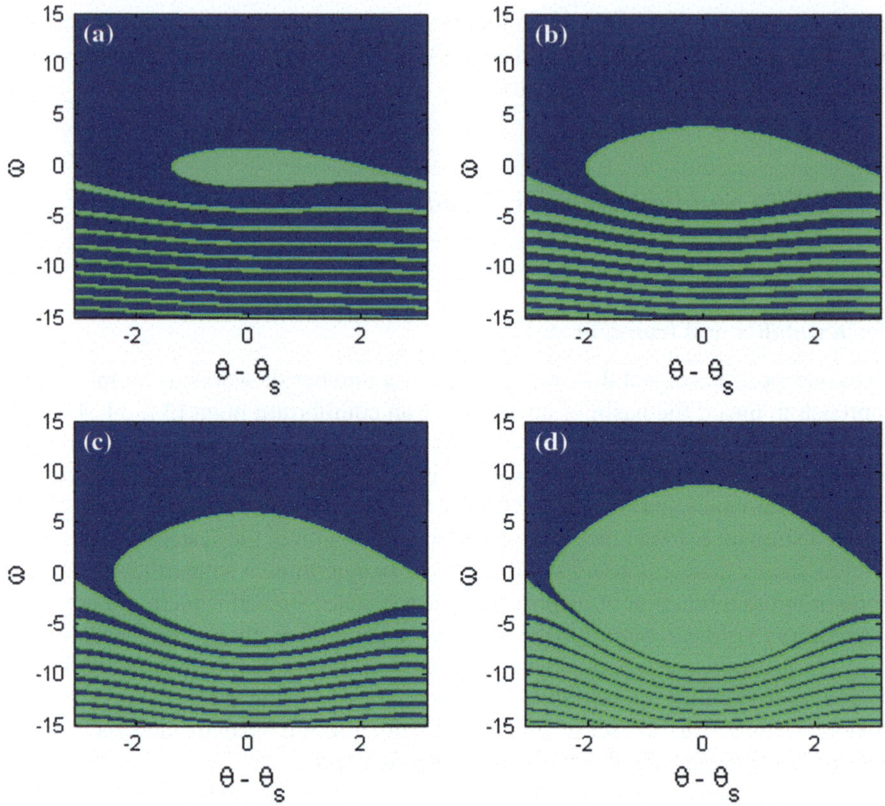

Fig. 4 Basin stability for a generator connected to an infinite bus for different values of maximum power transfer P_{max}, which stands for the coupling between the generator and the infinite bus: **a** $P_{max} = 2$, **b** $P_{max} = 5$, **c** $P_{max} = 10$, **d** $P_{max} = 20$

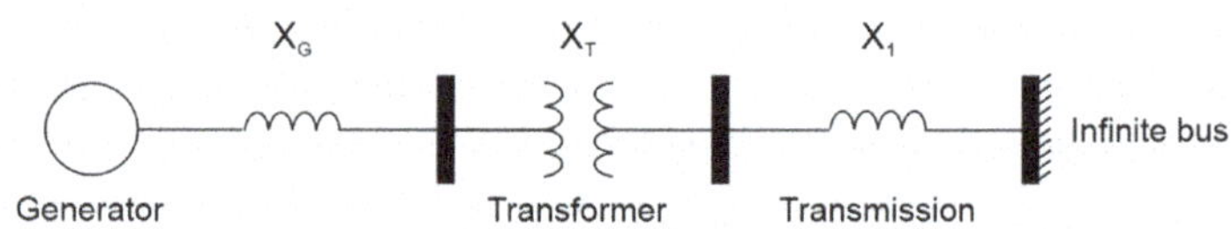

Fig. 5 Generator-to-infinite bus model of a power network. The infinite bus stands for a node with constant voltage, angular velocity and rotor angle deviation

out of it. In Fig. 4, the basin of attraction of the equilibrium point in the origin is represented in green. As it can be seen from the figure, the area of the basin of stability grows as P_{max} increases.

From Fig. 4, it can be noted that the concept of basin stability is very illustrative of the geometry of the basin of attraction of the a system's steady-state operation point. Further, as a inherent feature due to its numerical nature, the method unveils a tight geometric picture of the system's behavior within a set of perturbations $\left[\, \Delta\delta \;\, \Delta\omega \,\right]^T$. The stability basin relies on massive computation, since the method is based on

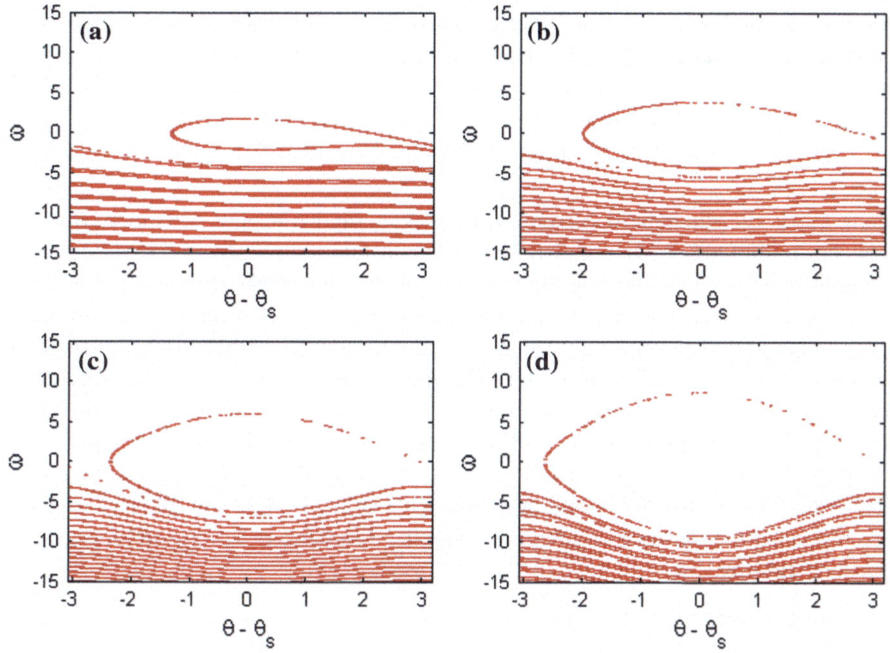

Fig. 6 Difference between the stability basin obtained by means of the classical model (Fig. 4) and the extended model considering voltage-angle interplay, as proposed in Ref. [52]

numerical integration of the system states. Nevertheless, as the integration of the equations for each set of perturbations is an independent process, the evaluation of the stability basin can be subjected to massive parallel computation. For instance, the simulations presented in Figs. 4 and 6 were performed using parallel algorithms tailored for General-Purpose Graphics Processing Unit (GPGPU) by means of the Compute Unified Device Architecture (CUDA) [41].

Coupling voltage dynamics into the generator equations

One of the main issues with studies considering the classical swing equation is that bus voltages are assumed constant all over the power network. In reality, this is not the case, since voltage magnitudes are related to and dependent on the power transfers and power angle among the nodes. As such, a more realistic and insightful model shall consider voltage stability as well. Considering a model for the sub-transient and transient voltage dynamics, an extended version of the second-order Kuramoto model which can account for voltage-angle stability interplay was studied in Ref. [52]. Such interplay plays a significant role in the stability of power networks, since the voltage magnitudes are directly related to the power transfer capability between two nodes. In turn, power transfer capability is closely related to the size of the basin of attraction of a node as it is hit by a perturbation (see Fig. 4). To this end, the oscillator model is formulated as a third order model in which the transient voltage dynamics is modeled by a first-order equation. From the power flow equations and the governing equations for the transient voltages, the node equations read

$$\begin{aligned} \ddot{\delta} &= -\alpha_i \dot{\delta} + P_{m,i} + \sum_{j=1}^{N} B_{ij} E_i E_j \sin\left(\delta_j - \delta_i\right) \\ \tau_i \dot{E}_i &= E_{f,i} - E_i + X_i \sum_{j=1}^{N} B_{ij} E_j \cos\left(\delta_j - \delta_i\right) \end{aligned} , \tag{18}$$

where $E_{f,i}$ is the stator field current, X_i is the generator transient reactance, B is the network susceptance matrix and τ_i is a time constant associated to the variation of voltage. As it turns out, the extended model was shown to make different predictions upon disturbance scenarios. The authors report that the transition to nonstationary states is facilitated in the extended model due to smaller stability region as compared to those of the classical model [52]. Indeed, we show in Fig. 6 the difference between the basin of stability of generator-to-infinite bus considering the swing equation (Eq. 12), as proposed in Ref. [33], and the swing equation with voltage-angle interplay (Eq. 18), as proposed in Ref. [52]. As pointed in Ref. [52], the resulting basin of attraction for the swing equation with voltage-angle interplay is smaller, thus reflecting the importance of considering the dynamics of voltage in the study of transient stability.

The role of dead tree gateways

Still regarding the concept of basin stability, the authors in Ref. [53] showed that the stability of a power network can be hinted by the measure of single-node basin stability [33, 34]. The idea was to identify nodes in the power network that are crucial for stability by means of statistical features of the network. In particular, the authors explore the role of network motifs in the stability of the power-network. As it turns out, the stability basin of a power network can be evaluated with respect to each of its nodes by disturbing each node in turn. As practical result, the study of the relation between network motifs and stability allowed to pinpoint nodes with poor stability without having to solve the network equations explicitly [53]. In particular, two kinds of nodes were found to be of significant influence upon single-node basin stability, namely the *detour nodes*[1] and the *dead tree gateways*.[2]Detour nodes were found to have either fair or high basin stability, while dead end gateways were found to be very likely to feature poor basin stability [53]. This is somewhat expected since detour nodes correspond to alternative or redundant paths which can be of great value in the case of power flow redistribution. On the other hand, dead tree gateways are expected to be associated with poor stability since no alternative paths exist in the case of link malfunctioning. This fact was further explored in Ref. [33] and, as a practical guideline, the authors hint that dead trees and dead tree gateways should be avoided in real power networks.

[1]Following the definition in Ref. [38], a detour at a node u belonging to the shortest path between nodes r and s, $P_G(r, s) = \{r, ..., u, v, ..., s\}$, is defined as the shortest path between u and s that does not go through link (u, v), that is, $P_{G-(u,v)}(u, s)$.

[2]Node adjacent to a dead tree.

Relating the swing equation to the non-uniform Kuramoto model

In Ref. [13], under the assumption of over-damped generators, the authors apply singular perturbation analysis to study transient stability in reduced network power system models in which generators are modeled by means of the swing equation

$$J_i\ddot{\theta}_i + D_i\dot{\theta}_i = \omega_i + \sum_{j=1}^{n} a_{ij}\sin\left(\theta_j - \theta_i + \varphi_{ij}\right), \tag{19}$$

where $\omega_i = P_{m_i} - E_i^2 Re\left(Y_{ii}\right)$, $a_{ij} = E_iE_j\left|Y_{ij}\right|$, $\varphi_{ij} = \arctan\left(\frac{Re(Y_{ij})}{Im(Y_{ij})}\right) \in [0, \pi 2)$ is the phase shift between generators i and j, Y is the network admittance matrix and E_i is the voltage at the node i. The authors provide algebraic conditions for synchronization as a function of network parameters. The results are derived on the basis of the so-called non-uniform Kuramoto model which results from Eq. 19 as one makes $J_i = 0$. The authors show that, on the long time scale and under the assumption of over-damped generators, the synchronization conditions for the transient stability of the coupled swing equations can be approximated with those of the non-uniform Kuramoto model. The expressions for the critical coupling leading to complete frequency synchronization are then developed on the basis of the Non-Uniform Kuramoto model as [13]

$$\Gamma_{crit} = \frac{1}{\cos\left(\varphi_{max}\right)}\left(\max_{i\neq j}\left|\frac{\omega_i}{D_i} - \frac{\omega_j}{D_j}\right| + 2\max_{i\in\{1,\ldots,n\}}\frac{a_{ij}}{D_i}\sin\left(\varphi_{ij}\right)\right) \tag{20}$$

From Eq. 20, the authors in Ref. [13] summarize that frequency synchronization requires that the network connectivity dominate the non-uniformity in power inputs and the power losses.

2.2 *Local Stability in Networks of Second-Order Kuramoto Oscillators*

Local stability in highly connected networks

In Ref. [4], the authors investigate local stability of highly connected power-grids. By considering the network equations linearized about the synchronous state, which reads

$$\begin{aligned} \delta\dot{\theta}_i &= \delta\omega_i \\ \delta\dot{\omega}_i &= -\alpha\delta\omega_i - P_i - P_{max}\delta\theta_i\textstyle\sum_{j=1}^{N} L_{ij}\cos\left(\theta_j - \theta_i\right) + P_{max}\sum_{j=1}^{N}\cos\left(\theta_j - \theta_i\right)\delta\theta_i\text{'} \end{aligned} \tag{21}$$

where L is the network Laplacian matrix, one is led to the simplification of the coupling terms as the summation of the couplings is a constant value in the synchronous state, that is,

$$\frac{\bar{P} - P_i}{P_{max}} = \sum_{j=1}^{N} A_{ij} \sin\left(\theta_j - \theta_i\right) \tag{22}$$

This allows the simplification

$$\cos\left(\theta_j - \theta_i\right) \approx 1 \tag{23}$$

since the phase differences become smaller as the number of terms being summed increases while the summation remains constant.

This reasoning allows Eq. 25 to be rewritten as

$$\begin{aligned} \delta\dot{\theta}_i &= \delta\omega_i \\ \delta\dot{\omega}_i &= -\alpha\delta\omega_i + P_{max}\delta\theta_i \sum L_{ij} - P_{max} \textstyle\sum_{j=1}^{N} L_{ij}\delta\theta_j \end{aligned} \tag{24}$$

As $\sum_j L_{ij} = 0$, Eq. 24 can be simplified to

$$\begin{aligned} \delta\dot{\theta}_i &= \delta\omega_i \\ \delta\dot{\omega}_i &= -\alpha\delta\omega_i - P_{max} \textstyle\sum_{j=1}^{N} L_{ij}\delta\theta_j \end{aligned} \tag{25}$$

As the states of the $i = 1, ..., N$ coupled systems are collected into the network state vector $\delta x = \left[\, \delta\theta_1 \; \delta\omega_1 \; \cdots \; \delta\theta_N \; \delta\omega_N \,\right]$, one obtains

$$\delta\dot{x} = (I \otimes J + P_{max} L \otimes E)\, \delta x, \tag{26}$$

where $J = \begin{bmatrix} 0 & 1 \\ 0 & -\alpha \end{bmatrix}$ is the Jacobian matrix of Eq. 25 and $E = \begin{bmatrix} 0 & 0 \\ 1 & 0 \end{bmatrix}$ is the inner coupling matrix of the individual systems. Put into its diagonal form, Eq. 26 yields N vibrational modes with individual dynamics given by

$$\delta\dot{\xi}_i = (J + \gamma_i P_{max} E)\, \delta\xi_i, \tag{27}$$

where γ_i are eigenvalues of the Laplacian matrix and $\gamma_N \geq \cdots \geq \gamma_2 \geq \gamma_1 = 0$. According to the MSF framework [45], the local stability of the synchronous state relies on that of the second mode, i.e., $\delta\xi_2$, whose eigenvalues correspond to its Lyapunov exponents due to the linearity of Eq. 27. As a consequence, one obtains the Lyapunov exponents as

$$\lambda_{1,2} = \frac{-\alpha \pm \sqrt{\alpha^2 - 4\gamma_2 P_{max}}}{2} \tag{28}$$

As observed in [4], since α, γ_2 and P_{max} are positive quantities, the Lyapunov exponents given by Eq. 28 are negative, which reveals that the synchronous state is locally stable for highly connected networks. Indeed, power networks are inherently stable to sufficiently small perturbations and that is a necessary condition for its functioning, otherwise it would not be able to handle small power fluctuations to which the grid is continuously subjected [25]. The results directly relate local stability to the value of the damping coefficient α. Although the damping parameter is represented in simplified form in the model (24), it is known to depend on a series of factors such as the presence of automatic voltage regulators and power system stabilizers [25]. Thus, the tuning of the damping coefficient can help enhance the stability of power networks, as shown in Ref. [36].

Optimizing generator parameters towards most stable steady states

A similar approach was adopted in Ref. [36], as the authors allow for heterogenous couplings among oscillators and consider a reduced network in which the loads are modeled as constant impedances. Under the reduced network model the generators are connected with all-to-all couplings and the values of the coupling between any two nodes is related to the nodes' voltages and to the admittance between them. The main contribution of the study is the derivation of algebraic expressions to the tuning of generators such that they operate in the point of maximum stability as a function of a regulation droop parameter, R_i and the damping parameter, D_i. The droop parameter regulates the output of the equation of the variation in mechanical power relatively to the angular velocity and it is defined as

$$\frac{\partial P_{mi}}{\partial \omega_i} = \frac{-1}{\omega_R R_i}, \tag{29}$$

where ω_R is the grid reference angular velocity. Such parameters are optimized on the basis of the real part of the smallest nonzero eigenvalue of the coupling matrix evaluated at the steady-state, which is denoted α_2, and H_i is the inertia constant of the *ith* generator, which is related to the total inertia of the *ith* node, J_i, by means of the expression $J_i = \frac{2H_i}{\omega_R}$. According to Ref. [36], tuning the droop parameter R_i, the optimal stability point is reached as

$$R_i = \frac{1}{4H_i\sqrt{\alpha_2} - D_i} \tag{30}$$

or, equivalently, tuning the damping parameter, it yields

$$D_i = 4H_i\sqrt{\alpha_2} - \frac{1}{R_i} \tag{31}$$

for each generator in the reduced network model. A hypothesis that allows the derivation of this result is that $\beta_i = (D_i + 1/R_i)/2H_i = \beta$ is the same for all generator nodes. The authors show by means of numerical simulations that the optimization of generator parameters improves stability against small perturbations for a model of the Northern Italy power grid [36]. Indeed, the importance of damping to stable synchronization is recognized in classical power systems stability theory [25] and was recognized to have influence upon the onset of synchronization [14, 21].

Although the hypothesis on uniform β might not hold in practice, the parameter tuning based on the results presented in Ref. [36] are significant in the sense that they help enhance the comprehension about the factors influencing local stability in power networks and even provide an approach to the derivation of near-optimal or sub-optimal parameter tuning in real power grids.

Algebraic expressions for local stability in the multi-rate Kuramoto model

The local stability of the synchronous state in power networks was also studied by means of the multi-rate Kuramoto model [14], defined as

$$\begin{aligned} J_i\ddot{\theta}_i + D_i\dot{\theta}_i &= P_{m_i} + P_{max}\textstyle\sum_{j=1}^{n} \sin\left(\theta_j - \theta_i\right), \ i \in \{1, ..., m\} \\ D_i\dot{\theta}_i &= P_{m_i} + P_{max}\textstyle\sum_{j=1}^{n} \sin\left(\theta_j - \theta_i\right), \ i \in \{m+1, ..., n\} \end{aligned}, \tag{32}$$

where equations 1 to m describe generators and $m+1$ to n load node equations. It can be noted that as $m = n$, the coupled equations are all second-order, while for $m = 0$, the model (32) becomes a set of coupled first-order equations. It can be noted also that this formulation assumes all-to-all coupling among the oscillators. Considering all positive inertia $J_i > 0$ and damping $D_i > 0$, the authors in Ref. [14] show that the local exponential stability of frequency synchronization depends on the damping coefficients D_i but not on the inertia coefficients, J_i. The authors achieve an algebraic expression that allows the evaluation of local stability as a function of the network and oscillator parameters.

However, since they are local in nature, the results do not apply in the case of transient stability studies, in which inertial terms are known to have profound influence [14, 25].

2.3 Onset of Complete Frequency Locking in Structure-Preserving Generator-Motor models

The existence of a synchronous manifold

For a power-grid with N nodes, the second-order model for electricity distribution developed in [17] and further studied in [4] is given by

$$\begin{aligned} \dot{\theta}_i &= \omega_i \\ \dot{\omega}_i &= -\alpha\omega_i + P_i - P_{max}\textstyle\sum_{j=1}^{N} A_{ij}\sin\left(\theta_j - \theta_i\right) \end{aligned} \tag{33}$$

for $i = 1, 2, ..., N$, where θ_i is the phase of the *ith* oscillator, ω_i is time-dependent frequency, α is a parameter of dissipation due to friction, P_i is the power consumed ($P_i < 0$) or generated ($P_i > 0$) in the *ith* node at time t. Further, A_{ij} is an entry in the *ith* row and *jth* column of the network adjacency matrix A, which is assumed symmetric since the network is undirected; if nodes i, j are connected, then $A_{ij} = A_{ji} = 1$, and

$A_{ij} = A_{ji} = 0$, otherwise. As the N oscillators operate in the same frequency ω_s, the network is synchronized. As pointed in [4], the synchronous solution implies

$$\begin{array}{l} \dot{\omega}_i = 0 \\ \omega_i = \omega_s \end{array}, \tag{34}$$

where ω_s is the synchronous frequency. Taking the conditions on Eq. 34 into the equations of the node dynamics (33) yields

$$\begin{array}{l} \dot{\theta}_i = \omega_s \\ 0 = -\alpha\omega_s + P_i - P_{max} \sum_{j=1}^{N} A_{ij} \sin\left(\Delta_{j,i}\right) \end{array}, \tag{35}$$

where $\Delta_{j,i}$ is constant in the synchronous solution. Following [4], define the degree of the *ith* node as

$$k_i = \sum_{j=1}^{N} A_{ij} \tag{36}$$

and the existence of a synchronous solution for system (33) is guaranteed by the condition

$$P_{max} \geq \left\{ \frac{|\alpha\omega_s - P_i|}{k_i} \right\}_{max} = \left\{ \frac{\left|\bar{P} - P_i\right|}{k_i} \right\}_{max} = P^*_{max} \tag{37}$$

Note that the condition given by the inequality (37) is satisfied once $\sum_{j=1}^{N} A_{ij} \sin\left(\theta_j - \theta_i\right) \leq k_i$.

The onset of frequency synchronization: a necessary condition

In Ref. [17], the authors consider the node equation

$$\begin{array}{l} \dot{\theta}_j = \omega_j \\ \dot{\omega}_j = -\alpha\omega + P_j + P_{max} \sum_{k=1}^{N} A_{jk} \sin\left(\theta_k - \theta_j\right) \end{array}, \tag{38}$$

where θ_j and ω_j are functions of time and denote the phase and angular velocity of the *jth* oscillator, respectively, α is a damping parameter, P_j is the normalized power delivered ($P_j > 0$) or consumed ($P_j < 0$) in the *jth* node, P_{max} is the maximum capacity of the transmission line between two nodes and A_{jk} is the entry of the symmetric adjacency matrix at row j and column k that is equal to one if nodes j and k are linked and to zero, otherwise. As one considers a network of coupled generators and loads, the synchronous state can be taken as $\omega_1(t) = \omega_2(t) = \cdots = \omega_N(t)$, which means that the angular velocities of all the oscillators are equal to the synchronous velocity $\omega_s(t)$. In this case, $\dot{\omega}_j - \dot{\omega}_k = 0$. Following [4], upon the assumption of the condition $\dot{\omega}_j = 0$, $\omega_j = \omega_s$, $j = 1, 2, ..., N$ for some constant synchronous angular velocity ω_s, Eq. 38 becomes

$$0 = -\alpha\omega_s + P_j + P_{max} \sum_{k=1}^{N} A_{jk} \sin\left(\Delta_{kj}\right) \tag{39}$$

where $\Delta_{kj} = \theta_k - \theta_j$ is a constant. As one recognizes that $\Delta_{j,i} = -\Delta_{i,j}$ then, for an undirected network,

$$\sum_{i=1}^{N} \sum_{j=1}^{N} A_{ij} \sin\left(\theta_j - \theta_i\right) = 0 \tag{40}$$

and the system of algebraic Eqs. 35 becomes

$$\sum_{i=1}^{N} (-\alpha\omega_s + P_i) = 0 \tag{41}$$

and, one step ahead,

$$-N\alpha\omega_s + \sum_{i=1}^{N} P_i = 0 \tag{42}$$

which finally yields the synchronous frequency of the synchronized network

$$\omega_s = \frac{\sum P_i}{\alpha N} = \frac{\bar{P}}{\alpha}, \tag{43}$$

where $\bar{P}$ is the average value of all the P_i. Under these conditions, it was shown in [4] that the summation of the equations of the N nodes yields the synchronous angular velocity

$$\omega_s = \sum \frac{P_j}{\alpha N} \tag{44}$$

In a power-grid, the balance between generators and consumers requires that $\sum_{j=1}^{N} P_j = 0$, which means that $\omega_s = 0$. Over the time, whenever these conditions do not hold, instabilities appear. Two indexes can be used to evaluate how effective a specific power grid is in handling instabilities: *synchronization quality* and *persistence over time*. The former refers to how tight the matching among the angular velocities of the oscillators is, while the latter refers to how long they remain synchronized given a reference time window. Actually, over the time, consumer's power request keep changing which requires continuous adjustments in the power supplied by generators. This scenario implies that in fact the synchronous angular velocity (frequency) keeps changing over the time.

Following this reasoning, it was shown in Ref. [4] that there is a lower bound for the existence of a synchronous manifold to coupled second-order Kuramoto oscillators is given by

$$P_{sm} = \max_{j} \left\{ \frac{\left|\alpha\omega_s - P_j\right|}{k_j} \right\}, \tag{45}$$

where $d_j = \sum_{k=1}^{N} A_{jk}$, is the degree of node j.

Consider the second-order Kuramoto model given by Eq. 38, with damping given by $\alpha > 0$, and coupling strength P_{max}. A necessary condition for complete frequency synchronization

$$P_{max} \geq P_{sm} \triangleq \max_{j} \left\{ \frac{\left|\alpha\omega_s - P_j\right|}{k_j} \right\}, \tag{46}$$

where P_j, $j = 1, ..., N$ is the power injected/consumed by node j and d_j stands for its degree.

This indicates that the variance in the distribution of P_j has to be compensated with stronger coupling or a high value of the node connectivity. This can also be seen as one looks for the coupling strength at the onset of synchronization for a fully connected network.

The onset of complete frequency synchronization: a tight condition

In Ref. [21], the authors consider the deviation equations $\ddot{\theta}_i - \ddot{\theta}_j$ given by

$$\dot{\omega}_i - \dot{\omega}_j = -\alpha \left(\omega_i - \omega_j\right) + \left(P_i - P_j\right) + P_{max} E \left(\theta_i, \theta_j, \theta_k\right) \tag{47}$$

Following [10], the maximum value for the function E, given by

$$E\left(\theta_i, \theta_j, \theta_k\right) = 2\sin\left(\theta_j - \theta_i\right) + \sum_{k=1, k\neq i,j}^{N} \left(\sin\left(\theta_k - \theta_i\right) + \sin\left(\theta_j - \theta_k\right)\right) \tag{48}$$

can be obtained by means of the calculation of the function extremes with respect to the phase angles $\theta_i, \theta_j, \theta_k$. There are two solutions and the maximum takes place for $\theta_i = \theta_j$ or $2\theta_k = \theta_i + \theta_j$. In case $\theta_i = \theta_j$, there is no power flow. For $2\theta_k = \theta_i + \theta_j$ the function (48) reaches a maximum. The optimal solution $\left(\theta_j - \theta_i\right)_{opt}$ provides the maximum of Eq. 48 as a function of the number of oscillators in the network. Thus, the optimal phase difference as found in Ref. [10] is given by

$$\left(\theta_j - \theta_i\right)_{opt} = 2 \arccos\left(\frac{-(N-2) + \sqrt{(N-2)^2 + 32}}{8} \right) \tag{49}$$

and it occurs when the function in Eq. 48 reaches

$$E_{max} = 2\sin\left(\theta_j - \theta_i\right)_{opt} + 2(N-2)\sin\left(\frac{\left(\theta_j - \theta_i\right)_{opt}}{2} \right) \tag{50}$$

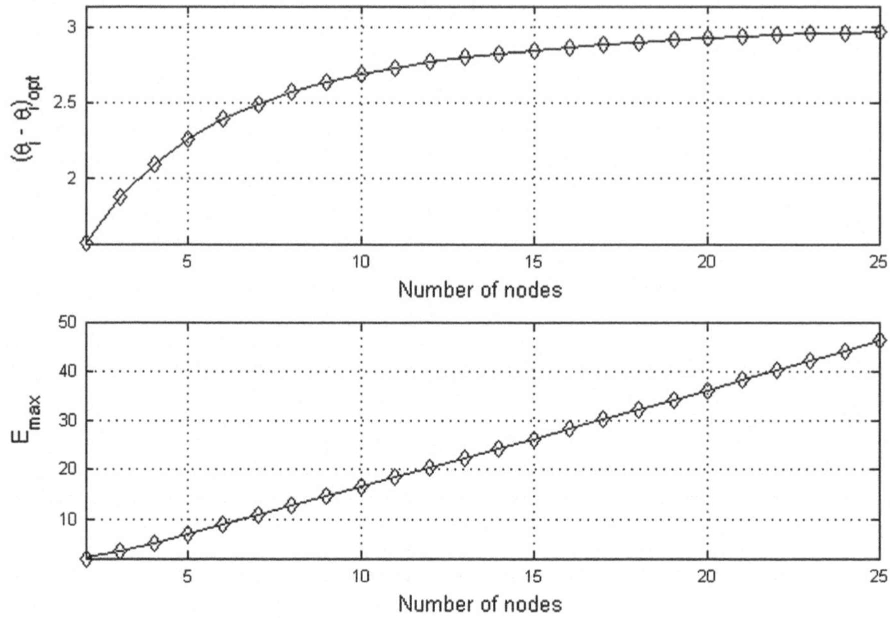

Fig. 7 Optimal phase angle $\left(\theta_j - \theta_i\right)_{opt}$ as a function of the number of nodes N and E_{max} as a function of the optimal phase angle and the number of nodes

The behavior of E_{max} as a function of the number of oscillators is shown in Fig. 7. Equation 50 provides a tight lower bound to the onset of complete frequency synchronization, as originally observed in Ref. [10] and recently refined in Ref. [16]. Following Refs. [10, 16], consider the optimal phase difference $\left(\theta_j - \theta_i\right)_{opt}$ and E_{max} as in Eqs. 49 and 50. For complete frequency synchronization of the second-order Kuramoto model in Eq. 38, $N \geq 2$, it is necessary that

$$P_{max} \geq P_{critical} \triangleq max \frac{\left|P_i - P_j\right|}{E_{max}}, \tag{51}$$

where $P_i, P_j,\ i, j = 1, ..., N$, are the injected/consumed power and $\left(\theta_j - \theta_i\right)_{opt} \in \left[\frac{\pi}{2}, \pi\right]$.

As the influence of angular velocities are taken into account, it is possible to provide a closer estimate to the onset of complete frequency synchronization for networks of second-order Kuramoto oscillators with all-to-all couplings.

The onset of frequency synchronization: a closer estimate

Following Ref. [21], as one takes the limit $\left(\theta_j - \theta_i\right) \rightarrow \left(\theta_j - \theta_i\right)_{opt}$ then $E \rightarrow E_{max}$. Consider Eq. 49 and admit $E = E_{max}$ for the purpose of finding P_{max} for the onset of synchronization, as proposed in [21]. Further, denote $\omega_i - \omega_j = \Delta\omega_{ij}$ in Eq. 47 to obtain the following linear ODE

$$\Delta\dot{\omega}_{ij} + \alpha\Delta\omega_{ij} = \left(P_i - P_j\right) + P_{max}E_{max}, \tag{52}$$

whose solution reads

$$\Delta\omega_{ij} = \frac{P_i - P_j}{\alpha} + \frac{P_{max}E_{max}}{\alpha} + \omega_0 e^{-\alpha t} \tag{53}$$

for some arbitrary integration constant ω_0. As $t \to +\infty$, the expression can be rewritten as

$$P_{max} > max\frac{\left|\alpha\Delta\omega_{ij} - \left(P_i - P_j\right)\right|}{E_{max}}, \; i,j = 1, \ldots, N \tag{54}$$

For the purpose of estimating the critical coupling for the onset of complete frequency synchronization, Eq. 54 gives a lower bound for P_{max}. Consider $P_{i,}\ P_j$ as constants and one can approximate P_{max} by means of the upper bound of the deviation in angular velocities, $\Delta\omega_{max}$. As one considers that the angular velocity for a given oscillator belongs to the interval $[\omega_{min}, \omega_{max}]$, where $\omega_{min} = min\,(\omega_i)$ and $\omega_{max} = max\,(\omega_i)$, $i = 1, \ldots, N$, one can define that interval by observing each oscillator will rotate with angular velocity corresponding to its natural frequency as $t \to +\infty$ in the case of weak coupling. Put in terms of an equation, this means that for weak coupling

$$\omega_i = \frac{P_i}{\alpha} \tag{55}$$

For strong coupling and $t \to +\infty$,

$$\omega_i = \omega_s = \sum_i^N \frac{P_i}{\alpha N} \tag{56}$$

such that one obtains

$$\Delta\omega_{max} = \omega_{max} - \omega_{min} = \max_i \left|\frac{\sum_i^N P_i - NP_i}{\alpha N}\right| \tag{57}$$

Therefore, the value of P_{max} from Eq. 54 can be estimated by means of

$$P_{max} = max\left\{\frac{\left|\alpha\Delta\omega_{max} - \left(P_i - P_j\right)\right|}{E_{max}}\right\}, \; i,j = 1, \ldots, N \tag{58}$$

Note that as $\omega_i = \omega_j$, Eq. 47 becomes

$$0 = (P_i - P_j) + P_{max}\left\{2\sin(\theta_j - \theta_i) + \sum_{k=1,k\neq i,j}^{N} (\sin(\theta_j - \theta_k) + \sin(\theta_j - \theta_k))\right\} \quad (59)$$

and one once again obtains the equation

$$P_{max} = max\frac{|P_i - P_j|}{E_{max}} \quad (60)$$

Equation 60 gives the value below which synchronization between nodes i and j does not occur. Thereby, a necessary condition for synchronization is that the coupling strength is larger than P_{max} given by Eq. 60. Recall that this criterion was developed under the assumption that the network is fully connected. To generalize the result to undirected Laplacian matrices not fully connected, one can consider the results in Ref. [4] to establish that, for highly connected networks, the scaling

$$P_{max} = \frac{\lambda_2^{full}}{\lambda_2} P_{max}^{full} \quad (61)$$

holds, where λ_2 is its smallest nonzero eigenvalue of the Laplacian matrix G of a highly connected undirected and unweighted network with N nodes and λ_2^{full} is the smallest nonzero eigenvalue of the Laplacian matrix of the analogous fully connected matrix [21].

2.4 Synchronizability of Power Networks with Distributed Generation

Generation decentralization and the robustness against topological failures

The study of networks of second-order Kuramoto oscillators with bimodal distribution in Ref. [49] focused on the impacts of successive decentralization of generator units upon dynamic stability, synchrony and robustness against structural failures. The methodology consisted of evaluating the phase ordering of the power grid model by means of the order parameter

$$r_\infty = \lim_{t_0\to\infty} \frac{1}{\Delta t}\int_{t_0}^{t_0+\Delta t} \sum_j \frac{e^{i\theta_j(t)}}{N} dt \quad (62)$$

for increasing decentralization of the generator units. The authors find that the maximum allowed perturbation to the steady state shrinks with decentralization. This indicates that a decentralized grid is more likely to get unstable under a perturbation than a centralized one of equivalent power under the same perturbation. The reduced inertia of smaller generator units seem to reduce their capacity to handle perturbations to synchronous dynamics, thus facilitating the emergence of rotor-angle instabilities. Regarding synchrony, it was found that the order parameter increases with decentralization. Similar results were found in Ref. [21] by means of a different methodology that shows that the critical coupling for the onset of complete frequency synchronization substantially decreases with decentralization. Another factor in favor of decentralization is the increased robustness against structural failures, as concluded in Ref. [49]. It is intuitive that decentralized generation tends to be less affected by the loss of a single transmission line, for example. Besides, the relative importance of a single generator or particular transmission line decreases in a decentralized power grid. However, it has to be considered that in a decentralized grid the number of occasional unanticipated failures tends to increase due to the larger number of nodes and interconnections.

Further conclusions regarding distributed generation (DG) are presented in Ref. [29] by means of the evaluation of diverse vulnerability measures against increasing penetration of DG. The vulnerability measures take into account both the equations of power flow and statistical measures from Complex Networks theory. The authors consider the steady-state solution of the power flow equations to assess the vulnerability measures. This approach seems to provide a closer approximation of the power system properties than early studies based only on system topology [12]. Among the vulnerability measures, the authors consider the ones based on efficiency of transmission of active power, such as the Operational Vulnerability Index (OVI). The OVI is lower to the proportion that the distance of transmission of active power increases and it was found that power grids with shorter electrical distances between sources and loads have improved reliability [29]. Shorter distances between generation and loads is a characteristic of power grids with distributed generation, as a fair share of generation can be located in the distribution network.

3 Perspectives and challenges

There are many challenges regarding the application of the concepts and tools of complex networks to the study and understanding of power networks. We summarize below the ones we consider to be the most salient.

Dropping of assumption on homogeneous node parameters and connectivity over the power network

There are ways in which the study of robustness and stability in power networks from the viewpoint of complex networks can be enhanced. They mainly involve relaxing assumptions made on the modeling of loads, transmission lines and network connectivity. There is a considerable overlap of theory developed in the context of the Kuramoto model that is applied in studies of power networks. It is common to have the assumption of homogeneous loads all over the network. This simplifies the calculations but may render the results little representative of the real power network dynamics. As another instance, the assumption on all-to-all connectivity holds for reduced-network models, in which a connected network of rotating generators and passive loads are boiled down to a network of rotating generators with all-to-all connections through links that represent both line impedances and load impedances. However, the assumption on passive loads (i.e., loads that do not have a dynamical behavior) is limited, since a fair share of loads do have a dynamical behavior that influence stability of real power networks.

Assessing and clarifying the limitations of complex network approach

Many of the techniques and tools of complex networks have been applied in the context of power networks. Important insights have been obtained that enhanced the understanding of underlying phenomena affecting stability and robustness. However, many of the discussions lack clarity regarding the limitations of the proposed methodologies and the consequences of the modelling assumptions upon the applicability of the results in real power networks.

The increase of renewable sources in power networks

The increasing addition of renewable generation such as solar and wind shall bring new dynamical features to the interconnected power networks [24, 54]. There are a number of reasons for that. First, solar and wind power are known to be intermittent in nature, in the sense that their output at any given time heavily depends upon the availability of solar radiation and sufficient wind velocity, respectively. As such, the output of a solar or wind power plant is quite uncertain. Second, the DC nature of solar and wind power outputs and the AC nature of power networks require the DC-AC conversion of power by means of electronic interfaces. In this process, a potential increase in the harmonics content of the power signal may affect power quality and system stability.

These issues are likely to influence the approaches currently in use for generator and load modeling and, further, the representation of power grids by means of its corresponding graph.

The impact of smart-grid technologies

The concept of smart-grid refers to power networks in which the subsystems intensively exchange information among them, which contributes to improve efficiency and reliability of operation and control. This is increasingly important as the share of renewable sources of intermittent nature in the grid increases. In order to simplify coordination and control of the overall power network, smart-grids with distributed generation have been thought of within the framework of *microgrids*. A microgrid is thought of as a cluster of generators and loads at the distribution level that are connected to the host power system by means of a single point of connection [42]. Thereby, each cluster would be designed to be responsive to control and coordination signals coming from the host power system. The matter of how to define such clusters and optimize them for synchronization seems a problem tailored for the complex networks framework (check studies in this topic in Ref. [3, 23, 30–32, 40, 46, 57, 60]).

References

1. Andersson, G., Donalek, P., Farmer, R., Hatziargyriou, N., Kamwa, I., Kundur, P., Martins, N., Paserba, J., Pourbeik, P., Sanchez-Gasca, J., Schulz, R., Stankovic, A., Taylor, C., Vittal, V.: Causes of the 2003 major grid blackouts in north america and europe, and recommended means to improve system dynamic performance. IEEE Trans. Power Syst. **20**(4), 1922–1928 (2005)
2. Bergen, A.R., Hill,D.J.: A structure preserving model for power system stability analysis. IEEE Trans. Power Appar. Syst PAS **100**(1), 25–35 (1981)
3. Bohm, C., Plant, C., Shao, J., Yang, Q.: Clustering by synchronization. In: Proceedings of the 16th ACM SIGKDD International Conference on Knowledge Discovery and Data Mining, Washington, DC, USA, pp. 25–28 (2010)
4. Carareto, R., Baptista, M.S., Grebogi, C.: Natural synchronization in power-grids with anti-correlated units. Commun. Nonlinear Sci. Numer. Simul. **18**(4), 1035–1046 (2013)
5. Carreras, B.A., Lynch, V.E., Dobson, I., Newman, D.E.: Complex dynamics of blackouts in power transmission systems. Chaos **14**, 643–652 (2004)
6. Chassin, D.P., Posse, C.: Evaluating north american electric grid reliability using the barabasi-albert network model. Phys. A Stat. Mech. Appl. **355**(2–4), 667–677 (2005)
7. Chen, X., Sun, K., Cao, Y., Wang, S.: Identification of vulnerable lines in power grid based on complex network theory. In: IEEE Power Engineering Society General Meeting, pp. 1–6. (2007)
8. Choi, Y.P., Li, Z., Ha, S.Y., Xue, X., Yun, S.B.: Complete entrainment of kuramoto oscillators with inertia on networks via gradient-like flow. J. Differ. Equ. **257**(7), 2591–2621 (2014)
9. Choi, Y-P., Ha, S-Y., Yun, S-B.: Complete synchronization of kuramoto oscillators with finite inertia. Physica D: Nonlinear Phenomena **240**(1), 32–44 (2011)
10. Chopra. N., Spong, M.W.: On synchronization of kuramoto oscillators. In: Proceedings of the 44th IEEE Conference on Decision and Control and the European Control Conference, Seville, Spain, pp. 3916–3922 (2005)
11. Crucitti, P., Latora, V., Marchiori, M.: Model for cascading failures in complex networks. Phys. Rev. E **69**, 045104 (2004)
12. Crucitti, P., Latora, V., Marchiori, M.: A topological analysis of the italian electric power grid. Physica A **338**, 92–97 (2004)
13. Dorfler. F., Bullo, F.: Synchronization and transient stability in power networks and non-uniform kuramoto oscillators. SIAM J. Control Optim. (2010)
14. Dorfler, F., Bullo, F.: On the critical coupling for kuramoto oscillators. SIAM J. Appl. Dyn. Syst. **10**(3), 1070–1099 (2011)
15. Dorfler, F., Bullo, F.: Kron reduction of graphs with applications to electrical networks. IEEE Trans. Circuits Syst. I Regul. pap. **60**, 150–163 (2013)
16. Dorfler, F., Bullo, F.: Synchronization in complex networks of phase oscillators: a survey. Automatica **50**(6), 1539–1564 (2014)
17. Filatrella, G., Nielsen, A.H., Pedersen, N.F.: Analysis of a power grid using a kuramoto-like model. Eur. Phys. J. B **61**, 485–491 (2008)
18. Union for the Coordination of Electricity Transmission (UCTE). Final Report of the Investigation Committee on the 28 September 2003 Blackout in Italy. Technical report (2004)
19. US-Canada Power System Outage Task Force. Final Report on the August 14, 2003 blackout in the United States and Canada. Technical report (2004)
20. Fortuna, L., Frasca, M., Fiore, A.S.: Analysis of the italian power grid based on kuramoto-like model. In: Physcon 2011 Leon Spain, 5-8 September 2008
21. Grzybowski, J.M.V., Macau, E.E.N., Yoneyama, T.: On synchronization in power-grids modelled as networks of second-order kuramoto oscillators. Chaos: an interdisciplinary. J. Nonlinear Sci. **26**(11), 113113 (2016)
22. Hines, P., Blumsack, S., Sanchez, E.C., Barrows, C.: The topological and electrical structure of power grids. In: Conference on 43rd Hawaii International System Sciences (HICSS), pp. 5-8. 2010

23. Huang, L., Lai, Y.C., Gatenby, R.A.: Optimization of synchronization in complex clustered networks. Chaos: an interdisciplinary. J. Nonlinear Sci. **18**(1), 013101 (2008)
24. Katoh, S., Ohara, S., Itoh, T.: Technologies for mitigating fluctuation caused by renewable energy sources. In: International Power Electronics Conference (IPEC-Hiroshima 2014 - ECCE ASIA) pp. 850–856 (2014)
25. Kundur, P.: Power System Stability and Control. McGraw-Hill (1994)
26. Kundur, P., Paserba, J., Ajjarapu, V., Andersson, G., Bose, A., Canizares, C., Hatziargyriou, N., Hill, D., Stankovic, A., Taylor, C., Van Cutsem, T., Vittal, V.: Definition and classification of power system stability IEEE/CIGRE joint task force on stability terms and definitions. IEEE Trans. Power Syst. **19**(3), 1387–1401 (2004)
27. Latora, V., Marchiori, M.: Efficient behavior of small-world networks. Phys. Rev. Lett. **87**, 198701 (2001)
28. Willis, H.L: Power Distribution Planning Reference Book. Marcel Dikker, Inc, (2004)
29. Liu, C., Xu, Q., Chen, Z., Bak, C.L.: Vulnerability evaluation of power system integrated with large-scale distributed generation based on complex network theory. In: 47th International Universities Power Engineering Conference (UPEC) pp. 1–5 (2012)
30. Liu, X. Chen, T.: Fixed-time cluster synchronization for complex networks via pinning control. arXiv:1509.03350 (2015)
31. Liu, X., Chen, T.: Finite-time and fixed-time cluster synchronization with or without pinning control. IEEE Trans. Cybern. **99**, 1–13 (2016)
32. McGraw, P.N., Menzinger, M.: Clustering and the synchronization of oscillator networks. Phys. Rev. E **72**, 015101 (2005)
33. Menck, P.J., Heitzig, J., Kurths, J., Schellnhuber, H.J.: How dead ends undermine power grid stability. Nat. Commun. **5**(3969), 1–8 (2014)
34. Menck, P.J., Heitzig, J., Marwan, N., Kurths, J.: How basin stability complements the linear-stability paradigm. Nat. Phys. **9**, 89–92 (2013)
35. Menck, P.J., Kurths, J.: Topological identification of weak points in power grids. In: Nonlinear Dynamics of Electronic Systems, Proceedings of the (NDES), Vol. 205, pp. 1–4 (2012)
36. Motter, A.E., Myers, S.A., Anghel, M., Nishikawa, T.: Spontaneous synchrony in power-grid networks. Nat. Phys. **9**, 191–197 (2013)
37. Carlotto, T., Onetta, M.S., Grzybowski, J.M.V.: Natural connectivity and the mitigation of cascading failures in a model of Eletrosul transmission system. In: Proceedings of the Congress of Applied and Computational Mathematics, CMAC-SUL SBMAC Curitiba, PR, Brazil (2014)
38. Nardelli, E., Proietti, G., Widmayer, P.: Finding the detour-critical edge of a shortest path between two nodes. Inf. Process. Lett. **67**(1), 51–54 (1998)
39. Nishikawa, T., Motter, A.E.: Comparative analysis of existing models for power-grid synchronization. New J. Phys. **17**(1), 015012 (2015)
40. Nixon, M., Friedman, M., Ronen, E., Friesem, A.A., Davidson, N., Kanter, I.: Synchronized cluster formation in coupled laser networks. Phys. Rev. Lett. **106**, 223901 (2011)
41. NVIDIA. CUDA technology (2007)
42. Olivares, D.E., Mehrizi-Sani, A., Etemadi, A.H., Canizares, C.A., Iravani, R., Kazerani, M., Hajimiragha, A.H., Gomis-Bellmunt, O., Saeedifard, M., Palma-Behnke, R., Jimenez-Estevez, G.A., Hatziargyriou, N.D.: Trends in microgrid control. IEEE Trans. Smart Grid **5**(4), 1905–1919 (2014)
43. Olmi, S., Navas, A., Boccaletti, S., Torcini, A.: Hysteretic transitions in the kuramoto model with inertia. Phys. Rev. E **90**, 042905 (2014)
44. Pahwa, S., Hodges, A., Scoglio, C., Wood, S.: Topological analysis of the power grid and mitigation strategies against cascading failures. In: IEEE International Systems Conference, pp. 272–276 (2010)
45. Pecora, L., Carroll, T.: Master stability functions for synchronized coupled systems. Phys. Rev. Lett. **80**, 2109–2112 (1998)
46. Pecora, L.M., Sorrentino, F., Hagerstrom, A.M., Murphy, T.E., Roy, R.: Cluster synchronization and isolated desynchronization in complex networks with symmetries. Nat. Commun. **5**, 1–8 (2014)

47. Peng, J., Kurths, J.: Basin stability of the kuramoto-like model in small networks. Eur. Phys. J. Spec. Top. **223**, 2483–2491 (2014)
48. Pinto, R.S., Saa, A.: Synchrony-optimized networks of kuramoto oscillators with inertia. Physica A Stat. Mech. Appl. **463**, 77–87 (2016)
49. Rohden, M., Sorge, A., Timme, M., Witthaut, D.: Self-organized synchronization in decentralized power grids. Phys. Rev. Lett. **109**, 064101 (2012)
50. Rosas-Casals, M., Corominas-Murtra, B.: Assessing european power grid reliability by means of topological measures. Trans. Ecol. Env. **121**, 515–525 (2009)
51. Rosas-Casals, M., Valverde, S., Sole, R.V.: Topological vulnerability of the European power grid under errors and attacks. Int. J Bifurc. Chaos **17**, 2465–2475 (2007)
52. Schmietendorf, K., Peinke, J., Friedrich, R., Kamps, O.: Self-organized synchronization and voltage stability in networks of synchronous machines. Eur. Phys. J. Spec. Top. **223**(12), 2577–2592 (2014)
53. Schultz, P., Heitzig, J., Kurths, J.: Detours around basin stability in power networks. New J. Phys. **16**, 125001 (2014)
54. Shafiullah, G.M., Oo, A.M.T.: Analysis of harmonics with renewable energy integration into the distribution network. In: IEEE Innovative Smart Grid Technologies–Asia (ISGT ASIA), pp. 1–6 (2015)
55. Solé, R.V., Rosas-Casals, M., Corominas-Murtra, B., Valverde, S.: Robustness of the european power grids under intentional attack. Phys. Rev. E **77**, 026102 (2008)
56. Thompson. M.J.: Fundamentals and advancements in generator synchronizing systems. In: Proceedings of the 65th Annual Conference for Protective Relay Engineers, 2-5 April , pp. 203–214 (2012)
57. Wang, K., Fu, X., Li, K.: Cluster synchronization in community networks with nonidentical nodes. Chaos **19**, 023106 (2009)
58. Xu, S., Zhou, H., Li, C., Yang, X.: Vulnerability assessment of power grid based on complex network theory. In: Asia-Pacific Power and Energy Engineering Conference, pp. 1–4 (2009)
59. Yang, Y., Nishikawa, T., Motter, A.E.: Vulnerability and cosusceptibility determine the size of network cascades. Phys. Rev. Lett. **118**, 048301 (2017)
60. Zemanova, C., Zhou, C., Kurths, J.: Structural and functional clusters of complex brain networks. Physica D **224**, 202–212 (2006)

The manufacturer's authorised representative in the EU is Springer Nature Customer Service Centre GmbH, Europaplatz 3, 69115 Heidelberg, Germany. If you have any concerns regarding our products, please contact ProductSafety@springernature.com

Printed and bound by CPI Group (UK) Ltd, Croydon, CR0 4YY
31/07/2026
02182873-0004